PROGRESS IN

Nucleic Acid Research and Molecular Biology

Volume 74

PROGRESS IN
Nucleic Acid Research and Molecular Biology

edited by

KIVIE MOLDAVE

Department of Molecular Biology an Biochemistry
University of California, Irvine
Irvine, California

Volume 74

ACADEMIC PRESS
An imprint of Elsevier

Amsterdam   Boston   Heidelberg   London   New York   Oxford
Paris   San Diego   San Francisco   Singapore   Sydney   Tokyo

This book is printed on acid-free paper.

Copyright © 2003, Elsevier (USA).

All Rights Reserved.
No part of this publication may be reproduced or transmitted in any form or by any means, electronic or mechanical, including photocopy, recording, or any information storage and retrieval system, without permission in writing from the Publisher.

The appearance of the code at the bottom of the first page of a chapter in this book indicates the Publisher's consent that copies of the chapter may be made for personal or internal use of specific clients. This consent is given on the condition, however, that the copier pay the stated per copy fee through the Copyright Clearance Center, Inc. (222 Rosewood Drive, Danvers, Massachusetts 01923), for copying beyond that permitted by Sections 107 or 108 of the U.S. Copyright Law. This consent does not extend to other kinds of copying, such as copying for general distribution, for advertising or promotional purposes, for creating new collective works, or for resale. Copy fees for pre-2003 chapters are as shown on the title pages. If no fee code appears on the title page, the copy fee is the same as for current chapters.
0079-6603/2003 $35.00

Permissions may be sought directly from Elsevier's Science & Technology Rights Department in Oxford, UK: phone: (+44) 1865 843830, fax: (+44) 1865 853333, e-mail: permissions@elsevier.com.uk. You may also complete your request on-line via the Elsevier homepage (http://elsevier.com), by selecting "Customer Support" and then "Obtaining Permissions."

Academic Press
An Elsevier Imprint.
525 B Street, Suite 1900, San Diego, California 92101-4495, USA
http://www.academicpress.com

Academic Press
84 Theobald's Road, London WC1X 8RR, UK
http://www.academicpress.com

International Standard Book Number: 0-12-540074-8

PRINTED IN THE UNITED STATES OF AMERICA
03 04 05 06 07 08   9 8 7 6 5 4 3 2 1

Contents

SOME ARTICLES PLANNED FOR FUTURE VOLUMES ix

Fine Tuning the Transcriptional Regulation of the CXCL1 Chemokine 1
Katayoun Izadshenas Amiri and Ann Richmond

I. Chemotactic Cytokines 2
II. Receptors of Chemokines 5
III. Chemokines in Wound Healing and Diseases 7
IV. Differential Expression of Chemokines 10
V. Transcriptional Regulation of CXC Chemokines 11
VI. Regulation of NF-κB Activation 13
VII. Role of CBP in CXCL1 Transcription 18
VIII. PARP as a Transcriptional Activator of the CXCL1 Gene 20
IX. CDP as a Transcriptional Repressor of the CXCL1 Gene 23
X. Reflection 25
XI. References 26

Enzymes That Cleave and Religate DNA at High Temperature: The Same Story with Different Actors 37
Marie-Claude Serre and Michel Duguet

I. Topoisomerases 38
II. Recombinases 59
III. Concluding Remarks 73
References 74

Molecular Mimicry in the Decoding of Translational Stop Signals. 83
Elizabeth S. Poole, Marjan E. Askarian-Amiri,
Louise L. Major, Kim K. McCaughan, Debbie-Jane
G. Scarlett, Daniel N. Wilson, and Warren P. Tate

I. Introduction 84
II. Atomic Level Structures of the Ribosome and Its Subunits 85

III. The Active Center of the Ribosome 87
IV. The Three-Dimensional Space of the Active Center 90
V. The Mimicry Complex in Termination 94
VI. The RF Structure: Does It Look Like a tRNA? 99
VII. The Paradox between the Structural and Functional Evidence
    for the Bacterial RF "Anticodon" 103
VIII. Specific Interactions within the Bacterial Termination Complex 106
IX. The Termination Mechanism 111
X. Summary 114
    References 115

Phylogenetics and Functions of the Double-Stranded RNA-Binding Motif: A Genomic Survey 123

Bin Tian and Michael B. Mathews

I. Introduction 124
II. Survey Methods 125
III. dsRBM Occurrence, Frequency, and Conservation 126
IV. Domain Structures of dsRBM Proteins 129
V. dsRBM Protein Families in Five Type Species 131
VI. dsRBMs in Other Taxa 146
VII. Concluding Remarks 148
    References 152

Mending the Break: Two DNA Double-Strand Break Repair Machines in Eukaryotes 159

Lumir Krejci, Ling Chen, Stephen Van Komen, Patrick Sung, and Alan Tomkinson

I. Introduction 160
II. Biological Relevance of the DSB 161
III. Genetic Pathways for Homologous Recombination 162
IV. Recombination Genes: the *RAD52* Epistasis Group 166
V. General Introduction to NHEJ 178
VI. Mechanisms and Function of NHEJ in Eukaryotes 179
    References 191

The Yeast and Plant Plasma Membrane H$^+$ Pump ATPase: Divergent Regulation for the Same Function ... 203

Benoit Lefebvre, Marc Boutry, and Pierre Morsomme

I. Introduction	204
II. PM H$^+$-ATPase Gene Organization and Expression	205
III. Structure and Activity	206
IV. Functions	209
V. Translational Regulation	212
VI. Posttranslational Regulation	212
VII. Trafficking	218
VIII. General Conclusions	225
References	228

The Genes Encoding Human Protein Kinase CK2 and Their Functional Links ... 239

Walter Pyerin and Karin Ackermann

I. Introduction	240
II. Protein Kinase CK2	241
III. The Human Protein Kinase CK2 Genes	245
IV. Expression of Human CK2 Genes	250
V. Functional Links of CK2 Genes	262
References	268

Heterochromatin, Position Effects, and the Genetic Dissection of Chromatin ... 275

Joel C. Eissenberg and Lori. Wallrath

I. Introduction	276
II. Heterochromatin and Gene Silencing	278
III. Heterochromatin and Gene Activation	278
IV. Genetic Strategies for Dissection of Heterochromatin	279
V. *cis*-Spreading *trans*-Inactivation, and Heterochromatic Associations	281
VI. Molecular Composition of Heterochromatin and the Regulation of Heterochromatin Silencing	283
VII. Setting Up Heterochromatin and Euchromatin Domains	291
VIII. Summary and Future Directions	292
References	293
Index	301

Some Articles Planned for Future Volumes

Tandem CCCH Zinc Finger Proteins in the Regulation of mRNA Turnover
　Perry Blackshear

Initiation and Recombination: Early and Late Events in the Replication of Herpes Simplex Virus
　Paul E. Boehmer

Molecular Regulation, Evolutionary and Functional Adaptations Associated with C to U Editing of Mammalian Apolipoprotein B mRNA
　Nicholas O. Davidson, Shrikant Anant, and Valerie Blanc

Conformational Polymorphism of d(A-G)n and Related Oligonucleotide Sequences
　Jacques Fresco and Nina G. Dolinnaya

DNA-Protein Interactions Involved in the Initiation and Termination of Plasmid Rolling Circle Replication
　Saleem A. Kahn, T.-L. Chang, M. G. Kramer, and M. Espinosa

FGF3: A Gene with a Finely Tuned Spatiotemporal Pattern of Expression During Development
　Christian Lavialle

Specificity and Diversity in DNA Recognition by E. Coli Cyclic AMP Receptor Protein
　James C. Lee

Steroid Signaling in Procaryotes
　Edmond Maser

Oxygen Sensing and Oxygen-regulated Gene Expression in Yeast
　Robert O. Poyton

Ribonucleases in Cancer Chemotherapy
　Robert T. Raines, P. A. Leland, M. C. Herbert, and K. E. Staniszewski

Broad Specificity of Serine/Arginine (SR)-rich Proteins Involved in the Regulation of Alternative Splicing of Premessnger RNA
　James Stevenin, Cyril Bourgeois, and Fabrice Lejune

DNA Double-Strand Break Repair in Eukaryotic Cells
　Patrick Sung, Lumir Krejci, and Alan Tomkinson

FOXO Forkhead Transcription Factors in Insulin and Growth Factor Action
　Terry Unterman, Shaodong Guo, and Xiaohui Zhang

Fine Tuning the Transcriptional Regulation of the CXCL1 Chemokine

KATAYOUN IZADSHENAS AMIRI[*,†]
AND ANN RICHMOND[*,‡]

*Department of Cancer Biology, Vanderbilt University School of Medicine, Nashville, Tennessee 37232
†Department of Microbiology, Meharry Medical College, Nashville, Tennessee 37203
‡Department of Veterans Affairs, Nashville, Tennessee 37232

I. Chemotactic Cytokines	2
II. Receptors of Chemokines	5
III. Chemokines in Wound Healing and Diseases	7
IV. Differential Expression of Chemokines	10
V. Transcriptional Regulation of CXC Chemokines	11
VI. Regulation of NF-κB Activation	13
A. Signaling Pathways Leading to Activation of NF-κB in Melanoma Cells	15
VII. Role of CBP in CXCL1 Transcription	18
VIII. PARP as a Transcriptional Activator of the CXCL1 Gene	20
IX. CDP as a Transcriptional Repressor of the CXCL1 Gene	23
X. Reflection	25
References	26

Constitutive activation of the transcription factor nuclear factor-κB (NF-κB) plays a major role in inflammatory diseases as well as cancer by inducing the endogenous expression of many proinflammatory proteins such as chemokines, and facilitating escape from apoptosis. The constitutive expression of chemokines such as CXCL1 has been correlated with growth, angiogenesis, and metastasis of cancers such as melanoma. The transcription of CXCL1 is regulated through interactions of NF-κB with other transcriptional regulatory molecules such as poly(ADP-ribose) polymerase-1 (PARP-1) and cAMP response element binding protein (CREB)-binding protein (CBP). It has been proposed that these two proteins interact with NF-κB and other enhancers to form an enhanceosome at the promoter region of CXCL1 and modulate CXCL1 transcription. In addition to these positive cofactors, a negative regulator, CAAT displacement protein (CDP), may also be involved in the transcriptional regulation of CXCL1. It has been postulated that the elevated expression of CXCL1 in melanomas is due to altered interaction between these molecules. CDP interaction with the promoter down-regulates transcription, whereas PARP and/or CBP interactions enhance transcription. Thus, elucidation of the interplay between components of the enhanceosome of this gene is important in finding more efficient and new therapies for conditions such as cancer as well as acute and chronic inflammatory diseases.

I. Chemotactic Cytokines

Chemokines are small, proinflammatory, inducible, secreted cytokines that are involved in trafficking, activation, and proliferation of many cell types such as myeloid, lymphoid, pigment epidermal, and endothelial cells (1). Chemokine proteins are encoded by 70–130 amino acids, which also include a signal peptide sequence of 20–25 amino acids. It is interesting to note that although chemokines share little homology in their primary sequence, their overall tertiary structure is similar (2). Chemokines have been observed to form dimers in concentrated solutions and on crystallization, however, these concentrations are much higher than the biological concentrations. It is now commonly accepted that chemokines act as monomers in biological systems (2–4). To date, over 50 chemokines have been identified and assigned to four classes according to their arrangement of the first two of four conserved cysteine residues: C, CC, CXC, and CX_3C chemokines (Table I). The C chemokines such as lymphotactin (XCL1) lack two of the four conserved cysteine residues, whereas CC chemokines have the first two cysteines adjacent to each other; examples are chemoattractant protein-1 (CCL2), macrophage inflammatory protein-1α (CCL3), and regulated upon activation of normal T cells expressed and secreted (CCL5). In CX_3C chemokines, three amino acid residues separate the first two cysteines, e.g., fractalkine (CX_3CL1), whereas in CXC chemokines such as melanocyte growth stimulatory activity/growth-related oncogene (CXCL1-3), interleukin 8 (CXCL-8), monokine induced by interferon-γ (CXCL9), interferon-γ-inducible protein-10 (CXCL10) and IFN-inducible T cell α-chemoattractant (CXCL11), only one amino acid residue separates the first two cysteines (1, 9). CXC chemokines may contain a Glu-Leu-Arg (ELR) motif at the amino terminus (e.g., CXCL1-3, CXCL8). Chemokines that contain the ELR motif are associated with angiogenesis, whereas chemokines lacking this motif are associated with angiostasis (e.g., CXCL9, CXCL10, and CXCL11) (5, 6).

TABLE I
CXC, C, CX₃C, AND CC CHEMOKINE AND RECEPTOR FAMILIES[a]

Systematic name	Chromosome	Human ligand	Mouse ligand	Chemokine receptor(s)
CXC chemokine/receptor family				
CXCL1	4q21.1	GROα/MGSAα	GRO/MIP2/KC?	CXCR2>CXCR1
CXCL2	4q21.1	GROβ/MGSAβ	GRO/MIP2/KC?	CXCR2
CXCL3	4q21.1	GROγ/MGSAγ	GRO/MIP2/KC?	CXCR2
CXCL4	4q21.1	PF4	PF4	Unknown
CXCL5	4q21.1	ENA7B	GCP2/LIX?	CXCR2
CXCL6	4q21.1	GCP2	GCP2/LIX?	CXCR1, CXCR2
CXCL7	4q21.1	NAP2	Unknown	CXCR2
CXCL8	4q21.1	IL-8	Unknown	CXCR1, CXCR2
CXCL9	4q21.1	MIG	MIG	CXCR3 (CD183)
CXCL10	4q21.1	IP10	IP10CRG2	CXCR3 (CD183)
CXCL11	4q21.1	ITAC	ITAC	CXCR3 (CD183)
CXCL12	10q11.21	SDF1α/β	SDF1/PBSF	CXCR4 (CD184)
CXCL13	4q21.1	BCA1	BLC	CXCR5
CXCL14	5q31.1	BRAK/Bolckine	BRAK	Unknown
(CXCL15)	Unknown	Unknown	Lungkine/WECHE	Unknown
CXCL16	17p13	Unknown	Unknown	CXCR6
C chemokine/receptor family				
XCL1	1q24.2	Lymphotactin/SCM1α/ATAC	Lymphotactin	XCR1
CX$_3$C chemokine/receptor family				
CX$_3$CL1	16q13	Fractalkine	Neurotactin/ABCD3	CX$_3$CR1
CC chemokine/receptor family				
CCL1	17q11.2	I309	TCA3/P500	CCR3
CCL2	17q11.2	MCP1/MCAF/TDCF	JE?	CCR2
CCL3	17q12	MIP1α/LD78α	MIP1α	CCR1, CCR5
CCL3L1	17q12	LD78β	Unknown	CCR1, CCR5
CCL4	17q12	MIP1β	MIP1β	CCR5 (CD195)
CCL5	17q12	RANTES	RANTES	CCR1, CCR3, CCR5 (CD195)
(CCL6)		Unknown	C10/MRP1	Unknown

(*Continues*)

TABLE I (Continued)

Systematic name	Chromosome	Human ligand	Mouse ligand	Chemokine receptor(s)
CCL7	17q11.2	MCP3	MARC?	CCR1, CCR2, CCR3
CCL8	17q11.2	MCP2	MCP2?	CCR3, CCR5 (CD195)
(CCL9/10)		Unknown	MRP2/CCF18/MIP1γ	CCR1
CCL11	17q11.2	Eotaxin	Eotaxin	CCR3
(CCL12)		Unknown	MCP5	CCR3
CCL13	17q11.2	MCP4	Unknown	CCR2, CCR3
CCL14	17q12	HCC1	Unknown	CCR1, CCR5
CCL15	17q12	HCC/LKN1/MP1δ	Unknown	CCR1, CCR3
CCL16	17q12	HCC4/LEC/LCC1	Unknown	CCR1, CCR2
CCL17	16q13	TARC	TARC/ABCO2	CCR4
CCL18	17q12	DC-CK1/PARC/AMAC1	Unknown	Unknown
CCL19	9p13.3	MIP3$_\beta$/ELC/exodus-3	MIP3$_\beta$/ELC/exodus-3	CCR7 (CD197)
CCL20	2q36.3	MIP3$_\alpha$/LARC/exodus-1	MIP3$_\alpha$/LARC/exodus-1	CCR6
CCL21	9p13.3	6Ckine/SLC/exodus-2	6Ckine/SLC/exodus-2/TCA4	CCR7 (CD197)
CCL22	16q13	MDC/STCP1	ABCD1	CCR4
CCL23	17q12	MPIF1/CK$_\beta$8/CK$_\beta$8-1	Unknown	CCR1
CCL24	7q11.23	Eotaxin-2/MPIF2	MPIF2	CCR3
CCL25	19p13.3	TECK	TECK	CCR9
CCL26	7q11.23	Eotaxin-3	Unknown	CCR3
CCL27	9p13.3	CTACK/ILC	ALP/CTACK/ILC/ESkine	CCR10
CCL28	5p12	MEC		CCR3/CCR10

[a]Reproduced with permission from International Union of Immunological Societies/World Health Organization Subcommittee on Chemokine Nomenclature. *J. Leukoc. Biol.* **70**, 465–466 (2001). The Society for Leukocyte Biology.

The expression of the chemokine superfamily is regulated through multiple pathways. Cytokines such as interleukin-1 (IL-1)[1] and tumor necrosis factor-α (TNF-α) induce their expression through activation of nuclear factor-κB (NF-κB) (7), whereas interferon-γ (IFN-γ) acts through the Janus kinases (JAK)/signal tranducer and activator of transcription proteins (STAT) pathway (8). Transforming growth factor-β (TGF-β) and glucocorticoids negatively regulate chemokine expression (9). The functions of chemokines are mediated through seven transmembrane domain, G protein-coupled cell surface chemokine receptors (9–11). The binding of chemokines to their receptors occurs through interactions of two regions with the receptor. The low-affinity binding of chemokines to the receptors is mediated by an exposed loop of the backbone between the second and third cysteines, whereas high-affinity binding to the receptors requires the N-terminus. The binding of the NH_2-terminal to the receptor is required for receptor signaling and its amino acid composition is important in determining the degree of chemokine binding to the receptor (2).

II. Receptors of Chemokines

The functions of chemokines are mediated through seven transmembrane G-protein-coupled cell surface receptors. Chemokine receptors are 340–370 amino acids in length with 25–80% amino acid identity (12). The structures of chemokine receptors have not yet been fully solved; however, they are believed to have common features such as four extracellular domains each with one cysteine residue, a conserved 10 amino acid sequence in the second

---

[1] Abbreviations: IL, interleukin; TNF-α, tumor necrosis factor-α; NF-κB, nuclear factor-κB; IFN-γ, interferon-γ; JAK, Janus kinases; STAT, signal tranducer and activator of transcription proteins; TGF-β, transforming growth factor-β; GTP, guanosine triphosphate; GDP, guanosine diphosphate; PLC^2, phospholipase Cβ2; IP$_3$, inositol 1,4,5-triphosphate; PIP$_2$, phosphatidylinositol 4,5-bisphosphate; PKC, protein kinase C; PI3K, phosphatidylinositol 3-kinase; MAPK, mitogen-activated protein kinases; HIP, Hsp70 interacting protein; ICAM-1, intercellular adhesion molecule-1; VCAM-1, vascular cell adhesion molecule-1; CTAPIII, connective tissue-activating peptide; MS, multiple sclerosis; AP-1, activator protein-1; NF-IL6, nuclear factor activated by interleukin 6; IRF, IFN-regulatory factor; Sp1, stimulating protein-1; HMGI/Y, high mobility group-I/Y; C/EBP, CAAT enhancer binding protein; IUR, immediate upstream region; NRF, NF-κB repressing factor; CBP, cAMP response element binding protein (CREB)-binding protein; HAT, histone acetyltransferase; HDAC, histone deacetylase; CDP, CAAT displacement protein; PARP, poly(ADP-ribose) polymerase; IκB, inhibitor of κB; NLS, nuclear localization signal; IKK, IκB kinase; PKA, protein kinase A; CKII, casein kinase II; PIP$_3$, phosphatidylinositol 3,4,5-trisphosphate; PDK, 3-phosphoinositide-dependent protein kinase; PTEN, phosphatase and tensin homologue deleted on chromosome 10; NIK, NF-κB-inducing kinase; TRAF2, TNF receptor-associated factor 2; ERK1/2, extracellular signal-regulated kinase 1 and 2; Cdk, cyclin-dependent kinases; NAD$^+$, β-nicotinamide adenine dinucleotide; BER, base excision repair; HD, Cut homeodomain; CYP7A1, human cholesterol-7 hydroxylase; HNF-1, hepatocyte nuclear factor-1.

intracellular loop, an acidic NH_2-terminus that contains N-linked glycosylation sites and is involved in the ligand binding, and an intracellular C-terminus containing serine and threonine phosphorylation sites (13, 14).

There is a wide variation in terms of chemokine and chemokine receptor selectivity. Certain chemokines bind only one receptor, whereas others may bind many different receptors. Receptors may also be exclusive for one or more chemokines (15, 16). In general chemokines from the same gene cluster tend to bind similar receptors (17). Once the receptor binds a chemokine, it gets phosphorylated by a G-protein receptor-coupled kinase, an event proposed to stabilize receptor desensitization. Subsequently, the receptor is internalized (18). The receptors may also undergo heterologous desensitization, where its serine residues are phosphorylated without ligand binding (19).

To date 19 human chemokine receptors have been identified among which 6 receptors selectively bind CXC chemokines and thus have been designated CXCR1 through 6, and 10 receptors, CCR1 to CCR10, all bind the CC chemokines. Receptors for fractalkine and lymphotactin were recently identified and named CX3CR1 (20) and XCR1 (21), respectively. The Duffy antigen receptor also includes the chemokine receptor family and binds promiscuously to both CXC and CC chemokines (22).

The biological functions of CXCL1 are known to be mediated through CXCR2. This receptor has been shown to be expressed on all granulocytes, monocytes, and mast cells, and on some $CD8^+$ T cells and $CD56^+$ natural killer cells as well as melanocytes. By binding and activating CXCR2, CXCL1 modulates inflammation, angiogenesis, wound healing, tumorigenesis, and cell motility (23–25).

The signal transduction mechanism of chemokine receptors is dependent on the trimeric G-protein that is coupled to the receptor. G-proteins function as molecular switches that can flip between two states: active when guanosine triphosphate (GTP) is bound, and inactive when guanosine diphosphate (GDP) is in place. When the ligand binds to the receptor, a conformational change in the receptor causes its association with the G-protein to facilitate the exchange of GDP for GTP. In this activated state, the G-protein dissociates into $G\alpha$ and $G\beta\gamma$ subunits, so that $G\beta\gamma$ is able to activate the membrane bound phospholipase $C\beta 2$ (PLC^2). PLC^2 in turn catalyzes the synthesis of 1,4,5-trisphosphate (IP_3) from phosphatidylinositol 4,5-bisphosphate (PIP_2). IP_3 mobilizes calcium, leading to activation of calcium-sensitive kinases such as protein kinase C (PKC), which in turn phosphorylate proteins that activate a series of signaling events leading to cellular responses (26, 27). Chemokine receptors can also activate several other intracellular effectors such as Ras and Rho (28), phospholipase A2, phosphatidylinositol-3-kinase (PI-3K) (29), tyrosine kinases (30, 31), and the mitogen-activated protein kinase (MAPK) pathway (32, 33).

Chemokine receptors become desensitized to repeated stimulation with the same or other ligands after activation. Although this process is not fully understood, it is thought to be stabilized by phosphorylation of serine and threonine residues in the C-terminus of the receptor by G-protein-coupled receptor kinases. Receptor phosphorylation in some instances facilitates receptor sequesteration and internalization (34, 35). Desensitization and receptor internalization are thought to be critical for maintaining the cell's capacity to sense a chemoattractant gradient. Recently, our laboratory demonstrated that CXCR2 associates with Hsc/Hsp 70-interacting protein (HIP), and this association seems to play an important role in receptor internalization (36). Subsequently, the receptors are targeted for degradation and/or recycling back to the membrane. Thus, desensitization and trafficking of chemokine receptors may also play important roles in the regulation of inflammatory processes and cancer.

III. Chemokines in Wound Healing and Diseases

Chemokines were originally described as mediators of leukocyte recruitment and activation. Chemotaxis of leukocytes is induced by early response mediators and cytokines such as histamine, C5a, TNF-α, and IL-1, which increase the expression of adhesion molecules on the vascular endothelial cells. This allows for localization and adherence of leukocyte populations to areas of inflammation. The early response mediators and cytokines help in slowing the circulatory flow of leukocytes by up-regulating molecules such as selectins on the endothelial surface. Once leukocytes have been localized to the inflamed vascular wall, the presence of adhesion molecules such as intercellular adhesion molecule-1 (ICAM-1) and vascular cell adhesion molecule-1 (VCAM-1) allows firm adhesion to the endothelium. After adherence to the vascular endothelium, leukocytes can migrate into the tissue following the chemotactic gradient in the inflamed tissue (37).

Leukocyte migration is one of the important components of wound healing. After acute injury in skin wound models, the destroyed blood vessels release platelets and neutrophils that serve as sources for growth factors and other mediators such as connective tissue-activating peptide III (CTAP-III). CTAPIII is processed to CXCL7 by proteases that are released by neutrophils, which in turn stimulates the migration and extravasation of neutrophils via the receptor CXCR2 (38). The diapedesis of neutrophils is further mediated by CXCL1, which is produced by both endothelial and dermal cells. A third major mediator of the neutrophil migration, CXCL5, is produced by mononuclear cells in the provisional matrix of the wound (39, 40). Furthermore, hypoxia and bacterial products as well as proinflammatory cytokines (TNF-α, IL-1) produced by neutrophils below the wound surface stimulate production of CXCL8, which stimulates the migration and proliferation of keratinocytes in addition to neutrophils.

The keratinocytes in turn express CCL2, which has been considered the main attractant of mast cells, monocytes, and lymphocytes to the wound area (39, 41, 42). Furthermore, CCL2 in addition to other angiogenic chemokines, e.g., CXCL1 and CXCL8, may also stimulate endothelial-cell locomotion during the angiogenesis phase of wound healing (39, 43, 44). CCL2 also may indirectly contribute to fibroblast proliferation by recruiting IL-4-producing mast cells. IL-4 stimulates fibroblast proliferation and helps to limit the inflammatory reaction by down-regulating the expression of chemokines, e.g., CCL2 and CXCL8 (45, 46). Thus, the attraction of resident and inflammatory cells appears to be tightly regulated by the complex and phase-specific expression of chemokines, placing these small molecules at center stage in the process of wound healing.

Almost any stimulus that alters cellular homeostasis may elicit the expression of inducible chemokines leading to an over 300-fold increase in the mRNA level within a few hours of activation (47). However, with a system that is easily inducible, there is also a greater potential for persistent expression, leading to diseases. Chemokines have been associated with various diseases such as the autoimmune disease multiple sclerosis (MS), bacterial and viral infections, atherosclerosis, asthma, graft rejection, as well as neoplasia. MS is a chronic relapsing neuroinflammatory disease in which myelinated nerve fibers are targeted by T lymphocytes and macrophages in the central nervous system. Studies have shown that the appearance of CCL5, CCL3, CCL4, CCL2, and CXCL10 mRNA and protein directly correlates with inflammatory lesions (48–53). Consistent with these data is the presence of the receptors for these chemokines, CCR2, CCR5, and CXCR3 on macrophages, activated microglia and T cell in lesions (54).

Chemokines also play an important role in bacterial and viral infections. Studies have shown increased MIP-2 production during lipopolysaccharide-induced endotoxemia in animal models, which may be responsible for the recruitment of neutrophils (55). Other chemokines such as CCL3, CCL5, and CCL2 have also been implicated in septic responses (56–58). In viral infections however, chemokine receptors seem to play an important role. Studies have shown that some chemokine receptors (CCR3, CCR2b, CCR5, and CXCR4) act as cofactors with CD4 for macrophage cell line-trophic HIV entry into monocytes and T cells (59, 60). Also, other viruses such as cytomegalovirus (CMV) have been shown to encode chemokine receptors (61).

The relevance of chemokines in atherosclerosis was demonstrated in animal models in which chemokines CXCL8, CXCL12, CXCL10, CCL2, and CCL5 were associated with the lesions. A model for involvement of chemokines in progression of atherosclerosis proposes that activated endothelial or arterial smooth muscle cells release chemokines to induce firm adhesion of monocytes to the vascular endothelium. This is followed by diapedesis into the subendothelium, where they take up lipid and become the foam cells within the fatty streak. The smooth muscle migration into the intima and thrombus formation over the plaque may also involve some chemokines (47).

In asthma, a chronic inflammatory disease of small airways, mononuclear, eosinophil, and mast cells infiltrate the submucosa leading to mucous gland hyperplasia and subepithelial fibrosis. Asthma is characterized by airway hyperresponsiveness (smooth muscle contraction) to nonspecific stimuli. This is thought to be due to the involvement of chemokines such as CCL2, CCL12, CCL24, and CCL5. CCL2 was able to induce changes in airway physiology when instilled into the lungs of normal mice, resulting in increased levels of histamine mast cell degranulation (62).

Chemokines may also influence allograft biology through their involvement in immune suppression, inflammatory responses in acute and chronic rejection, as well as recruitment of leukocytes to the allograft leading to ischemia-reperfusion (63). Studies in heart and skin transplants have demonstrated the presence of CXCL1 and CCL2 in the early response and CXCL10, CXCL9, CCL5, and CCL4 at the later stages. The presence of these chemokines may be necessary for the movement of $CD4^+$ and $CD8^+$ T lymphocytes, macrophages, natural killer cells, and antigen-presenting cells that are involved in the acute rejection (64–66).

Interestingly, elevated expression of chemokines has been observed in many tumor cell types, implicating a role for chemokines in neoplasia. The potential of chemokine involvement in tumors has sparked a new interest in the field of cancer biology. Studies of a variety of tumor cell types indicate the role of chemokines as tumor growth factors as well as stimulators of angiogenesis and metastasis. Murine models have shown that chemokine secretion by tumor cells may influence angiogenesis as well as tumor growth. The metastatic and angiogenic abilities of a number of tumors have been attributed to elevated levels of ELR^+, angiogenic chemokines (5, 6). Constitutively expressed chemokines have been shown to transform melanocytes and high levels of endogenous CXCL1 and CXCL8 have been detected in melanomas (24, 25, 67–69). The role of these chemokines in tumor growth has also been implicated in many other tumor types such as pancreas, head and neck, and non-small-cell lung (70–72).

The directional metastasis of malignant tumors has also been attributed to the presence of chemokines and their receptors in tumors (73–75). A recent publication on the involvement of CXCR4 and CCR7 in directing breast cancer tumors to their secondary sites further supports this hypothesis (76). The two chemokine receptors CXCR4 and CCR7 were shown to be expressed in high levels in breast tumors compared to the normal breast tissue. Interestingly enough, the ligands for these receptors, CXCL12 and CCL21, respectively, showed high expression in such organs as lung, liver, bone marrow, and regional lymph nodes, which are associated with the distinct breast cancer metastasis pattern. Moreover, the high expression of CXCR4 and CCR7 induced actin polymerization and pseudopodia formation, resulting in enhanced invasiveness of the breast tumors. Thus, chemokines appear to play a significant role in tumor development due to their involvement in tumor growth, angiogenesis, and metastasis.

Thus, with the association of chemokines in a wide variety of diseases, it is imperative to study the mechanisms underlying their regulation in order to develop more efficient therapies. The focus in our laboratory lies on the chemokine CXCL1 in a melanoma model, since CXCL1 appears to contribute to the growth and proliferation of melanoma. Studies in our laboratory have shown that the level of this chemokine is elevated in many types of melanoma cell lines and that this elevation in the expression may be due to deregulation in the transcriptional regulation of the CXCL1 gene. In this review, the possible mechanism of deregulation will be discussed.

IV. Differential Expression of Chemokines

Cytokines are not stored intracellularly and as such their stimulus-dependent secretion heavily relies on *de novo* protein synthesis. Hence, cytokine expression is regulated primarily at the initial phase of protein synthesis—transcription. This regulation appears to be stimulus and cell type specific and is mediated through inducible transcription factors such as NF-κB and activator protein (AP-1) (77–80). Following a stimulus, early response cytokines such as TNF-α and IL-1 are rapidly produced, which in turn induce macrophages and other cell types to secrete additional cytokines such as CXCL8 and IL-6. The promoters of many of these cytokine genes contain binding sites for NF-κB, which mediates the up-regulation of their transcription in response to the stimulus and is in fact required for their maximal transcription (81–84).

The cause of the differential expression of the NF-κB-dependent cytokines is uncertain, but recent studies support the speculation that involvement of other transcription factors may be important in determining the transcription rates of cytokine genes. Although cytokine promoters have the NF-κB binding site in common, each individual promoter also binds other transactivators as well as repressors that interact with NF-κB to direct transcription. The binding sites for the transcription factors AP1 and nuclear factor activated by interleukin 6 (NF-IL6), for example, are present in the promoters of CXCL8, IL-6, and CCL5 genes, where the positive cooperation between these transcription factors and NF-κB enhances the transcription of these genes (85–87). In addition to these transcription factors, CCL5 gene expression requires IFN-regulatory factor (IRF)-1, 3, and 7 transcription factors as well (88, 89). Other transcription factors common in cytokine gene expression include stimulating protein-1 (Sp1) and high mobility group-I/Y (HMGI/Y) (90, 91). There are additional factors that may lead to differential expression of cytokines. These may include differences in chromatin and transcription factor accessibility, posttranscriptional RNA processing, mRNA stability, as well as differences in translational efficiency (92–94). Thus, the differential activation of the many inducible transcription factors and their accessibility to the binding sites as well as posttranscriptional events may explain the cell type-specific and stimulus-specific expression of cytokines.

V. Transcriptional Regulation of CXC Chemokines

The expression of the CXC chemokines is thought to be NF-κB dependent, thus indicating a disregulation in the activation of the transcription factor NF-κB may be involved in the up-regulation of these chemokines in inflammatory diseases and cancer. However, as mentioned previously, the transcriptional regulation of these chemokines is more complex and involves factors other than NF-κB. In CXCL8 transcription, NF-κB interacts with either AP1, NF-IL6, or CAAT enhancer binding protein (C/EBP) elements (95–97), whereas in CXCL1 transcription, the NF-κB interacting elements are Sp1 (HMG I/Y) and immediate upstream region (IUR) (90) (Fig. 1). It is postulated that the factors binding to these elements may form an enhanceosome-like transcriptional response element in modulating the gene expression (98, 99).

The transactivation of CXCL8 transcription occurs through the interaction between the NF-κB element of CXCL8 and either the AP1 or the NF-IL6 element in a cell type-specific manner (100–102). Repression of CXCL8 transcription is, on the other hand, regulated by NRF, an NF-κB repressing factor. In the absence of stimulation, NRF inhibits transcription of CXCL8, but after stimulation with IL-1, NRF is necessary for full induction of CXCL8 transcription (103). Upon the binding of NF-κB to its binding site, the p65 subunit activates the promoter via recruitment of CBP to the site. The intrinsic histone acetylase activity (HAT) of CBP/p300 will then stabilize the transcription from the promoter. This stabilization becomes imperiled in the presence of CDP, which has been shown to be capable of recruiting the histone deacetylase activity (HDAC) that would counteract the HAT activity of CBP (104). Thus, it is plausible that similar interactions among NF-κB, CBP, and CDP are involved in the transcriptional regulation of CXCL1.

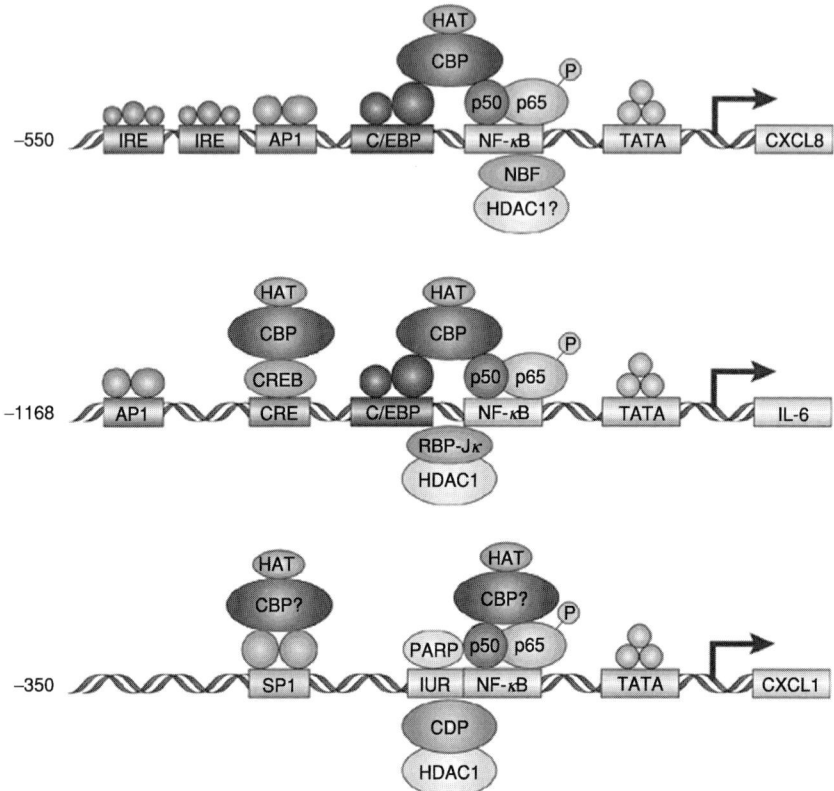

FIG. 1. Modulation of transcription of cytokines involves similar regulatory elements. Reprinted by permission from A. Richmond, *Nat. Rev. Immunol.* **2**, 664–674 (2002), ©2002 Macmillan Publishers Ltd.

Within the CXCL1 promoter, the IUR element is thought to regulate the transcription of CXCL1 both positively and negatively. It contains the consensus sequence GGGATCGATC, which binds the negative modulator CDP (*105*), as well as the TCGATC sequence that binds the positive modulator, PARP-1 (*106*). Thus, PARP-1 may induce NF-κB-activated transcription of CXCL1, whereas CDP would repress this activity. However, the intricate mechanism of regulation by these molecules is yet to be identified in the context of the CXCL1 promoter. Thus, efforts to shed light on the matter would greatly contribute to the efficacy of therapeutic means used today in inflammatory diseases and cancer.

VI. Regulation of NF-κB Activation

The major known player in the transcription induction of CXCL1 is the transcription factor NF-κB. In the healthy human, NF-κB regulates the expression of genes involved in normal immunological responses (e.g., generation of immunoregulatory molecules such as antibody light chains) in response to proinflammatory cytokines and byproducts of microbial and viral infections (*107*). However, increased activation of NF-κB results in enhanced expression of proinflammatory mediators, leading to acute inflammatory injury to lungs and other organs, development of multiple organ dysfunction, as well as angiogenesis and tumor growth (*107–109*).

The NF-κB protein is composed of two subunits, which may vary, affecting the transcriptional activity of the protein. There are five known mammalian NF-κB subunits, each characterized by ankyrin repeat elements: Rel (c-Rel), p65 (RelA), RelB, p50, and p52. The transcription regulatory action of NF-κB depends on the composition of the NF-κB dimers, p50 homodimers lack strong transactivation domains and can actually inhibit gene expression by competing with p65/p50 or other transactivating complexes for the κB sites (*107, 108*).

In the absence of activation, NF-κB is sequestered in the cytoplasm by being associated with IκB (an inhibitor of κB) protein. The IκB protein binds to the nuclear localization signal (NLS) of NF-κB, inhibiting its translocation into the nucleus (*107, 108, 110*). When the cell is exposed to activating signals, such as TNF-α, the IκB protein is phosphorylated, ubiquitinated, and then broken down in the 26 S proteasome (*111*). This frees the NF-κB to translocate into the nucleus where it binds to κB sites in the promoter/enhancer regions of specific genes, including the promoter for IκB, to transactivate transcription. There are five known IκB proteins: IκB-α,β,ε,γ, and Bcl-3. In addition to these five, p105 and p100, the precursors of p50 and p52, respectively, possess domains that act as IκBs. The IκB subtypes show different affinity for the different NF-κB dimers in a cell-specific manner. For example, in endothelial cells, IκB-α has similar inhibitory activity for p50/RelA, p50/RelB, and p50/c-Rel, whereas IκB-β more strongly inhibits p50/RelA than the other two (*107, 108*).

IκBs are also active in the nucleus and interact with NF-κB dimers as they do in the cytoplasm. IκB-α binds p50/RelA and inhibits its transactivating activity in the nucleus as well as freeing the heterodimer from κB sites on DNA to induce their transport from the nucleus to the cytoplasm. On the contrary, Bcl-3 can act as a transcriptional activator in the nucleus by binding to p50/p50 homodimers, which can act as transcriptional repressors by occupying NF-κB binding sites and preventing the transactivating NF-κB heterodimers such as p50/RelA or p50/c-Rel from binding to these sites (*112*). The inducible degradation of IκB is controlled by three large multiprotein complexes: IκB kinase (IKK), IκB ubiquitin ligase, and 26 S proteasome. IKK activity is induced upon various stimulation, leading to phosphorylation of two N-terminal serine residues of IκB, Ser-32 and Ser-36 (*113, 114*). The phosphorylation of IκB targets it for ubiquitination by IκB-ubiquitin ligase and subsequent degradation by the 26 S proteasome (*107, 108, 111*).

The transactivation function of NF-κB is also regulated in the nucleus through interaction with HDAC corepressor proteins. Ashburner *et al.* (*115*) demonstrate in their study a direct interaction between the HDAC1 and NF-κB p65 subunit by which HDAC1 exerts its corepressor function. Overexpression of HDAC1 and HDAC2 was shown to repress TNF-α-induced NF-κB-regulated gene expression. HDAC2 does not interact with NF-κB directly, but can probably regulate NF-κB activity through its association with HDAC1. In accordance with this, chemical inhibitors of HDAC activity such as trichostatin A increased expression of an NF-κB-dependent reporter gene as well as an endogenous IL-8 gene. Thus, the association of NF-κB with the HDAC1 and HDAC2 corepressor proteins functions to repress basal expression of NF-κB-dependent genes as well as to control the induced level of expression of these genes.

Activation of NF-κB may be induced by a variety of pathogenic stimuli, including bacterial products, such as metalloproteases (MMP-3 and 9), viral proteins, cytokines such as IL-1 and TNF-α, growth factors such as PDGF, radiation, ischemia/reperfusion, and oxidative stress through multiple pathways (*107, 108*). The activation of NF-κB occurs within minutes of stimulation since *de novo* protein synthesis is not required and the activated NF-κB in turn mediates expression of more than 150 genes involved in inflammatory and immune responses (*116*). It is important to note that the promoters and enhancers of NF-κB-dependent genes also contain binding sites for other transcription factors and interplay of these factors can potentiate or repress the ability of NF-κB to initiate transcription. However, it is in events of abnormal, constitutive activation of NF-κB that major problems arise resulting in many chronic inflammatory diseases as well as cancer. Persistent activation of NF-κB inhibits apoptosis and promotes proliferation leading to hyperplasia (*107–109, 117, 118*).

There have been reports on NF-κB activation through IKK-independent pathways as well. Protein kinase A (PKA), casein kinase II (CKII), and p38 MAPK have all been implicated in the phosphorylation of NF-κB Rel A, leading to enhanced interaction of NF-κB with transcriptional coactivators and components of basal transcriptional machinery (*119–122*). PKA phosphorylation of the NF-κB p65 subunit on serine 276 was shown to weaken the interaction between the N- and C-terminal of p65 and create an additional site for interaction with the coactivator protein CBP (*119*). Based on these data, it is feasible that there is an interaction between NF-κB and the coactivator protein CBP within the CXCL1 enhanceosome and this interaction among others may be enhanced in melanomas and inflammatory diseases leading to constitutive expression of CXCL1.

A. Signaling Pathways Leading to Activation of NF-κB in Melanoma Cells

The hallmark of cancer cells lies in their ability to escape apoptosis by overactivation of growth and survival pathways. There are two major pathways that are associated with survival and proliferation: PI3K/Akt and Ras/MAPK pathways. Stimulation of any one of these pathways leads to activation of antiapoptotic and prosurvival proteins mediated through NF-κB (Fig. 2).

Akt, originally identified as a homologue of the viral oncogene v-Akt, is closely related to protein kinase A and C and was thus named protein kinase B (PKB). There are three mammalian Akt genes that encode proteins containing a pleckstrin homology (PH) domain in the N-terminus, a central kinase domain, and a regulatory carboxy terminus. There are two regulatory phosphorylation sites within Akt: threonine 308 and serine 473. In unstimulated cells, Akt exists in an unphosphorylated state in the cytoplasm. Upon growth factor stimulation and PI3K activation and subsequent production of PIP_3, Akt is recruited to the plasma membrane and is phosphorylated at T308 and S473 by 3-phosphoinositide-dependent protein kinase 1 (PDK1) and PDK2, respectively (123). Fully activated Akt then becomes available to phosphorylate its substrates within the same basic motif, R-X-R-X-X-S/T (124). Thus far, 13 substrates of Akt have been identified, which may be grouped according to their functions in cell survival, cell cycle, glucose metabolism, and protein synthesis. The Akt substrates involved in cell survival regulation include Bad, the forkhead family of transcription factors, FLICE inhibitory protein, and IKK. Thus, activation of Akt may regulate NF-κB activity through IKK. Indeed, NF-κB activation in correlation with Akt activation has been shown in several different carcinomas such as ovarian, breast, pancreatic, and melanoma (125, 126). A model for NF-κB activation through Akt proposed by Madrid et al. (127) suggests Akt utilizes IKK-β in a p38-dependent manner that requires serines 529 and 536 of RelA/p65 to directly stimulate NF-κB activity and Akt signaling in response to IL-1 exposure stimulates NF-κB by activating p38 in a manner dependent on IKK. Whether Akt acts only through IKK to activate NF-κB or directly interacts with NF-κB to activate the transcription factor remains elusive.

Up-regulation of Akt in cancers may be a consequence of mutation or deletion of the phosphatase and tensin homologue deleted on chromosome 10 (PTEN) tumor suppressor gene. PTEN is a lipid phosphatase that plays a

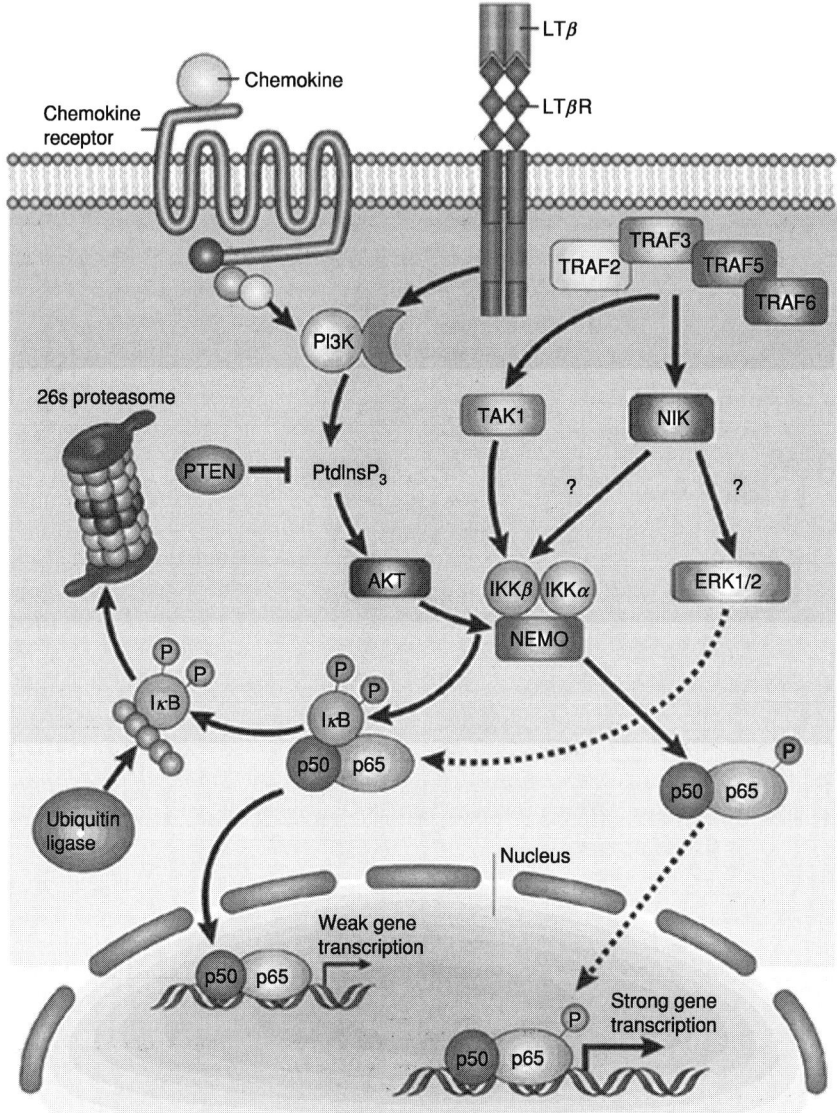

FIG. 2. The signal transduction pathways involved in activation of the transcription factor NF-κB. Reprinted by permission from A. Richmond, *Nat. Rev. Immunol.* **2,** 664–674 (2002), ©2002 Macmillan Publishers Ltd.

crucial role in deactivation of AKT, since one of its primary targets is the direct product of PI3K, PIP3. Loss of PTEN function in cancer cell lines results in accumulation of PIP_3 and activation of Akt and escape from apoptosis. Indeed, overexpression of wild-type PTEN sensitizes them to apoptosis probably through inhibition of the PI3K/Akt pathway (128). As in many cancers, PTEN mutations have been reported in melanomas, suggesting a mechanism for overactivation of Akt in this cancer (129).

The second major pathway leading to NF-κB activation involves the Ras family members, which play important roles in cell differentiation, growth, transformation, and apoptosis (130–132). There are four Ras genes in the mammalian genome, expressing the very homologous proteins N-Ras, H-Ras, M-Ras, and K-Ras of 21 kDa. Ras proteins have been shown to activate the PI3K, Rac, RhoA, coupled with activation of the Raf/MAPK pathway to promote oncogenic transformation (133). The ultimate targets of these pathways are transcription factors such as NF-κB (134).

Work in our laboratory has shown that the CXCL1 protein is endogenously expressed in almost 70% of the melanoma cell lines and tumors, but not in normal melanocytes. Overexpression of human CXCL1, 2, or 3 in immortalized murine melanocytes (melan-a cells) enables these cells to form tumors in SCID and nude mice. Differential display examination of the CXCL1 effect on melanocyte transformation revealed overexpression of the Ras genes. One of the mRNAs identified in the screen as overexpressed in CXCL1 transformed melan-a clones was the newly described M-Ras gene. Overexpression of CXCL1 up-regulates M-Ras expression at both the mRNA and protein levels, and this induction requires an intact ELR motif in the CXCL1 protein. K- and N-Ras proteins are also elevated in CXCL1-expressing melan-a clones, leading to an overall increase in the amount of activated Ras. CXCL1-expressing melan-a clones also exhibited enhanced AP-1 activity. Overexpression of wild-type M-Ras or a constitutively activated M-Ras mutant in control melan-a cells as monitored by an AP-1-luciferase reporter also showed enhanced AP-1 activity, whereas expression of a dominant negative M-Ras blocked AP-1-luciferase activity in CXCL1-transformed melan-a clones. In the *in vitro* transformation assay, overexpression of M-Ras mimicked the effects of CXCL1 by inducing cellular transformation in control melan-a cells, whereas overexpression of dominant negative M-Ras blocked transformation (118). These data suggest that CXCL1-mediated transformation may require Ras activation in melanocytes.

Another NF-κB-regulating kinase that is overexpressed in melanoma is NF-κB-inducing kinase (NIK). NIK was first identified by way of its association with TNF receptor-associated factor 2 (TRAF2) and shares homology with mitogen-activated protein kinase kinase kinases (135). NIK physically associates with and activates both IKK-α and IKK-β (136, 137) and overexpression of NIK has been shown to activate NF-κB (*135*). Interestingly, our laboratory found that NIK is overexpressed in melanoma cells and the IKK-associated NIK activity is enhanced compared to normal cells. When a catalytically inactive form of NIK was expressed, the constitutive activation of NF-kB and CXCL1 promoter was blocked in Hs294T melanoma cells, but not in normal human epidermal melanocytes. The NIK-induced activation of NF-kB was shown to be through the MAPK signaling kinases extracellular signal-regulated kinase 1 and 2 (ERK1/2), since overexpression of dominant-negative ERK leads to a decrease in NF-kB promoter activity (*138*). Thus, these data suggest that NIK is involved in up-regulation of NF-kB activity and hence CXCL1 transcription through an NIK/MEKK-IKK-IκB signaling pathway.

VII. Role of CBP in CXCL1 Transcription

Many inducible transcription factors such as NF-κB are activated through interactions with cellular coactivators (*139*). The transcriptional activity of NF-κB appears to be optimized through interactions with the coactivator CBP as discussed above. CBP is a nuclear protein belonging to a family of highly homologous proteins such as p300 and p270 that are involved in many physiological responses such as proliferation, differentiation, and apoptosis (*140*). CBP is thought to regulate transcription through its HAT activity, since *in vitro* transcription experiments have shown that transcription is induced only on the template with chromatin structure, but not on the naked DNA (*141, 142*). Thus, through its intrinsic HAT activity, CBP activates transcription by transferring an acetyl group to the ε-amino group of a lysine residue of the histone and neutralizes the positive charge of the molecule and thus loosens the interaction between the histone and the negatively charged DNA, leading to chromatin remodeling (*143*). Furthermore, the HAT activity of CBP is restricted not only to histones, but also to other targets such as transcription factors and the transcription apparatus (*144, 145*).

The human CBP locus is located in chromosomal region 16p13.3 and codes for a protein containing (1) the bromodomain, which is found in mammalian HATs; (2) an ADA2-homology domain, which is homologous to the yeast transcriptional coactivator, Ada2p; (3) a KIX domain; and (4) three cysteine-histidine (CH)-rich domains referred to as CH1, CH2, and CH3. It has been suggested that the bromodomain recognizes acetylated residues and could function in identification of different acetylated domains. The CH domains and the KIX domain are thought to mediate protein–protein interaction (*146*). The KIX domain is involved in the interaction between CBP and the p65 (Rel A) subunit of NF-κB, as well as CBP binding to the transcriptionally active serine-133-phosphorylated form of CREB (142, 147).

CBP also appears to be under cell cycle control. Many kinases such as PKA, calcium/calmodulin-dependent kinase, MAP kinase, and cyclin-dependent kinases Cdk2 and Cdc2 have been implicated in the positive and negative phosphorylation control of CBP. The cyclinE–Cdk2 complex has been shown to bind to the C-terminal region of CBP and negatively regulate CBP-mediated coactivation of NF-κB, whereas the other mentioned kinases aid in CBP-mediated transcriptional activation (146). The phosphorylation sites on CBP have not been identified yet, but appear to be critical for its regulation.

There are multiple proposed models for the mode of action of CBP in the regulation of transcription. In the bridging model, CBP acts as a connecting bridge between transcription factors and the transcription apparatus. Data showing the interaction of CBP with a variety of transcription factors as well as components of the basal transcriptional machinery such as TATA box-binding protein (TBP), TFIIB, TFIIE, and TFIIF are supportive of this model (146, 148). In a second model, CBP plays a role in transcriptional activation by assembling a diverse group of cofactor proteins into multicomponent coactivator complexes. Thus, CBP serves as a scaffold for the assembly of transcription cofactors, increases the relative concentrations of these factors in the transcription area, and allows for protein–protein and protein–DNA interactions. Yet another model proposes that CBP may take advantage of either its intrinsic HAT activity or other HATs assembled in multicomponent complexes to target the chromatin and/or transcription factors for transcription activation. Even though targets of CBP *in vivo* are yet to be identified, *in vitro* studies have shown that CBP can acetylate all four core histones (149).

More recently, acetylation of transcription factors such as p53, E2F-1, 2, and 3, MYB, MyoD, GATA-1, CREB, and NF-κB (142, 147, 150, 151) by CBP has been reported and, in almost all cases, the acetylation led to enhanced DNA-binding activity. It is possible that acetylation regulates protein–protein interaction or facilitates protein–DNA binding by changing the conformation of the protein and introducing a DNA-recognition surface (146). These findings also indicate that there may be competitive interactions between the transcription factors for association with CBP, thus providing an additional regulatory event. A good example of this competitive behavior is among CBP, NF-κB, and CREB. CBP recognizes the phosphorylated serine 276 on the p65 subunit of NF-κB through an S domain in the KIX region. The KIX region is also responsible for recognizing the phosphorylated serine 133 form of CREB. Thus phosphorylation of these two proteins promotes their interaction with CBP. Activation of protein kinase A was shown to increase amounts of phosphorylated CREB and decrease NF-κB-mediated transcription (*152, 153*). Thus, the activity of PKA is a determinant in the binding of CBP to its interacting proteins.

A recent report by Chen *et al.* (*154*) demonstrated that the RelA subunit of NF-κB is subject to inducible acetylation by CBP/p300. Acetylation of NF-κB resulted in weak interaction with IκB in the nucleus, which in turn led to decreased IκB-dependent nuclear export of the complex and thus increased NF-κB transactivation. This acetylation was reversed by HDAC3. Thus, CBP/p300 may be involved in the regulation of CXCL1 transcription in more than one way. It could acetylate not only the histones, but also NF-κB itself, and contribute to the prolonged and/or constitutive activation of the transcription factor.

VIII. PARP as a Transcriptional Activator of the CXCL1 Gene

Another putative coactivator of NF-κB in CXCL1 transcription is PARP. The mammalian poly(ADP-ribose) polymerase (PARP)-1, the major isoform of the PARP family, is a nuclear protein that is heavily associated with chromatin. PARP is composed of 1014 amino acids (114 kDa) and is continuously expressed in eukaryotes. It has a 46-kDa DNA-binding domain at the N-terminus that contains two Zn fingers, facilitating the binding of PARP to DNA strand breaks, as well as bipartite nuclear localization signal. The two regions of nuclear localization signals are separated by the DEVD sequence, which is the target of caspase-3 during apoptosis and subsequently gets cleaved, and thus inactivated, resulting in the formation of two proteolytic fragments of PARP, a 29-kDa amino terminus and an 85-kDa carboxyl terminus (*155, 156*). A 54-kDa domain of PARP located in the carboxyl terminus represents the β-nicotinamide adenine dinucleotide (NAD^+)-binding domain containing a highly conserved "PARP signature" sequence comprising the catalytically crucial amino acid residue Glu-988. Between the DNA-binding domain and the NAD^+-binding domain lies a 22-kDa "automodification domain," which facilitates the homodimerization and/or heterodimerization of PARP with other chromatin proteins (*156*).

The catalytic activity of PARP is stimulated 500-fold by noncovalent contact of the DNA-binding domain with DNA strand breaks and results in the transfer of successive units of the ADP-ribose moiety from NAD^+ to itself and other nuclear protein acceptors such as topoisomerase I and II, histones, HMG proteins, p53, and NF-κB (*155, 157–159*). It is thought that PARP mediates stress-induced signaling by binding to damaged DNA containing single-strand breaks and nucleotide excisions. Automodification upon DNA binding releases PARP from DNA due to the highly negatively charged poly(ADP-ribose), rendering DNA more accessible to the DNA repair machinery (159). PARP has been shown to associate *in vivo* with XRCC1, a DNA repair protein involved in base excision repair (BER) of DNA and PARP-1-deficient cells display a severe DNA repair defect that appears to be a primary cause of the observed genomic instability and cytotoxicity of DNA damaging agents that induce BER (160, 161). The catalytic activity of PARP appears to be of importance since in the absence of NAD^+, PARP inhibits DNA repair by irreversible binding to damaged DNA (161). When the damage is too extensive, overactivation of PARP leads to depletion of NAD^+ and ATP. As apoptosis is ATP dependent, inactivation of PARP by cleavage prevents energy depletion and enables completion of apoptosis. Thus, PARP also serves as a marker for the onset of apoptosis due to its cleavage by caspase-3 into smaller, inactive fragments (162, 163).

Studies with PARP-1-deficient cells and animals have revealed other functions of PARP including roles in genomic recombination and instability, DNA replication, regulation of telomere function, and transcriptional regulation (155, 164). The PARP-1 knockout lines exhibit tetraploidy, increased frequency of sister chromatid exchanges, and micronucleus formation, all markers of genomic instability (164, 165). They also show shorter telomere length, which is associated with aging. Tankyrase1, a telomeric PARP localized at the telomeres, inhibits telomeric-repeat binding factor 1 (TRF1), an inhibitor of telomere elongation, through its ADP-ribosylation activity and thus promotes telomere elongation (158, 166, 167). Microarray analysis of the primary PARP-1 gene, Adprt1, null fibroblasts, revealed down-regulation of several genes involved in the regulation of cell cycle progression, mitosis, DNA replication, chromosomal assembly, or processing. On the other hand, some cytoskeletal and extracellular matrix genes implicated in cancer or in normal or premature aging were up-regulated (158). These data (along with more recent publications) suggest a role for PARP in the regulation of gene expression.

Soldatenkov *et al.* (168) report a negative role for PARP in transcription regulation, where the direct interaction of the PARP protein with its own gene promoter results in suppression of transcription. However, in response to DNA damage, PARP catalytic activity is stimulated and automodification of the protein will subsequently prevent its interaction with the promoter. This relieves the PARP-mediated block on the promoter and allows for transcription of PARP and other genes suppressed by PARP. Upon DNA damage repair, the DNA binding activity and thus repressor function of PARP are restored. Although in this instance the catalytic activity of PARP relieves suppression, it can in other instances promote suppression. It has been demonstrated that poly(ADP-ribosyl)ation of transcription factors and members of the basal transcriptional machinery prevents binding of these proteins to DNA, thus interrupting formation of new transcription complexes. However, it is important to note that once transcription preinitiation complexes have been formed, the transcription factors are inaccessible to ADP-ribosylation (169). Thus, poly(ADP-ribosyl)ation prevents binding of transcription factors to DNA, whereas binding to DNA prevents their modification.

Conversely, others have reported a positive role for PARP in the formation of the preinitiation complex (PIC). Meisterernst et al. (159) report that the presence of PARP is required during assembly of RNA polymerase II (RNAPII) and other general transcription factors with a preformed complex consisting of TFIID and possibly TFIIA in vitro. They also propose that PARP activation of the PIC is achieved through recruitment of general transcription factors or conformational changes within the components of the basal transcription machinery and that PARP effects on supercoiled templates are DNA concentration dependent and do not require damaged DNA. Furthermore, PARP-1 has been implicated as a transcriptional coactivator in activation of some genes such as B-MYB, reg, and cTNT (170–172).

PARP-1 has also been reported to interact directly with transcription factors such as NF-κB (168, 173, 174). Studies on PARP-$1^{-/-}$ cells revealed that they are defective in NF-κB-dependent transcriptional activation in response to TNF-α and lipopolysaccharide (LPS), indicative of a functional link between PARP-1 and NF-κB. Furthermore, these cells were unable to induce the expression of NF-κB itself after exposure to TNF-α (175, 176). This would suggest a role of PARP-1 as a signaling molecule, controlling the expression of a transcription factor. Moreover, Hassa et al. (177) report that PARP-1 is required for specific NF-κB transcriptional activation, which occurs through interaction of PARP-1 with both subunits of NF-κB, where each subunit binds to a different PARP-1 domain. They also demonstrate that this interaction is independent of PARP-1 activity and/or DNA binding. However, the published literature regarding PARP-1 and NF-κB interaction is somewhat controversial. Inhibition of the enzymatic activity of PARP has been shown not to affect NF-κB DNA binding, which indicates a lack of a direct role for the catalytic activity of PARP-1 in activating NF-κB (173), reinforcing the proposed interaction by Hassa et al. (177). On the other hand Chang and Alvarez-Gonzalez (178) report that direct PARP-1 interaction with NF-κB inhibits NF-κB from binding to its site and this inhibition is relieved by the auto-poly(ADP-ribosyl)ation of PARP-1. Taken together, the data demonstrate that PARP-1 may be a cofactor in the activation of NF-κB through its direct interaction with the transcription factor. However, the nature of this interaction is not yet defined and the question of whether this interaction could be stimulated or inhibited by PARP activity remains to be answered.

IX. CDP as a Transcriptional Repressor of the CXCL1 Gene

Although CBP and PARP are potential positive coactivators in CXCL1 transcription, the negative regulation of CXCL1 transcription is thus far attributed to CDP (*10*). CDP/Cux/Cut proteins are members of a unique family of homeoproteins that is conserved among higher eukaryotes and contain a Cut homeodomain as well as one or more "Cut repeat" DNA-binding domain(s) (*179–181*). The CDP, Cux, and Cut proteins contain three Cut repeats and a Cut homeodomain and are referred to as the classic CDP/Cux/Cut proteins. CDP, CCAAT displacement protein, was first identified in the testis of the sea urchin *Psammechinus miliaris* and later in mammals and was characterized as a transcriptional repressor. There are six evolutionarily conserved domains in mammalian and *Drosophila* CDP/Cut proteins: a region that forms a coiled-coil structure, three regions of ~70 amino acids, three Cut repeats (CR1, CR2, and CR3), and a Cut homeodomain (HD) (*179, 180*).

The cut repeats are believed to function as specific DNA-binding domains, in addition to the HD. These DNA-binding domains enable CDP/Cut to bind to a wide range of DNA sequences including sequences related to CCAAT, ATCGAT, AT-rich matrix attachment regions, and Sp1-sites. Several other types of sequences diverging from the consensus sequences have been isolated as well by way of PCR-mediated site selection using GST-fusion proteins containing various CDP/Cut DNA-binding domains (*180*). These reports are indicative of the flexibility of the CDP/Cut proteins in choosing their DNA targets.

In vitro studies of CDP/Cut DNA-binding modes have shown several combinations of domains such as CR1CR2, CR1HD, CR2HD, CR3HD, and CR2CR3 implementing this function. However, *in vivo* experiments in mammalian cells display only two modes of DNA binding. The CCAAT-displacement activity involves primarily CR1 and CR2, whereas binding to ATCGAT involves CR3HD. CR1CR2 exhibits fast "on" and "off" rates, whereas CR3HD displays slow on and off rates. Thus, although CR1CR2 binds to DNA rapidly but only transiently, CR3HD makes a stable interaction with the DNA (*179, 180*).

The DNA-binding activity of CDP/Cut is regulated in a cell-cycle-dependent manner, where strong binding is observed in S phase as a result of both an increase in CDP/Cut expression and dephosphorylation of CDP/Cut by the Cdc25A phosphatase (*180*). In G_2 phase, however, its DNA-binding activity is inhibited through cyclinA-Cdk1. Santaguida *et al.* (*181*) report that the cyclinA–Cdk1 complex activated through the action of the Cdk-activating kinase phosphorylates CDP/Cut on two serine residues around the Cut homeodomain and inhibits its DNA binding. Phosphorylation of conserved sites in the Cut repeats by PKC and CKII or pCAF-mediated acetylation of the Cut homeodomain has also been implicated in the inhibition of CDP/Cut DNA-binding activity (*180, 182, 183*). These data indicate that down-modulation of CDP/Cut activity may be very important for cell cycle progression in late S and in G_2.

The exact mechanism of repression by CDP/Cut is unknown, however, two mechanisms have been proposed: passive repression in which CDP/Cut competes with activators for occupancy of binding sites, hence the name "CCAAT displacement protein," and active repression involving direct interaction of CDP/Cut with HDAC1 (*182*). In support of the active repression theory, two active repression domains within the carboxy-terminal domain of CDP/Cut have been identified that are able to interact with HDAC1.

Mammalian CDP/Cut proteins have been shown to repress genes in proliferating precursor cells that are turned on as cells become terminally differentiated and CDP/Cut activity ceases. Thus, CDP/Cut proteins are transcriptional repressors that inhibit terminally differentiated gene expression during early stages of differentiation. CDP/Cut was shown to repress the p21/waf1 gene, a cyclin inhibitor, as well as bind to promoters of various histone genes, which are regulated in a cell-cycle-dependent manner (*184*). In accordance with this, CDP/Cut DNA-binding activity oscillates during the cell cycle and reaches a maximum at the end of G_1 and during the S phase. Interestingly, CDP/Cut regulates histone genes positively, where the peak of expression of these genes coincides with or closely precedes the DNA replication phase (*180, 183*). Thus, the regulatory effect of CDP/Cut on transcription might vary depending on the proteins with which it interacts and that these CDP/Cut-interacting proteins may be cell and tissue specific.

However, there is limited insight into the function of CDP/Cut in mammals *in vivo*. To address this question, a few mutational studies have been conducted in which either the CR1 or the Cut HD was mutated. The phenotypes observed in CR1 homozygous mutant mice include curly vibrissae, wavy hair, and high postnatal lethality among the litters, whereas homozygous Cut HD mutants displayed high postnatal lethality, growth retardation, nearly complete hair loss, and severely reduced male fertility (*182*). Thus absence of fully functional CDP/Cut results only in limited disturbances of normal tissue development.

The repression activity of CDP/Cut has also been observed in non-cell-cycle-related genes such as human cholesterol-7 hydroxylase (CYP7A1), CXCL1, and mouse mammary tumor virus (MMTV) expression. CYP7A1 is expressed only in the liver and is regulated by several liver-enriched transcription factors. CDP/Cut was found to be a major negative regulator of the expression of this gene, where it mediates its repressor activity by displacing two hepatic transcriptional activators, hepatocyte nuclear factor-1 (HNF-1) and C/EBP, from their binding sites within intron 1 of the gene (*185*). The expression of MMTV in the mammary glands has also been shown to be repressed by CDP/Cut, since mutations in the putative CDP/Cut binding domain elevated reporter gene expression by six-fold, whereas overexpression of CDP/Cut was able to suppress reporter gene expression up to 20-fold (*186*). The same phenomenon was observed in the regulation of CXCL1 transcription by Nirodi *et al.* (*105*). The authors showed that CDP mediates its suppressive function by binding to the GCATCGATC sequence within the IUR element in the promoter of the CXCL1 gene. However, further investigation is needed to establish if the repression activity of CDP/Cut is due to its displacement activity or direct interaction with HDAC1 or both. It is possible that CDP/Cut exerts its transcriptional repression simply by displacing the coactivator PARP-1.

X. Reflection

Thus, based on the above discussion, alterations in the interactions of the CXCL1 enhanceosome proteins may be the cause for overexpression of this chemokine as seen in many malignancies and inflammatory diseases. Based on the available data, these alterations may affect the interactions of CDP with the promoter negatively, but have a positive effect on PARP and NF-κB. Thus, overexpression of CXCL1 may be attributed to a concomitant decrease/loss of CDP binding and enhanced PARP binding to the promoter as well as to NF-κB in melanomas. The question then becomes: How does PARP achieve enhanced interactions with the promoter and NF-κB, whereas CDP interaction is at a loss? To answer this question, it is necessary to explore the explicit role of PARP and CDP in the regulation of CXCL1 transcription in "normal" models that can be compared with melanoma models. The significance of these proteins in the regulation of CXCL1 transcription may be investigated through ablation of their expression as well as mutational studies. It is also important to establish the exact sequences that CDP and PARP bind within the IUR element and to look for any differences in these sequences between normal and melanoma models. The effect of PARP catalytic activity on NF-κB and IUR interaction should also be investigated, as it may be of great importance for therapeutic inventions. Thus, by altering the interaction of these proteins in the CXCL1 enhanceosome complex, we may reduce the inflammatory and tumor growth responses in tissues.

Acknowledgments

We are appreciative of the support for the work described herein from the Department of Veterans Affairs for a SRCS award to Ann Richmond and to the NIH for CA56704 and CA34590.

References

1. Baggiolini, M., Dewald, B., and Moser, B. (1994). Interleukin-8 and related chemotactic cytokines—CXC and CC chemokines. *Adv. Immunol.* **55,** 97–179.
2. Clark-Lewis, I., Kim, K. S., Rajarathnam, K., Gong, J. H., Dewald, B., Moser, B., Baggiolini, M., and Sykes, B. D. (1995). Structure-activity relationships of chemokines. *J. Leukoc. Biol.* **57,** 703–711.
3. Clore, G. M., Appella, E., Yamada, M., Matsushima, K., and Gronenborn, A. M. (1990). Three-dimensional structure of interleukin 8 in solution. *Biochemistry* **29,** 1689–1696.
4. Baldwin, E. T., Weber, I. T., St. Charles, R., Xuan, J. C., Appella, E., Yamada, M., Matsushima, K., Edwards, B. F., Clore, G. M., Gronenborn, A. M. *et al.* (1991). Crystal structure of interleukin 8: Symbiosis of NMR and crystallography. *Proc. Natl. Acad. Sci. USA* **88,** 502–506.
5. Strieter, R. M., Polverini, P. J., Kunkel, S. L., Arenberg, D. A., Burdick, M. D., Kasper, J., Dzuiba, J., Van Damme, J., Walz, A., Marriott, D. *et al.* (1995). The functional role of the ELR motif in CXC chemokine-mediated angiogenesis. *J. Biol. Chem.* **270,** 27348–27357.
6. Arenberg, D. A., Kunkel, S. L., Polverini, P. J., Morris, S. B., Burdick, M. D., Glass, M. C., Taub, D. T., Iannettoni, M. D., Whyte, R. I., and Strieter, R. M. (1996). Interferon-gamma-inducible protein 10 (IP-10) is an angiostatic factor that inhibits human non-small cell lung cancer (NSCLC) tumorigenesis and spontaneous metastases. *J. Exp. Med.* **184,** 981–992.
7. Beg, A. A., Finco, T. S., Nantermet, P. V., and Baldwin, A. S., Jr. (1993). Tumor necrosis factor and interleukin-1 lead to phosphorylation and loss of I kappa B alpha: A mechanism for NF-kappa B activation. *Mol. Cell. Biol.* **13,** 3301–3310.
8. Kotenko, S. V., and Pestka, S. (2000). Jak-Stat signal transduction pathway through the eyes of cytokine class II receptor complexes. *Oncogene* **19,** 2557–2565.
9. Richmond, A., Luan, J., Du, J., and Haghnegahdar, H. (1999). The role of ELR+– CXC chemokines in wound healing and melanoma biology. *In* "Chemokines in Disease: Biology and Clinical Research" (C. A. Herbert, Ed.). *In* "Chemokines in Disease: Biology and Clinical Research" p. 191–214. Humana Press Inc, Totowa, NJ.
10. Ahuja, S. K., and Murphy, P. M. (1996). The CXC chemokines growth-regulated oncogene (GRO) alpha, GRObeta, GROgamma, neutrophil-activating peptide-2, and epithelial cell-derived neutrophil-activating peptide-78 are potent agonists for the type B, but not the type A, human interleukin-8 receptor. *J. Biol. Chem.* **271,** 20545–20550.
11. Combadiere, C., Ahuja, S. K., Tiffany, H. L., and Murphy, P. M. (1996). Cloning and functional expression of CC CKR5, a human monocyte CC chemokine receptor selective for MIP-1 (alpha), MIP-1(beta), and RANTES. *J. Leukoc. Biol.* **60,** 147–152.
12. Balkwill, F. (1998). The molecular and cellular biology of the chemokines. *J. Viral Hepat.* **5,** 1–14.
13. Murdoch, C., and Finn, A. (2000). Chemokine receptors and their role in inflammation and infectious diseases. *Blood* **95,** 3032–3043.
14. Olson, T. S., and Ley, K. (2002). Chemokines and chemokine receptors in leukocyte trafficking. *Am. J. Physiol. Regul. Integr. Comp. Physiol.* **283,** R7–28.
15. Neote, K., DiGregorio, D., Mak, J. Y., Horuk, R., and Schall, T. J. (1993). Molecular cloning, functional expression, and signaling characteristics of a C-C chemokine receptor. *Cell* **72,** 415–425.
16. Luster, A. D. (1998). Chemokines—chemotactic cytokines that mediate inflammation. *N. Engl. J. Med.* **338,** 436–445.
17. Murphy, P. M., Baggiolini, M., Charo, I. F., Hebert, C. A., Horuk, R., Matsushima, K., Miller, L. H., Oppenheim, J. J., and Power, C. A. (2000). International union of pharmacology. XXII. Nomenclature for chemokine receptors. *Pharmacol. Rev.* **52,** 145–176.

18. Ali, H., Richardson, R. M., Haribabu, B., and Snyderman, R. (1999). Chemoattractant receptor cross-desensitization. *J. Biol. Chem.* **274**, 6027–6030.
19. Le, Y., Li, B., Gong, W., Shen, W., Hu, J., Dunlop, N. M., Oppenheim, J. J., and Wang, J. M. (2000). Novel pathophysiological role of classical chemotactic peptide receptors and their communications with chemokine receptors. *Immunol. Rev.* **177**, 185–194.
20. Imai, T., Hieshima, K., Haskell, C., Baba, M., Nagira, M., Nishimura, M., Kakizaki, M., Takagi, S., Nomiyama, H., Schall, T. J., and Yoshie, O. (1997). Identification and molecular characterization of fractalkine receptor CX3CR1, which mediates both leukocyte migration and adhesion. *Cell* **91**, 521–530.
21. Yoshida, T., Imai, T., Kakizaki, M., Nishimura, M., Takagi, S., and Yoshie, O. (1998). Identification of single C motif-1/lymphotactin receptor XCR1. *J. Biol. Chem.* **273**, 16551–16554.
22. Szabo, M. C., Soo, K. S., Zlotnik, A., and Schall, T. J. (1995). Chemokine class differences in binding to the Duffy antigen-erythrocyte chemokine receptor. *J. Biol. Chem.* **270**, 25348–25351.
23. Devalaraja, R. M., Nanney, L. B., Du, J., Qian, Q., Yu, Y., Devalaraja, M. N., and Richmond, A. (2000). Delayed wound healing in CXCR2 knockout mice. *J. Invest Dermatol.* **115**, 234–244.
24. Haghnegahdar, H., Du, J., Wang, D., Strieter, R. M., Burdick, M. D., Nanney, L. B., Cardwell, N., Luan, J., Shattuck-Brandt, R., and Richmond, A. (2000). The tumorigenic and angiogenic effects of MGSA/GRO proteins in melanoma. *J. Leukoc. Biol.* **67**, 53–62.
25. Owen, J. D., Strieter, R., Burdick, M., Haghnegahdar, H., Nanney, L., Shattuck-Brandt, R., and Richmond, A. (1997). Enhanced tumor-forming capacity for immortalized melanocytes expressing melanoma growth stimulatory activity/growth-regulated cytokine beta and gamma proteins. *Int. J. Cancer.* **73**, 94–103.
26. Bokoch, G. M. (1995). Chemoattractant signaling and leukocyte activation. *Blood* **86**, 1649–1660.
27. Wu, D., LaRosa, G. J., and Simon, M. I. (1993). G protein-coupled signal transduction pathways for interleukin-8. *Science* **261**, 101–103.
28. Bokoch, G. M., Vlahos, C. J., Wang, Y., Knaus, U. G., and Traynor-Kaplan, A. E. (1996). Rac GTPase interacts specifically with phosphatidylinositol 3-kinase. *Biochem. J.* **315**, 775–779.
29. Turner, S. J., Domin, J., Waterfield, M. D., Ward, S. G., and Westwick, J. (1998). The CC chemokine monocyte chemotactic peptide-1 activates both the class I p85/p110 phosphatidylinositol 3-kinase and the class II PI3K-C2alpha. *J. Biol. Chem.* **273**, 25987–25995.
30. Huang, R., Lian, J. P., Robinson, D., and Badwey, J. A. (1998). Neutrophils stimulated with a variety of chemoattractants exhibit rapid activation of p21-activated kinases (Paks): Separate signals are required for activation and inactivation of paks. *Mol. Cell. Biol.* **18**, 7130–7138.
31. Mellado, M., Rodriguez-Frade, J. M., Aragay, A., del Real, G., Martin, A. M., Vila-Coro, A. J., Serrano, A., Mayor, F., Jr., and Martinez-A, C. (1998). The chemokine monocyte chemotactic protein 1 triggers Janus kinase 2 activation and tyrosine phosphorylation of the CCR2B receptor. *J. Immunol.* **161**, 805–813.
32. Brill, A., Hershkoviz, R., Vaday, G. G., Chowers, Y., and Lider, O. (2001). Augmentation of RANTES-induced extracellular signal-regulated kinase mediated signaling and T cell adhesion by elastase-treated fibronectin. *J. Immunol.* **166**, 7121–7127.
33. Wain, J. H., Kirby, J. A., and Ali, S. (2002). Leucocyte chemotaxis: Examination of mitogen-activated protein kinase and phosphoinositide 3-kinase activation by monocyte chemoattractant proteins-1, -2, -3 and -4. *Clin. Exp. Immunol.* **127**, 436–444.
34. Aramori, I., Ferguson, S. S., Bieniasz, P. D., Zhang, J., Cullen, B., and Cullen, M. G. (1997). Molecular mechanism of desensitization of the chemokine receptor CCR-5: Receptor signaling and internalization are dissociable from its role as an HIV-1 co-receptor. *EMBO J.* **16**, 4606–4616.

35. Richmond, A., Mueller, S., White, J. R., and Schraw, W. (1997). C-X-C chemokine receptor desensitization mediated through ligand-enhanced receptor phosphorylation on serine residues. *Methods Enzymol.* **288,** 3–15.
36. Fan, G. H., Yang, W., Sai, J., and Richmond, A. (2002). Hsc/Hsp70 interacting protein (hip) associates with CXCR2 and regulates the receptor signaling and trafficking. *J. Biol. Chem.* **277,** 6590–6597.
37. Imhof, B. A. (1995). Leukocyte migration and adhesion. *Adv. Immunol.* **58,** 345–416.
38. Brandt, E., Petersen, F., Ludwig, A., Ehlert, J. E., Bock, L., and Flad, H. D. (2000). The beta-thromboglobulins and platelet factor 4: Blood platelet-derived CXC chemokines with divergent roles in early neutrophil regulation. *J. Leukoc. Biol.* **67,** 471–478.
39. Engelhardt, E., Toksoy, A., Goebeler, M., Debus, S., Brocker, E. B., and Gillitzer, R. (1998). Chemokines IL-8, GROalpha, MCP-1, IP-10, and Mig are sequentially and differentially expressed during phase-specific infiltration of leukocyte subsets in human wound healing. *Am. J. Pathol.* **153,** 1849–1860.
40. Gillitzer, R., and Goebeler, M. (2001). Chemokines in cutaneous wound healing. *J. Leukoc. Biol.* **69,** 513–521.
41. Gibran, N. S., Ferguson, M., Heimbach, D. M., and Isik, F. F. (1997). Monocyte chemoattractant protein-1 mRNA expression in the human burn wound. *J. Surg. Res.* **70,** 1–6.
42. Trautmann, A., Toksoy, A., Engelhardt, E., Brocker, E. B., and Gillitzer, R. (2000). Mast cell involvement in normal human skin wound healing: Expression of monocyte chemoattractant protein-1 is correlated with recruitment of mast cells which synthesize interleukin-4 in vivo. *J. Pathol.* **190,** 100–1066.
43. Goede, V., Brogelli, L., Ziche, M., and Augustin, H. G. (1999). Induction of inflammatory angiogenesis by monocyte chemoattractant protein-1. *Int. J. Cancer* **82,** 765–770.
44. Weber, K. S., Nelson, P. J., Grone, H. J., and Weber, C. (1999). Expression of CCR2 by endothelial cells: Implications for MCP-1 mediated wound injury repair and in vivo inflammatory activation of endothelium. *Arterioscler. Thromb. Vasc. Biol.* **19,** 2085–2093.
45. Leonard, E. J., Skeel, A., Yoshimura, T., and Rankin, J. (1993). Secretion of monocyte chemoattractant protein-1 (MCP-1) by human mononuclear phagocytes. *Adv. Exp. Med. Biol.* **351,** 55–64.
46. Trautmann, A., Krohne, G., Brocker, E. B., and Klein, C. E. (1998). Human mast cells augment fibroblast proliferation by heterotypic cell-cell adhesion and action of IL-4. *J. Immunol.* **160,** 5053–5057.
47. Gerard, C., and Rollins, B. J. (2001). Chemokines and disease. *Nat. Immunol.* **2,** 108–115.
48. Ransohoff, R. M. (1997). Chemokines and chemokine receptors in model neurological pathologies: Molecular and immunocytochemical approaches. *Methods Enzymol.* **287,** 319–348.
49. Karpus, W. J., and Kennedy, K. J. (1997). MIP-1 and MCP-1 differentially regulate acute and relapsing autoimmune encephalomyelitis as well as Th1/Th2 lymphocyte differentiation. *J. Leukoc. Biol.* **62,** 681–687.
50. Glabinski, A. R., Tani, M., Tuohy, V. K., Tuthill, R. J., and Ransohoff, R. M. (1995). Central nervous system chemokine mRNA accumulation follows initial leukocyte entry at the onset of acute murine experimental autoimmune encephalomyelitis. *Brain Behav. Immunol.* **9,** 315–330.
51. Karpus, W. J., Lukacs, N. W., McRae, B. L., Strieter, R. M., Kunkel, S. L., and Miller, S. D. (1995). An important role for the chemokine macrophage inflammatory protein-1 in the pathogenesis of the T cell-mediated autoimmune disease, experimental autoimmune encephalomyelitis. *J. Immunol.* **155,** 5003–5010.
52. Godiska, R., Chantry, D., Dietsch, G. N., and Gray, P. W. (1995). Chemokine expression in murine experimental allergic encephalomyelitis. *J. Neuroimmunol.* **58,** 167–176.

53. Ransohoff, R. M., Hamilton, T. A., Tani, M., Stoler, M. H., Shick, H. E., Major, J. A., Estes, M. L., Thomas, D. M., and Tuohy, V. K. (1993). Astrocyte expression of mRNA encoding cytokines IP-10 and JE/MCP-1 in experimental autoimmune encephalomyelitis. *FASEB J.* **7**, 592–600.
54. Simpson, J., Rezaie, P., Newcombe, J., Cuzner, M. L., Male, D., and Woodroofe, M. N. (2000). Expression of the chemokine receptors CCR2, CCR3, and CCR5 in multiple sclerosis central nervous system tissue. *J. Neuroimmunol.* **108**, 192–200.
55. Standiford, T. J., Strieter, R. M., Lukacs, N. W., and Kunkel, S. L. (1995). Neutralization of IL-10 increases lethality in endotoxemia. Cooperative effects of macrophage inflammatory protein-2 and tumor necrosis factor. *J. Immunol.* **155**, 2222–2229.
56. Standiford, T. J., Kunkel, S. L., Lukacs, N. W., Greenberger, M. J., Danforth, J. M., Kunkel, R. G., and Strieter, R. M. (1995). Macrophage inflammatory protein-1 alpha mediates lung leukocyte recruitment, lung capillary leak, and early mortality in murine endotoxemia. *J. Immunol.* **155**, 1515–1524.
57. VanOtteren, G. M., Strieter, R. M., Kunkel, S. L., Paine, R., 3rd, Greenberger, M. J., Danforth, J. M., Burdick, M. D., and Standiford, T. J. (1995). Compartmentalized expression of RANTES in a murine model of endotoxemia. *J. Immunol.* **154**, 1900–1908.
58. Zisman, D. A., Kunkel, S. L., Strieter, R. M., Tsai, W. C., Bucknell, K., Wilkowski, J., and Standiford, T. J. (1997). MCP-1 protects mice in lethal endotoxemia. *J. Clin. Invest.* **99**, 2832–2836.
59. Choe, H., Farzan, M., Sun, Y., Sullivan, N., Rollins, B., Ponath, P. D., Wu, L., Mackay, C. R., LaRosa, G., Newman, W., Gerard, N., Gerard, C., and Sodroski, J. (1996). The beta-chemokine receptors CCR3 and CCR5 facilitate infection by primary HIV-1 isolates. *Cell* **85**, 1135–1148.
60. Feng, Y., Broder, C. C., Kennedy, P. E., and Berger, E. A. (1996). HIV-1 entry cofactor: Functional cDNA cloning of a seven-transmembrane, G protein-coupled receptor. *Science* **272**, 872–877.
61. Gao, J. L., and Murphy, P. M. (1994). Human cytomegalovirus open reading frame US28 encodes a functional beta chemokine receptor. *J. Biol. Chem.* **269**, 28539–28542.
62. Lukacs, N. W., and Kunkel, S. L. (1998). Chemokines and their role in disease. *Int. J. Clin. Lab. Res.* **28**, 91–95.
63. Melter, M., McMahon, G., Fang, J., Ganz, P., and Briscoe, D. M. (1999). Current understanding of chemokine involvement in allograft transplantation. *Pediatr. Transplant.* **3**, 10–21.
64. Yun, J. J., Fischbein, M. P., Laks, H., Fishbein, M. C., Espejo, M. L., Ebrahimi, K., Irie, Y., Berliner, J., and Ardehali, A. (2000). Early and late chemokine production correlates with cellular recruitment in cardiac allograft vasculopathy. *Transplantation.* **69**, 2515–2524.
65. Belperio, J. A., Burdick, M. D., Keane, M. P., Xue, Y. Y., Lynch, J. P., 3rd, Daugherty, B. L., Kunkel, S. L., and Strieter, R. M. (2000). The role of the CC chemokine, Rantes, in acute lung allograft rejection. *J. Immunol.* **165**, 461–472.
66. Kapoor, A., Morita, K., Engeman, T. M., Vapnek, E. M., Hobart, M., Novick, A. C., and Fairchild, R. L. (2000). Intragraft expression of chemokine gene occurs early during acute rejection of allogeneic cardiac grafts. *Transplant. Proc.* **32**, 793–795.
67. Schadendorf, D., Fichtner, I., Makki, A., Alijagic, S., Kupper, M., Mrowietz, U., and Henz, B. M. (1996). Metastatic potential of human melanoma cells in nude mice—characterisation of phenotype, cytokine secretion and tumour-associated antigens. *Br. J. Cancer* **74**, 194–199.
68. Singh, R. K., Varney, M. L., Bucana, C. D., and Johansson, S. L. (1999). Expression of interleukin-8 in primary and metastatic malignant melanoma of the skin. *Melanoma Res.* **9**, 383–387.

69. Scheibenbogen, C., Mohler, T., Haefele, J., Hunstein, W., and Keilholz, U. (1995). Serum interleukin-8 (IL-8) is elevated in patients with metastatic melanoma and correlates with tumour load. *Melanoma Res.* **5,** 179–181.
70. Takamori, H., Oades, Z. G., Hoch, O. C., Burger, M., and Schraufstatter, I. U. (2000). Autocrine growth effect of IL-8 and GROalpha on a human pancreatic cancer cell line, Capan-1. *Pancreas* **21,** 52–56.
71. Richards, B. L., Eisma, R. J., Spiro, J. D., Lindquist, R. L., and Kreutzer, D. L. (1997). Coexpression of interleukin-8 receptors in head and neck squamous cell carcinoma. *Am. J. Surg.* **174,** 507–512.
72. Olbina, G., Cieslak, D., Ruzdijic, S., Esler, C., An, Z., Wang, X., Hoffman, R., Seifert, W., and Pietrzkowski, Z. (1996). Reversible inhibition of IL-8 receptor B mRNA expression and proliferation in non-small cell lung cancer by antisense oligonucleotides. *Anticancer Res.* **16,** 3525–3530.
73. Wakabayashi, H., Cavanaugh, P. G., and Nicolson, G. L. (1995). Purification and identification of mouse lung microvessel endothelial cell-derived chemoattractant for lung-metastasizing murine RAW117 large-cell lymphoma cells: Identification as mouse monocyte chemotactic protein 1. *Cancer Res.* **55,** 4458–4464.
74. Youngs, S. J., Ali, S. A., Taub, D. D., and Rees, R. C. (1997). Chemokines induce migrational responses in human breast carcinoma cell lines. *Int. J. Cancer* **71,** 257–266.
75. Wang, J. M., Deng, X., Gong, W., and Su, S. (1998). Chemokines and their role in tumor growth and metastasis. *J. Immunol. Methods.* **220,** 1–17.
76. Muller, A., Homey, B., Soto, H., Ge, N., Catron, D., Buchanan, M. E., McClanahan, T., Murphy, E., Yuan, W., Wagner, S. N., Barrera, J. L., Mohar, A., Verastegui, E., and Zlotnik, A. (2001). Involvement of chemokine receptors in breast cancer metastasis. *Nature* **410,** 50–56.
77. Cogswell, J. P., Godlevski, M. M., Wisely, G. B., Clay, W. C., Leesnitzer, L. M., Ways, J. P., and Gray, J. G. (1994). NF-kappa B regulates IL-1 beta transcription through a consensus NF-kappa B binding site and a nonconsensus CRE-like site. *J. Immunol.* **153,** 712–723.
78. Roebuck, K. A., Carpenter, L. R., Lakshminarayanan, V., Page, S. M., Moy, J. N., and Thomas, L. L. (1999). Stimulus-specific regulation of chemokine expression involves differential activation of the redox-responsive transcription factors AP-1 and NF-kappaB. *J. Leukoc. Biol.* **65,** 291–298.
79. Stylianou, E., Nie, M., Ueda, A., and Zhao, L. (1999). c-Rel and p65 trans-activate the monocyte chemoattractant protein-1 gene in interleukin-1 stimulated mesangial cells. *Kidney Int.* **56,** 873–882.
80. Shi, M. M., Chong, I., Godleski, J. J., and Paulauskis, J. D. (1999). Regulation of macrophage inflammatory protein-2 gene expression by oxidative stress in rat alveolar macrophages. *Immunology* **97,** 309–315.
81. Takaya, H., Andoh, A., Shimada, M., Hata, K., Fujiyama, Y., and Bamba, T. (2000). The expression of chemokine genes correlates with nuclear factor-kappaB activation in human pancreatic cancer cell lines. *Pancreas* **21,** 32–40.
82. Merola, M., Blanchard, B., and Tovey, M. G. (1996). The kappa B enhancer of the human interleukin-6 promoter is necessary and sufficient to confer an IL-1 beta and TNF-alpha response in transfected human cell lines: Requirement for members of the C/EBP family for activity. *J. Interferon Cytokine Res.* **16,** 783–798.
83. Garcia, G. E., Xia, Y., Chen, S., Wang, Y., Ye, R. D., Harrison, J. K., Bacon, K. B., Zerwes, H. G., and Feng, L. (2000). NF-kappaB-dependent fractalkine induction in rat aortic endothelial cells stimulated by IL-1beta, TNF-alpha, and LPS. *J. Leukoc. Biol.* **67,** 577–584.
84. Sugita, S., Kohno, T., Yamamoto, K., Imaizumi, Y., Nakajima, H., Ishimaru, T., and Matsuyama, T. (2002). Induction of macrophage-inflammatory protein-3alpha gene expression by TNF-dependent NF-kappaB activation. *J. Immunol.* **168,** 5621–5628.

85. Mastronarde, J. G., He, B., Monick, M. M., Mukaida, N., Matsushima, K., and Hunninghake, G. W. (1996). Induction of interleukin (IL)-8 gene expression by respiratory syncytial virus involves activation of nuclear factor (NF)-kappa B and NF-IL-6. *J. Infect. Dis.* **174**, 262–267.
86. Akira, S., and Kishimoto, T. (1992). IL-6 and NF-IL6 in acute-phase response and viral infection. *Immunol. Rev.* **127**, 25–50.
87. Casola, A., Garofalo, R. P., Haeberle, H., Elliott, T. F., Lin, R., Jamaluddin, M., and Brasier, A. R. (2001). Multiple cis regulatory elements control RANTES promoter activity in alveolar epithelial cells infected with respiratory syncytial virus. *J. Virol.* **75**, 6428–6439.
88. Lee, A. H., Hong, J. H., and Seo, Y. S. (2000). Tumour necrosis factor-alpha and interferon-gamma synergistically activate the RANTES promoter through nuclear factor kappaB and interferon regulatory factor 1 (IRF-1) transcription factors. *Biochem. J.* **350**, 131–138.
89. Genin, P., Algarte, M., Roof, P., Lin, R., and Hiscott, J. (2000). Regulation of RANTES chemokine gene expression requires cooperativity between NF-kappa B and IFN-regulatory factor transcription factors. *J. Immunol.* **164**, 5352–5361.
90. Wood, L. D., Farmer, A. A., and Richmond, A. (1995). HMGI(Y) and Sp1 in addition to NF-kappa B regulate transcription of the MGSA/GRO alpha gene. *Nucleic Acids Res.* **23**, 4210–4219.
91. Ueda, A., Okuda, K., Ohno, S., Shirai, A., Igarashi, T., Matsunaga, K., Fukushima, J., Kawamoto, S., Ishigatsubo, Y., and Okubo, T. (1994). NF-kappa B and Sp1 regulate transcription of the human monocyte chemoattractant protein-1 gene. *J. Immunol.* **153**, 2052–2063.
92. Schook, L. B., Albrecht, H., Gallay, P., and Jongeneel, C. V. (1994). Cytokine regulation of TNF-alpha mRNA and protein production by unprimed macrophages from C57B1/6 and NZW mice. *J. Leukoc. Biol.* **56**, 514–520.
93. Lin, G., Pearson, A. E., Scamurra, R. W., Zhou, Y., Baarsch, M. J., Weiss, D. J., and Murtaugh, M. P. (1994). Regulation of interleukin-8 expression in porcine alveolar macrophages by bacterial lipopolysaccharide. *J. Biol. Chem.* **269**, 77–85.
94. Villarete, L. H., and Remick, D. G. (1996). Transcriptional and post-transcriptional regulation of interleukin-8. *Am. J. Pathol.* **149**, 1685–1693.
95. Wolf, J. S., Chen, Z., Dong, G., Sunwoo, J. B., Bancroft, C. C., Capo, D. E., Yeh, N. T., Mukaida, N., and Van Waes, C. (2001). IL (interleukin)-1alpha promotes nuclear factor-kappaB and AP-1-induced IL-8 expression, cell survival, and proliferation in head and neck squamous cell carcinomas. *Clin. Cancer Res.* **7**, 1812–1820.
96. Wu, G. D., Lai, E. J., Huang, N., and Wen, X. (1997). Oct-1 and CCAAT/enhancer-binding protein (C/EBP) bind to overlapping elements within the interleukin-8 promoter. The role of Oct-1 as a transcriptional repressor. *J. Biol. Chem.* **272**, 2396–2403.
97. Roux, P., Alfieri, C., Hrimech, M., Cohen, E. A., and Tanner, J. E. (2000). Activation of transcription factors NF-kappa B and NF-IL-6 by human immunodeficiency virus type 1 protein R (Vpr) induces interleukin-8 expression. *J. Virol.* **74**, 4658–4665.
98. Luan, J., Shattuck-Brandt, R., Haghnegahdar, H., Owen, J. D., Strieter, R., Burdick, M., Nirodi, C., Beauchamp, D., Johnson, K. N., and Richmond, A. (1997). Mechanism and biological significance of constitutive expression of MGSA/GRO chemokines in malignant melanoma tumor progression. *J. Leukoc. Biol.* **62**, 588–597.
99. Wood, L. D., and Richmond, A. (1995). Constitutive and cytokine-induced expression of the melanoma growth stimulatory activity/GRO gene requires both NF-B and novel constitutive factors. *J. Biol. Chem.* **270**, 30619–30626.
100. Mukaida, N., Okamoto, S., Ishikawa, Y., and Matsushima, K. (1994). Molecular mechanism of interleukin-8 gene expression. *J. Leukoc. Biol.* **56**, 554–558.

101. Roger, T., Out, T., Mukaida, N., Matsushima, K., Jansen, H., and Lutter, R. (1998). Enhanced AP-1 and NF-kappaB activities and stability of interleukin 8 (IL-8) transcripts are implicated in IL-8 mRNA superinduction in lung epithelial H292 cells. *Biochem. J.* **330,** 429–435.
102. Matsusaka, T., Fujikawa, K., Nishio, Y., Mukaida, N., Matsushima, K., Kishimoto, T., and Akira, S. (1993). Transcription factors NF-IL6 and NF-kappa B synergistically activate transcription of the inflammatory cytokines, interleukin 6 and interleukin 8. *Proc. Natl. Acad. Sci. USA* **90,** 10193–10197.
103. Nourbakhsh, M., Kalble, S., Dorrie, A., Hauser, H., Resch, K., and Kracht, M. (2001). The NF-kappa b repressing factor is involved in basal repression and interleukin (IL)-1-induced activation of IL-8 transcription by binding to a conserved NF-kappa b-flanking sequence element. *J. Biol. Chem.* **276,** 4501–4508.
104. Vanden Berghe, W., De Bosscher, K., Boone, E., Plaisance, S., and Haegeman, G. (1999). The nuclear factor-kappaB engages CBP/p300 and histone acetyltransferase activity for transcriptional activation of the interleukin-6 gene promoter. *J. Biol. Chem.* **274,** 32091–32098.
105. Nirodi, C., Hart, J., Dhawan, P., Nepveu, A., and Richmond, A. (2001). The role of CDP in the negative regulation of CXCL1 gene expression. *J. Biol. Chem.* **276,** 26122–26131.
106. Nirodi, C., NagDas, S., Gygi, S. P., Olson, G., Aebersold, R., and Richmond, A. (2001). A role for poly(ADP-ribose) polymerase in the transcriptional regulation of the melanoma growth stimulatory activity (CXCL1) gene expression. *J. Biol. Chem.* **276,** 9366–9374.
107. Abraham, E. (2000). NF-κB activation. *Crit. Care Med.* **28,** N100–N104.
108. Makarov, S. S. (2000). NF-kappaB as a therapeutic target in chronic inflammation: Recent advances. *Mol. Med. Today.* **6,** 441–448.
109. Huang, S., DeGuzman, A., Bucana, C. D., and Fidler, I. J. (2000). Nuclear factor-kappaB activity correlates with growth, angiogenesis, and metastasis of human melanoma cells in nude mice. *Clin. Cancer Res.* **6,** 2573–2581.
110. Beg, A. A., Ruben, S. M., Scheinman, R. I., Haskill, S., Rosen, C. A., and Baldwin, A. S., Jr. (1992). I kappa B interacts with the nuclear localization sequences of the subunits of NF-kappa B: A mechanism for cytoplasmic retention. *Genes Dev.* **6,** 1899–1913.
111. DiDonato, J., Mercurio, F., Rosette, C., Wu-Li, J., Suyang, H., Ghosh, S., and Karin, M. (1996). Mapping of the inducible IkappaB phosphorylation sites that signal its ubiquitination and degradation. *Mol. Cell. Biol.* **16,** 1295–1304.
112. Nolan, G. P., Fujita, T., Bhatia, K., Huppi, C., Liou, H. C., Scott, M. L., and Baltimore, D. (1993). The bcl-3 proto-oncogene encodes a nuclear I kappa B-like molecule that preferentially interacts with NF-kappa B p50 and p52 in a phosphorylation-dependent manner. *Mol. Cell. Biol.* **13,** 3557–3566.
113. Zandi, E., Rothwarf, D. M., Delhase, M., Hayakawa, M., and Karin, M. (1997). The IkappaB kinase complex (IKK) contains two kinase subunits, IKKalpha and IKKbeta, necessary for IkappaB phosphorylation and NF-kappaB activation. *Cell* **91,** 243–252.
114. DiDonato, J. A., Hayakawa, M., Rothwarf, D. M., Zandi, E., and Karin, M. (1997). A cytokine-responsive IkappaB kinase that activates the transcription factor NF-kappaB. *Nature* **388,** 548–554.
115. Ashburner, B. P., Westerheide, S. D., and Baldwin, A. S., Jr. (2001). The p65 (RelA) subunit of NF-kappaB interacts with the histone deacetylase (HDAC) corepressors HDAC1 and HDAC2 to negatively regulate gene expression. *Mol. Cell. Biol.* **21,** 7065–7077.
116. Pahl, H. L. (1999). Activators and target genes of Rel/NF-kappaB transcription factors. *Oncogene* **18,** 6853–6866.
117. Bakker, T. R., Reed, D., Renno, T., and Jongeneel, C. V. (1999). Efficient adenoviral transfer of NF-kappaB inhibitor sensitizes melanoma to tumor necrosis factor-mediated apoptosis. *Int. J. Cancer* **80,** 320–323.

118. Wang, D., Yang, W., Du, J., Devalaraja, M. N., Liang, P., Matsumoto, K., Tsubakimoto, K., Endo, T., and Richmond, A. (2000). MGSA/GRO-mediated melanocyte transformation involves induction of Ras expression. *Oncogene* **19**, 4647–4659.
119. Zhong, H., Voll, R. E., and Ghosh, S. (1998). Phosphorylation of NF-kappa B p65 by PKA stimulates transcriptional activity by promoting a novel bivalent interaction with the coactivator CBP/p300. *Mol. Cell* **1**, 661–671.
120. Wang, D., Westerheide, S. D., Hanson, J. L., and Baldwin, A. S., Jr. (2000). Tumor necrosis factor alpha-induced phosphorylation of RelA/p65 on Ser529 is controlled by casein kinase II. *J. Biol. Chem.* **275**, 32592–32597.
121. Berra, E., Diaz-Meco, M. T., Lozano, J., Frutos, S., Municio, M. M., Sanchez, P., Sanz, L., and Moscat, J. (1995). Evidence for a role of MEK and MAPK during signal transduction by protein kinase C zeta. *EMBO J.* **14**, 6157–6163.
122. Vanden Berghe, W., Plaisance, S., Boone, E., De Bosscher, K., Schmitz, M. L., Fiers, W., and Haegeman, G. (1998). p38 and extracellular signal-regulated kinase mitogen-activated protein kinase pathways are required for nuclear factor-kappaB p65 transactivation mediated by tumor necrosis factor. *J. Biol. Chem.* **273**, 3285–3290.
123. Shiojima, I., and Walsh, K. (2002). Role of Akt signaling in vascular homeostasis and angiogenesis. *Circ. Res.* **90**, 1243–1250.
124. Datta, S. R., Brunet, A., and Greenberg, M. E. (1999). Cellular survival: A play in three Akts. *Genes Dev.* **13**, 2905–2927.
125. Ozes, O. N., Mayo, L. D., Gustin, J. A., Pfeffer, S. R., Pfeffer, L. M., and Donner, D. B. (1999). NF-kappaB activation by tumour necrosis factor requires the Akt serine-threonine kinase. *Nature* **401**, 82–85.
126. Romashkova, J. A., and Makarov, S. S. (1999). NF-kappaB is a target of AKT in anti-apoptotic PDGF signalling. *Nature* **401**, 86–90.
127. Madrid, L. V., Mayo, M. W., Reuther, J. Y., and Baldwin, A. S., Jr. (2001). Akt stimulates the transactivation potential of the RelA/p65 Subunit of NF-kappa B through utilization of the Ikappa B kinase and activation of the mitogen-activated protein kinase p38. *J. Biol. Chem.* **276**, 18934–18940.
128. Li, J., Simpson, L., Takahashi, M., Miliaresis, C., Myers, M. P., Tonks, N., and Parsons, R. (1998). The PTEN/MMAC1 tumor suppressor induces cell death that is rescued by the AKT/protein kinase B oncogene. *Cancer Res.* **58**, 5667–5672.
129. Guldberg, P.Thor Straten, P., Birck, A., Ahrenkiel, V., Kirkin, A. F., and Zeuthen, J. (1997). Disruption of the MMAC1/PTEN gene by deletion or mutation is a frequent event in malignant melanoma. *Cancer Res.* **57**, 3660–3663.
130. Gomez, J., Martinez-A, C., Fernandez, B., Garcia, A., and Rebollo, A. (1996). Critical role of Ras in the proliferation and prevention of apoptosis mediated by IL-2. *J. Immunol.* **157**, 2272–2281.
131. Joneson, T., and Bar-Sagi, D. (1999). Suppression of Ras-induced apoptosis by the Rac GTPase. *Mol. Cell. Biol.* **19**, 5892–5901.
132. Crespo, P., and Leon, J. (2000). Ras proteins in the control of the cell cycle and cell differentiation. *Cell. Mol. Life Sci.* **57**, 1613–1636.
133. Khosravi-Far, R., Solski, P. A., Clark, G. J., Kinch, M. S., and Der, C. J. (1995). Activation of Rac1, RhoA, and mitogen-activated protein kinases is required for Ras transformation. *Mol. Cell. Biol.* **15**, 6443–6453.
134. Norris, J. L., and Baldwin, A. S., Jr. (1999). Oncogenic Ras enhances NF-kappaB transcriptional activity through Raf-dependent and Raf-independent mitogen-activated protein kinase signaling pathways. *J. Biol. Chem.* **274**, 13841–13846.
135. Malinin, N. L., Boldin, M. P., Kovalenko, A. V., and Wallach, D. (1997). MAP3K-related kinase involved in NF-κB induction by TNF, CD95 and IL-1. *Nature* **385**, 540–544.

136. Woronicz, J. D., Gao, X., Cao, Z., Rothe, M., and Goeddel, D. V. (1997). IκB kinase-beta: NF-κB activation and complex formation with IκB kinase-α and NIK. *Science* **278**, 866–869.
137. Delhase, M., Hayakawa, M., Chen, Y., and Karin, M. (1999). Positive and negative regulation of IkappaB kinase activity through IKKbeta subunit phosphorylation. *Science* **284**, 309–313.
138. Dhawan, P., and Richmond, A. (2002). A novel NF-κB-inducing kinase-MAPK signaling pathway up-regulates NF-κB activity in melanoma cells. *J. Biol. Chem.* **277**, 7920–7928.
139. Gerritsen, M. E., Williams, A. J., Neish, A. S., Moore, S., Shi, Y., and Collins, T. (1997). CREB-binding protein/p300 are transcriptional coactivators of p65. *Proc. Natl. Acad. Sci. USA* **94**, 2927–2932.
140. Yao, T. P., Oh, S. P., Fuchs, M., Zhou, N. D., Ch'ng, L. E., Newsome, D., Bronson, R. T., Li, E., Livingston, D. M., and Eckner, R. (1998). Gene dosage-dependent embryonic development and proliferation defects in mice lacking the transcriptional integrator p300. *Cell* **93**, 361–372.
141. Michael, L. F., Asahara, H., Shulman, A. I., Kraus, W. L., and Montminy, M. (2000). The phosphorylation status of a cyclic AMP-responsive activator is modulated via a chromatin-dependent mechanism. *Mol. Cell. Biol.* **20**, 1596–1603.
142. Yuan, L. W., and Gambee, J. E. (2001). Histone acetylation by p300 is involved in CREB-mediated transcription on chromatin. *Biochim. Biophys. Acta.* **1541**, 161–169.
143. Lodish, H., Berk, A., Zipursky, L. S., Matsudaira, P., Baltimore, D., and Darnell, J. E. (1999). Role of deacetylation and hyperacetylation of histone N-terminal tails in yeast transcription control. In "Molecular Cell Biology," 4th ed. p. 384–390. Freeman W. H. & Co., New York.
144. Nakajima, T., Uchida, C., Anderson, S. F., Lee, C. G., Hurwitz, J., Parvin, J. D., and Montminy, M. (1997). RNA helicase A mediates association of CBP with RNA polymerase II. *Cell* **90**, 1107–1112.
145. Bannister, A. J., Oehler, T., Wilhelm, D., Angel, P., and Kouzarides, T. (1995). Stimulation of c-Jun activity by CBP: c-Jun residues Ser63/73 are required for CBP induced stimulation in vivo and CBP binding in vitro. *Oncogene* **11**, 2509–2514.
146. Chan, H. M., and La Thangue, N. B. (2001). p300/CBP proteins: HATs for transcriptional bridges and scaffolds. *J. Cell Sci.* **114**, 2363–2373.
147. Shenkar, R., Yum, H. K., Arcaroli, J., Kupfner, J., and Abraham, E. (2001). Interactions between CBP, NF-kappaB, and CREB in the lungs after hemorrhage and endotoxemia. *Am. J. Physiol. Lung Cell. Mol. Physiol.* **281**, L418–426.
148. Imhof, A., Yang, X. J., Ogryzko, V. V., Nakatani, Y., Wolfe, A. P., and Ge, H. (1997). Acetylation of general transcription factors by histone acetyltransferases. *Curr. Biol.* **7**, 689–692.
149. Bannister, A., and Konzarides, T. (1996). The CBP co-activator is a histone acetyl transferase. *Nature* **384**, 641–643.
150. Felzien, L. K., Farrell, S., Betts, J. C., Mosavin, R., and Nabel, G. J. (1999). Specificity of cyclin E-Cdk2, TFIIB, and E1A interactions with a common domain of the p300 coactivator. *Mol. Cell. Biol.* **19**, 4241–4246.
151. Boyes, J., Byfield, P., Nakatani, Y., and Ogryzko, V. (1998). Regulation of activity of the transcription factor GATA-1 by acetylation. *Nature* **396**, 594–598.
152. Parry, G. C., and Mackman, N. (1997). Role of cyclic AMP response element-binding protein in cyclic AMP inhibition of NF-kappaB-mediated transcription. *J. Immunol.* **1**, 5450–5456.
153. Ollivier, V., Parry, G. C., Cobb, R. R., de Prost, D., and Mackman, N. (1996). Elevated cyclic AMP inhibits NF-kappaB-mediated transcription in human monocytic cells and endothelial cells. *J. Biol. Chem.* **271**, 20828–20835.
154. Chen, L. F., Fischle, W., Verdin, E., and Greene, W. C. (2001). Duration of nuclear NF-kappaB action regulated by reversible acetylation. *Science* **293**, 1653–1657.
155. Smulson, M. E., Simbulan-Rosenthal, C. M., Boulares, A. H., Yakovlev, A., Stoica, B., Iyer, S., Luo, R., Haddad, B., Wang, Z. Q., Pang, T., Jung, M., Dritschilo, A., and Rosenthal, D. S.

(2000). Roles of poly(ADP-ribosyl)ation and PARP in apoptosis, DNA repair, genomic stability and functions of p53 and E2F-1. *Adv. Enzyme Regul.* **40**, 183–215.
156. Alvarez-Gonzalez, R., Spring, H., Muller, M., and Burkle, A. (1999). Selective loss of poly(ADP-ribose) and the 85-kDa fragment of poly(ADP-ribose) polymerase in nucleoli during alkylation-induced apoptosis of HeLa cells. *J. Biol. Chem.* **274**, 32122–32126.
157. Hassa, P. O., and Hottiger, M. O. (1999). A role of poly (ADP-ribose) polymerase in NF-kappaB transcriptional activation. *Biol. Chem.* **380**, 953–959.
158. Burkle, A. (2001). Physiology and pathophysiology of poly(ADP-ribosyl)ation. *BioEssays* **23**, 795–806.
159. Meisterernst, M., Stelzer, G., and Roeder, R. G. (1997). Poly(ADP-ribose) polymerase enhances activator-dependent transcription in vitro. *Proc. Natl. Acad. Sci. USA* **94**, 2261–2265.
160. Conde, C., Mark, M., Oliver, F. J., Huber, A., de Murcia, G., and Menissier-de Murcia, J. (2001). Loss of poly(ADP-ribose) polymerase-1 causes increased tumour latency in p53-deficient mice. *EMBO J.* **20**, 3535–3543.
161. D'Amours, D., Germain, M., Orth, K., Dixit, V. M., and Poirier, G. G. (1998). Proteolysis of poly(ADP-ribose) polymerase by caspase 3: Kinetics of cleavage of mono(ADP-ribosyl)ated and DNA-bound substrates. *Radiat. Res.* **150**, 3–10.
162. Davidovic, L., Vodenicharov, M., Affar, E. B., and Poirier, G. G. (2001). Importance of poly(ADP-ribose) glycohydrolase in the control of poly(ADP-ribose) metabolism. *Exp. Cell Res.* **268**, 7–13.
163. Boulares, A. H., Yakovlev, A. G., Ivanova, V., Stoica, B. A., Wang, G., Iyer, S., and Smulson, M. (1999). Role of poly(ADP-ribose) polymerase (PARP) cleavage in apoptosis. Caspase 3-resistant PARP mutant increases rates of apoptosis in transfected cells. *J. Biol. Chem.* **274**, 22932–22940.
164. Cayuela, M. L., Carrillo, A., Ramirez, P., Parrilla, P., and Yelamos, J. (2001). Genomic instability in a PARP-1(-/-) cell line expressing PARP-1 DNA-binding domain. *Biochem. Biophys. Res. Commun.* **285**, 289–294.
165. Simbulan-Rosenthal, C. M., Ly, D. H., Rosenthal, D. S., Konopka, G., Luo, R., Wang, Z. Q., Schultz, P. G., and Smulson, M. E. (2000). Misregulation of gene expression in primary fibroblasts lacking poly(ADP-ribose) polymerase. *Proc. Natl. Acad. Sci. USA* **97**, 11274–11279.
166. Seimiya, H., and Smith, S. (2002). The telomeric poly(ADP-ribose) polymerase, tankyrase 1, contains multiple binding sites for telomeric repeat binding factor 1 (TRF1) and a novel acceptor, 182-kDa tankyrase-binding protein (TAB182). *J. Biol. Chem.* **277**, 14116–14126.
167. Ziegler, M., and Oei, S. L. (2001). A cellular survival switch: Poly(ADP-ribosyl)ation stimulates DNA repair and silences transcription. *BioEssays* **23**, 543–548.
168. Soldatenkov, V. A., Chasovskikh, S., Potaman, V. N., Trofimova, I., Smulson, M. E., and Dritschilo, A. (2002). Transcriptional repression by binding of poly(ADP-ribose) polymerase to promoter sequences. *J. Biol. Chem.* **277**, 665–670.
169. Oei, S. L., Griesenbeck, J., Schweiger, M., and Ziegler, M. (1998). Regulation of RNA polymerase II-dependent transcription by poly(ADP-ribosyl)ation of transcription factors. *J. Biol. Chem.* **273**, 31644–31647.
170. Akiyama, T., Takasawa, S., Nata, K., Kobayashi, S., Abe, M., Shervani, N. J., Ikeda, T., Nakagawa, K., Unno, M., Matsuno, S., and Okamoto, H. (2001). Activation of Reg gene, a gene for insulin-producing beta-cell regeneration: Poly(ADP-ribose) polymerase binds Reg promoter and regulates the transcription by autopoly(ADP-ribosyl)ation. *Proc. Natl. Acad. Sci. USA* **98**, 48–53.
171. Butler, A. J., and Ordahl, C. P. (1999). Poly(ADP-ribose) polymerase binds with transcription enhancer factor 1 to MCAT1 elements to regulate muscle-specific transcription. *Mol. Cell. Biol.* **19**, 296–306.

172. Cervellera, M. N., and Sala, A. (2000). Poly(ADP-ribose) polymerase is a B-MYB coactivator. *J. Biol. Chem.* **275**, 10692–10696.
173. Ha, H. C., Hester, L. D., and Snyder, S. H. (2002). Poly(ADP-ribose) polymerase-1 dependence of stress-induced transcription factors and associated gene expression in glia. *Proc. Natl. Acad. Sci. USA* **99**, 3270–3275.
174. Ullrich, O., Diestel, A., Eyupoglu, I. Y., and Nitsch, R. (2001). Regulation of microglial expression of integrins by poly(ADP-ribose) polymerase-1. *Nat. Cell Biol.* **3**, 1035–1042.
175. Shall, S., and de Murcia, G. (2000). Poly(ADP-ribose) polymerase-1: What have we learned from the deficient mouse model? *Mutat. Res.* **460**, 1–15.
176. Oliver, F. J., Menissier-de Murcia, J., Nacci, C., Decker, P., Andriantsitohaina, R., Muller, S., de la Rubia, G., Stoclet, J. C., and de Murcia, G. (1999). Resistance to endotoxic shock as a consequence of defective NF-kappaB activation in poly (ADP-ribose) polymerase-1 deficient mice. *EMBO J.* **18**, 4446–4454.
177. Hassa, P. O., Covic, M., Hasan, S., Imhof, R., and Hottiger, M. O. (2001). The enzymatic and DNA binding activity of PARP-1 are not required for NF-kappa B coactivator function. *J. Biol. Chem.* **276**, 45588–45597.
178. Chang, W. J., and Alvarez-Gonzalez, R. (2001). The sequence-specific DNA binding of NF-kappa B is reversibly regulated by the automodification reaction of poly (ADP-ribose) polymerase 1. *J. Biol. Chem.* **276**, 47664–47670.
179. Moon, N. S., Berube, G., and Nepveu, A. (2000). CCAAT displacement activity involves CUT repeats 1 and 2, not the CUT homeodomain. *J. Biol. Chem.* **275**, 31325–31334.
180. Nepveu, A. (2001). Role of the multifunctional CDP/Cut/Cux homeodomain transcription factor in regulating differentiation, cell growth and development. *Gene* **270**, 1–15.
181. Santaguida, M., Ding, Q., Berube, G., Truscott, M., Whyte, P., and Nepveu, A. (2001). Phosphorylation of the CCAAT displacement protein (CDP)/Cux transcription factor by cyclin A-Cdk1 modulates its DNA binding activity in G(2). *J. Biol. Chem.* **276**, 45780–45790.
182. Luong, M. X., van der Meijden, C. M., Xing, D., Hesselton, R., Monuki, E. S., Jones, S. N., Lian, J. B., Stein, J. L., Stein, G. S., Neufeld, E. J., and van Wijnen, A. J. (2002). Genetic ablation of the CDP/Cux protein C terminus results in hair cycle defects and reduced male fertility. *Mol. Cell. Biol.* **22**, 1424–1437.
183. Coqueret, O., Martin, N., Berube, G., Rabbat, M., Litchfield, D. W., and Nepveu, A. (1998). DNA binding by cut homeodomain proteins is down-modulated by casein kinase II. *J. Biol. Chem.* **273**, 2561–2566.
184. Coqueret, O., Berube, G., and Nepveu, A. (1998). The mammalian Cut homeodomain protein functions as a cell-cycle-dependent transcriptional repressor which downmodulates p21WAF1/CIP1/SDI1 in S phase. *EMBO J.* **17**, 4680–4694.
185. Antes, T. J., Chen, J., Cooper, A. D., and Levy-Wilson, B. (2000). The nuclear matrix protein CDP represses hepatic transcription of the human cholesterol-7alpha hydroxylase gene. *J. Biol. Chem.* **275**, 26649–26660.
186. Zhu, Q., Gregg, K., Lozano, M., Liu, J., and Dudley, J. P. (2000). CDP is a repressor of mouse mammary tumor virus expression in the mammary gland. *J. Virol.* **74**, 6348–6357.

Enzymes That Cleave and Religate DNA at High Temperature: The Same Story with Different Actors

MARIE-CLAUDE SERRE AND
MICHEL DUGUET

Laboratoire d'Enzymologie des Acides
Nucléiques, Institut de Génétique et
Microbiologie, Université Paris-Sud, 91405
Orsay Cedex, France

I. Topoisomerases 38
   A. Thermophilic Topoisomerase I (Type IA, Eubacteria) 39
   B. Hyperthermophilic Topoisomerase III (Type IA, Archaea) 45
   C. Reverse Gyrase 46
   D. Topoisomerase V (Type IB, Archaea) 51
   E. Gyrase (Type IIa, Eubacteria and Some Archaea) 53
   F. Topoisomerase VI (Type IIb, Archaea) 57
II. Recombinases 59
   A. Holliday Junction Resolvases 61
   B. Viral Tyrosine Recombinases: The Fuselloviridae Integrases 68
   C. Cellular Tyrosine Recombinases: The Thermophilic Xer Proteins 69
III. Concluding Remarks 73
   References 74

DNA transactions such as replication, recombination, and transcription produce various changes in the structure and topology of the DNA molecule. These changes are triggered by specialized enzymes that produce transient or permanent breaks in the DNA backbone and are classified as topoisomerases and recombinases. The role of these enzymes is especially crucial in thermophilic organisms, since the integrity of the double helical structure must be maintained as replication/recombination or other DNA transactions take place at high temperatures. Indeed, thermophilic organisms exhibit a remarkable variety of DNA topoisomerases and recombinases. Some of them have been discovered through studies of hyperthermophiles, and they could represent either enzymes specialized in the functionning of DNA at high temperature (i.e., reverse gyrase) or remnants of the early evolution of living organisms (i.e., Topo VI and the unique Xer recombinase found in thermophiles). The study of these proteins is interesting, not only for phylogenic reasons, but also from a structural and mechanistic point of view: these enzymes may represent either simplified versions of recombinases and topoisomerases conserved in higher organisms or entirely new enzymes with new mechanistic solutions, albeit based on the same chemistry, to the problems faced by the DNA double helix at high temperature.

Thermophilic organisms face the problems of stability of their macromolecules at high temperature. In particular, the DNA molecule, which is in charge of the genetic information, must maintain its double helical structure near its melting point (T_m). The difficulty is that, at the same time, transmission of the genetic information requires at least local separation of the two DNA strands. This is true for DNA replication, transcription, repair, or recombination. In this context, enzymes in charge of the DNA structure and topology in thermophilic organisms have two apparent antagonizing roles: to favor DNA transactions as in mesophiles and to prevent uncontrolled double helix opening at high temperature. In this review, we focus on two main classes of enzymes working on DNA, topoisomerases and recombinases. To draw a picture of these enzymes, we have exploited the data of the 16 complete genomes from thermophilic organisms (3 from eubacteria, 13 from archaea) available at present, but the exceedingly fast release of new complete sequences from a number of genomes may change this picture.

I. Topoisomerases

Topoisomerases are classified into two types. Type I topoisomerases make a transient break in DNA one strand at a time and promote passage of the other strand, whereas type II topoisomerases cleave both strands of a double helix and can promote passage of another double helix. Both types use the same general chemistry to break DNA: a tyrosyl group of the enzyme attacks a phosphodiester bond on DNA and remains covalently linked to one side of the break. Attack of this phosphotyrosine link by the hydroxyl of the free DNA strand rejoins the broken DNA strand and releases the enzyme for the next cycle. Both types I and II topoisomerases can be further divided into two families (see Table I).

Type I topoisomerases are subdivised into families IA and IB on the basis of their biochemical properties. Type IA enzymes form a transient covalent bond to the 5'-phosphoryl of the broken strand, whereas type IB enzymes form a 3'-end covalent bond. These two families do not share any phylogenic relationship. Until recently, Topo IB was thought to be specific for eukaryotes and their viruses. The discovery of Topo V in *Methanopyrus kandleri* (1) and of Topo IB in some eubacteria (2) changed this. However, to date the presence of type IB topoisomerases in prokaryotes remains exceptional. However, the type IA family is widespread in all three kingdoms (eukarya, archaea, eubacteria) and includes three subfamilies, all containing thermophilic representatives: topoisomerases I, topoisomerases III, as well as reverse gyrase, an enzyme specific to hyperthermophilic organisms (Table I).

Type II topoisomerases are also subdivided into two families, IIa and IIb, that share little similarity. Type IIa represents the classic type II topoisomerases that cleave duplex DNA, leaving 4-nucleotides 5' overhangs on each side of the break. They are found in eukaryal (Topo II) and eubacterial (DNA gyrases and Topo IV) kingdoms. However, thermophilic eubacteria do not seem to possess Topo IV (Table I). In archaea, type IIb enzymes that cleave duplex DNA with 2-nucleotides 5' overhangs replace the type IIa. Exceptions are *Archaeoglobus fulgidus*, which contains type IIa and IIb topoisomerases, and the *thermoplasmales*, where no type IIb is present (Table I).

A. Thermophilic Topoisomerase I (Type IA, Eubacteria)

1. PHYLOGENIC POSITION AND SEQUENCE ORGANIZATION

This class belongs to the family of eubacterial topoisomerases related to the *top*A gene product found in *Escherichia coli*. As shown in Fig. 1, this family forms a rather homogeneous group, with the five thermophilic members so far sequenced [from *Thermotoga maritima* (3, 4), *Fervidobacterium islandicum* (5), *Aquifex aeolicus* (6), *Thermoaerobacter tengcongensis* (7), and *Thermobifida fusca*[1]] clearly integrated within the family. The two first members, *T. maritima* and *F. islandicum*, which both belong to the order Thermotogales, are more closely related. Although their sequences are conserved, the gene environment of the thermophilic eubacterial topoisomerases I in the different genomes is not conserved.

Sequence organization is schematically described in Fig. 2. The five thermophilic topoisomerases I appear to retain all the conserved motifs of the eubacterial topoisomerases I found in the 67-kDa amino-terminal fragment of *E. coli* TopA, suggesting a common basic architecture for the whole family. This was further confirmed by structural data (8). Moreover, the region of the active site tyrosine has a particular structure, called the CAP DNA binding domain, also found in the CAP (*c*atabolite gene *a*ctivator *p*rotein) transcriptional regulator and in type II topoisomerases (9). Despite these common features, the size of thermophilic topoisomerases I differs considerably, from 540 residues for *A. aeolicus* up to 996 for *T. fusca*. On the amino-terminal side, a prominent sequence of 112 amino acids is found in *T. fusca* upstream of the first highly conserved motif (Fig. 2), whereas a 33 amino acid prominent sequence appears in *F. islandicum* (5). This feature is rather unusual, since thermophilic proteins are often smaller than their mesophilic equivalents. As several mesophilic enzymes,

---

[1] Unfinished genome. Contigued sequences are available at the NCBI. The *T. fusca* Topo IA sequence was obtained by a BLAST using *T. maritima* Topo IA as the search sequence.

TABLE I
INVENTORY OF TOPOISOMERASES FROM THERMOPHILIC ORGANISMS[a]

	Optimal growth temperature (°C)	Type I Topoisomerases			Type II Topoisomerases			
			Type IA		Type IB	Type IIa		Type IIb
		Topo I	Topo III	Reverse gyrase	Topo V	Topo IIA (GyrA)	Topo IIB (GyrB)	Topo VIA	Topo VIB
Complete archaeal genomes									
Crenarchaea									
Aeropyrum pernix	90		APE1794	APE1376 APE1340				APE0703	APE0706
Pyrobaculum aerophilum	98		PAE2993	PAE1108				PAE2219	PAE2217
Sulfolobus solfataricus	80		SSO0907	SSO0963 SSO0420				SSO0969	SSO0968
Sulfolobus tokodaii	80		ST1216	ST1290 ST0374				ST1444	ST1294
Euryarchaea									
Archaeoglobus fulgidus	78–85		AF1506	AF1024		AF0465	AF0530	AF0940	AF0652
Methanobacterium thermoautotrophicum	60–65		MTH1624					MTH1008	MTH1007
Methanococcus jannaschii	80–85		MJ1652	MJ1512				MJ0369	MJ1028
Methanopyrus kandleri	98		MK1604	MK0289[b] MK0049	MK1436			MK0512	MK0921
Pyrococcus abyssi	96		PAB1430	PAB2423				PAB2411	PAB0407

Organism	%						
Pyrococcus horikoshii	98	PH0622	PH0800			PH1563	PH1564
Pyrococcus furiosus	97–100	PF0494	PF0495			PF1578	PF1579
Thermoplasma acidophilum	55–60	Ta0063		Ta1054	Ta1055		
Thermoplasma volcanium	60	TVN0019		TVN0542	TVN0541		
Complete bacterial genomes							
Aquifex aeolicus	90	aq_657	aq_886	aq_980	aq_1026		
			aq_1159				
Thermoanaerobacter tengcongenris	75	TTE1449	TTE1745	TTE0011[c]	TTE0010		
Thermotoga maritima	80	TM0258	TM0173	TM1084	TM0833		
Genes[d]							
Sulfolobus acidocaldarius (A)	70		Q08582				
Sulfolobus shibatae B12 (A)	80	U97022	X98420			O05208	O05207
Fersidobacterium islandicum (B)	65	U97022					
Thermus thermophilus (B)	60			AAF89614	AAF89615		

[a] For sequenced genomes, the gene numbers correspond to the notations available at http://www.ncbi.nlm.nih.gov/PMGifs/Genomes/micr.html.
[b] The M. kandleri reverse gyrase is composed of two subunits encoded by separate genes.
[c] The status of TTE0011 and TTE0010 as GyrA/GyrB or ParC/ParE is unclear (see Fig. 5).
[d] Accession numbers of unsequenced thermophilic archaeal (A) or eubacterial (B) genomes. Sequencing of T. thermophilus and S. acidocaldarius is in progress.

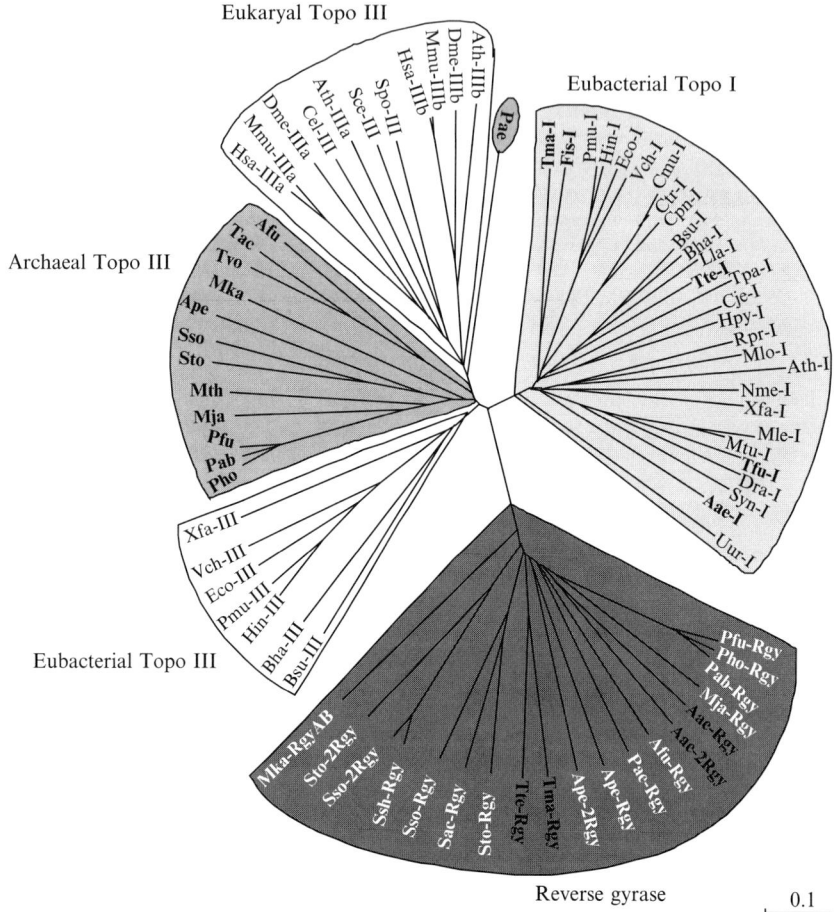

FIG. 1. Phylogenetic tree of topoisomerases IA. The phylogenetic tree was deduced from topoisomerases IA sequence alignments produced by ClustalX. Each group is identified as follow: eubacterial Topo I (topA), light gray shading; eukaryal Topo III and eubacterial Topo III, circled black, no shading; archaeal Topo III, medium gray shading; reverse gyrase, dark gray shading. Thermophilic organisms are indicated by bold lettering. Within the reverse gyrase group, eubacteria are black labeled and archaea white labeled. Species abbreviations are the following: Eukarya: Ath, *Arabidopsis thaliana*; Cel, *Caenorhabditis elegans*; Dme, *Drosophila melanogaster*; Hsa, *Homo sapiens*; Mmu, *Mus musculus*; Sce, *Saccharomyces cerevisiae*; Spo: *Schizosaccharomyces pombe*. Archaea: Afu, *Archaeoglobus fulgidus*; Ape, *Aeropyrum pernix*; Mja, *Methanococcus jannaschii*; Mka, *Methanopyrus kandleri*, Mth, *Methanobacterium thermoautotrophicum*; Pab, *Pyrococcus abyssi*; Pae, *Pyrobaculum aerophilum*; Pho, *Pyrococcus horikoshii*; Pfu, *Pyrococcus furiosus*; Sac, *Sulfolobus acidocaldarius*; Ssh, *Sulfolobus shibatae*; Sso, *Sulfolobus solfataricus*; Sto, *Sulfolobus tokodaii*; Tac, *Thermoplasma acidophilum*; Tvo, *Thermoplasma volcanium*. Eubacteria: Aae, *Aquifex aeolicus*; Bha, *Bacillus halodurans*; Bsu, *Bacillus subtilis*; Cje, *Campylobacter jejuni*; Cmu,

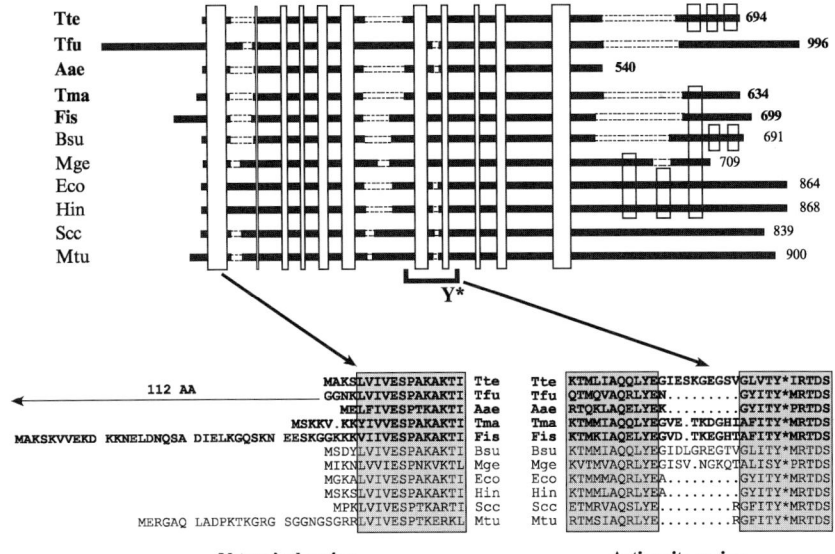

FIG. 2. Domain structure organization of eubacterial Topo IA. (Top) Domanial alignment of several eubacterial Topo IA. Species abbreviations are as in Fig. 1, with the following additions: Mge, *Mycoplasma genitalium*; Scc, *Synechococcus* sp. Thermophilic species are labeled in bold. Dotted lines represent gaps. Foreground white boxes correspond to the topoisomerases IA conserved motifs and background white boxes to zinc finger motifs. The position of the catalytic tyrosine is identified by a Y. The size of each topoisomerase is indicated on the right of the figure. (Bottom) Corresponding sequence alignments of the N-terminal and active site regions.

three of the thermophilic proteins have an 8–9 amino acid additional sequence in the active site region (Fig. 2). This insertion separates two highly conserved motifs in the active site without apparent effect on catalytic activity.

As commonly observed for eubacterial Topo I, the carboxy-terminal part of the proteins is largely variable in length and in sequence (Fig. 2). This region is totally absent in the enzyme from *A. aeolicus*, while a long C-terminal tail (238 residues) is present in *T. fusca*. The existence of putative zinc fingers, often found in the region, is also very different among the five thermophilic

---

Chlamydia muridanum; Cpn, *Chlamydophila pneumoniae*; Ctr, *Chlamydia trachomatis*; Dra, *Deinococcus radiodurans*; Eco, *Escherichia coli*; Fis, *Fervidobacterium islandicum*; Hin, *Haemophilus influenzae*; Hpy, *Helicobacter pylori* 26695; Lla, *Lactococcus lactis*; Mle, *Mycobacterium leprae*; Mlo, *Mesorhizobium loti*; Mtu, *Mycobacterium tuberculosis*; Nme, *Neisseria meningitidis*; Pmu, *Pasteurella multocida*; Rpr, *Rickettsia prowazekii*; Syn, *Synechocystis* sp.; Tfu, *Thermobifida fusca*; Tma, *Thermotoga maritima*; Tpa, *Treponema pallidum*; Tte, *Thermoanaerobacter tengcongensis*; Uur, *Ureaplasma urealyticum*; Vch, *Vibrio cholerae*; Xfa, *Xylella fastidiosa*.

topoisomerases: a total absence in *A. aeolicus* and *T. fusca*, a unique zinc motif in *T. maritima* and *F. islandicum*, and three motifs in *T. tengcongensis*. As shown below, the unique motif found in *T. maritima* does bind zinc, but is dispensable for topoisomerase activity.

Finally, attempts to define sequence features specific to the five thermophilic species presently known failed. One can simply notice the low cystein content of these enzymes and the higher proportion of isoleucine, glutamic acid, and lysine residues. Determination of the three-dimensional (3D) structure of topoisomerase I from *T. maritima*, which is currently under investigation, may shed light on the residues needed to stabilize the enzyme at high temperature.

2. ENZYMATIC PROPERTIES

Only two topoisomerases I from thermophilic eubacteria have been investigated for their enzymatic activities. In an early work, the native enzyme from *F. islandicum* was purified to near homogeneity (*10*). It shares the main characteristics of the *E. coli* enzyme, with an ATP-independent DNA relaxing activity at an optimal temperature of 70 °C, close to the growing temperature of its eubacterial host. An unexpected property is that the activity is inhibited by nucleosides triphosphates, possibly by chelation of the magnesium ions required for activity.

Recently, topoisomerase I from *T. maritima* has been overproduced in *E. coli* (*11*). The recombinant enzyme has been easily purified to near homogeneity by including a heating step that removed the majority of *E. coli* proteins. Although synthesized at 37 °C, the optimal activity of this recombinant enzyme was observed at 75 °C, the normal growth temperature of *T. maritima*, confirming the correct folding of the protein.

Peculiar to this topoisomerase is its exceptionally high DNA relaxation activity, 100- to 200-fold that of the *E. coli* enzyme (*12*): one molecule can catalyze 500–1000 strand passage events per minute, a rate close to that of topoisomerase IB (*13*). The biological reasons for such a high activity are unknown, but it may be necessary *in vivo* for rapid removing of negative supercoils locally occurring behind a transcription complex to prevent helix opening at high temperature.

Another interesting property of *T. maritima* topoisomerase I is its capacity to decatenate kinetoplastic DNA networks of minicircles, provided that these are nicked (*11*). Whether this property can be extended to other type Ia enzymes remains an open question. The decatenation activity is more reminiscent of topoisomerase III, which is present in *E. coli* but not in *T. maritima*. It is possible that in this organism, topoisomerase I replaces topoisomerase III for the resolution of intertwined daughter chromosomes at the end of replication.

As for *E. coli* topoisomerase I and for reverse gyrases, DNA cleavage occurs preferentially with a cytosine in position −4 of the break point

(*14*, *15*). Interestingly, the cleavage efficiency appears to be independent of temperature, suggesting that this step is not rate limiting (*11*).

The most striking difference with the *E. coli* enzyme concerns the carboxy-terminal region. In *T. maritima* topoisomerase I, disruption of the unique Zn motif by mutation of two cysteins has no influence on its enzymatic activity, relaxation, or decatenation (*11*). Moreover, neither its cleavage efficiency and specificity nor its thermal stability are affected, suggesting that, contrary to the case of *E. coli* (*16*), the Zn motif is not involved in enzyme activity, at least *in vitro*.

Recently, the relaxation and binding properties of the thermophilic topoisomerase have been investigated using a single DNA molecule bound at each extremity (*17*, *17a*). The experimental device allows us to adjust both the pulling force and the winding of DNA strands. *T. maritima* topoisomerase I is not active on plectonemic negative or positive supercoils produced at low pulling force, since it does not bind these structures. However, the enzyme recognizes and binds bubbles produced at a higher pulling force in an underwound DNA. Once bound, this topoisomerase is able to act indifferently on positively or negatively constrained DNA. The observation of individual steps in the relaxation of a single DNA molecule clearly confirms the enzyme-bridged strand passage model supported by biochemical and structural data and likely extendable to the whole topoisomerase IA family (*17*).

B. Hyperthermophilic Topoisomerase III (Type IA, Archaea)

Another group of topoisomerases IA is formed by archaeal topoisomerases III (Fig. 1). Our present knowledge of these enzymes essentially derives from sequence comparisons. Remarkably, in these comparisons, the group appears distinct from both eubacterial and eukaryal topoisomerases III that are present uniquely in mesophilic organisms. In particular, hyperthermophilic eubacteria do not contain a topoisomerase III. The only biochemical data on these enzymes concern the topoisomerase III from *Desulfurococcus amylolyticus* (*18*). As the other topoisomerases IA, it is able to relax negatively, but not positively supercoiled DNA in a reaction that is dependent on Mg^{2+} but not on ATP. Several interesting observations suggest that the critical point that governs the activity of this topoisomerase is its binding to the DNA substrate, in particular the availability of single-stranded regions in the DNA. Thus, at moderate temperatures (i.e., 60 °C), the relaxation rate is slow and critically dependent on the level of negative supercoiling, and not all supercoils are removed. At higher temperatures (i.e., 80 °C), relaxation is complete and fast, presumably because more single-stranded regions are present. Interestingly, the thermostability of *D. amylolyticus* topoisomerase III allowed experiments to be tried at temperatures exceeding 90 °C, a range in which DNA begins to melt. In this case, the enzyme no longer distinguishes between negative and positive supercoils, and a different activity occurs, i.e., the unwinding of the two DNA strands. In this reaction, the DNA linking number (Lk) decreases, contrary to the normal activity of the topoisomerases IA (relaxation of negative supercoils) for which Lk increases. This observation again suggests that the direction of strand passage for type IA topoisomerases is primarily directed by the structure of the DNA substrate.

Whether the properties of this topoisomerase may be extended to the other archaeal topoisomerases III is not known. Moreover, there is presently little information on the *in vivo* functions of these enzymes in hyperthermophilic archaea, and the unwinding reaction described *in vitro* may not be biologically relevant.

C. Reverse Gyrase

Among thermophilic topoisomerases IA, reverse gyrases constitute a group of original protein machines working on DNA, with their unique property to produce positive supercoils in a circular DNA (*19, 20*). Initially described in crenarchaeotae, they were found in all hyperthermophilic organisms, including archaea and eubacteria (*21, 22*).

1. Sequence Organization

Reverse gyrases are composed of a single polypeptide of relatively constant length (around 1200 amino acids), with the exception of the two subunits in *M. kandleri*. The first reverse gyrase sequence, obtained from the crenarchaeon *S. acidocaldarius*, provided some information on the way this enzyme could function (*23*). Two regions (each of them spanning half of the protein) can be delimited (Fig. 3): the N-terminal region contains conserved motifs found in helicases, whereas the C-terminal domain exhibits a clear similarity to type IA topoisomerases, confirming previous biochemical results showing single-strand cleavage and covalent link to the 5′ DNA ends (*24*). This organization is a general feature of the reverse gyrases family. As shown in Fig. 1, reverse gyrases form a well-delimited group, independently of their belonging to eubacterial or archaeal domains of life.

More precise examination of the N-terminal part in a dozen reverse gyrases indicates that it belongs to the helicase superfamily SF2 comprising recombination/repair proteins RecQ, Rad3, UvrB, the primosomal protein PriA, and the eukaryotic translation initiation factor eIF4A (*25*). Motifs I, Ia, II, III, V, and VI of the SF2 family are present and well conserved among reverse gyrases (*26*). These motifs are mainly devoted to ATP binding and hydrolysis (I, II, III, VI), and to single-strand DNA binding (Ia, V). Motifs denoted as IV or IVa are barely detectable in reverse gyrases. Peculiar to reverse gyrases is a putative Zn finger in the very N-terminal end of the enzyme, which resembles Zn fingers found in GATA transcription factors (*27*). Its function is presently unknown.

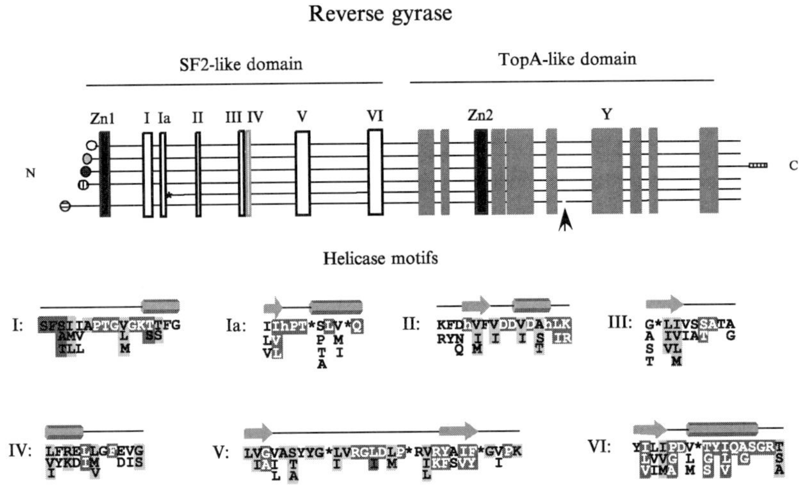

Reverse gyrase

FIG. 3. Domain organization of reverse gyrases. (Top) The two domains comprising reverse gyrases are indicated. Each domain has several conserved motifs. Helicase motifs (SF2-like family) are represented by white boxes labeled from I to VI. A hypothetical motif IV (shaded and italicized) was found on the bases of sequence homologies. TopA motifs are represented by gray boxes, and the position of the active site tyrosine is indicated (Y). Two potential zinc binding domains are represented by black boxes. Reverse gyrases can be grouped in terms of their sizes (abbreviations are as in Fig. 1). ⊙, Aae (aq_886), Afu, Mja, Pab, Pae, Pho, Pfu, Ssh, Sso (SSO0963), Sto (ST1290) Tma, Tte. ○, Ape (APE1376), Sac. ●, Ape (APE1340), Sso (SSO0420). These two reverse gyrases have a C-terminal extention. The longest is found in Ape and is signaled by ▭. ⊕, Aae (aq_1159). ⊖, Mka. This reverse gyrase is composed of two subunits. The subunit encoded by gene MK0049 corresponds to the N-terminal part of other reverse gyrases. Gene MK0289 encodes the subunit corresponding to the C-terminal part. The arrow indicates the separation between the two subunits. ★, Sto (ST0374). The lack of motifs 1 and 1a in the helicase domain makes it unlikely that this reverse gyrase is functional. (Bottom) Consensus sequences of the reverse gyrase helicase motifs. The consensus sequences were obtained by aligning 19 reverse gyrase sequences. Residue conservation is indicated by the following code: white letter, dark gray box, 100% similarity; black letter, light gray box, 80 to <100% similarity; black letter unboxed, 60 to <80% similarity. h, Hydrophobic; *, any (including gaps). At positions where several residues are listed, the occurrence is from the most (top) to the least (bottom) represented in the sequences. Above each motif, the secondary structure of *Archaeoglobus fulgidus* reverse gyrase is indicated. Gray arrow, β-strand; gray cylinder, α helix; black line, loop.

The C-terminal domain of reverse gyrases exhibits evident similarity to all type IA topoisomerases, with the totality of the 10 motifs conserved (26). Recent experiments demonstrating the ATP-independent topoisomerase I activity of this domain confirm this view (28). A characteristic of this domain in reverse gyrases is the presence of a zinc finger between motifs 2 and 3 (Fig. 3) that is in the first fourth of the enzyme, contrasting with the other type IA topoisomerase in which Zn fingers are exclusively at the C-terminus.

A few examples of reverse gyrases exhibit a different structure, while conserving the organisation in two domains. One is the enzyme from *M. kandleri*, made of two subunits (29, 30). These do not correspond to the limits of the previously defined domains. The large subunit, RgyB (138 kDa), spans the N-terminal (helicase-like) domain and nearly half of the topoisomerase domain. The small subunit (42 kDa) contains the rest of the topoisomerase domain and has no activity by itself (31). In the case of *M. jannaschii* and *P. horikoshii*, sequence data indicate that their reverse gyrases contain inteins. Remarkably, the supplementary sequence is inserted precisely into the region of the active site tyrosine, in the topoisomerase domain, suggesting that the precursor protein is inactive. The point of insertion is the same as that of the 8–9 amino acids insertion found in some eubacterial topoisomerases I, suggesting that this region of topoisomerase IA sequence/structure is especially sensitive to rearrangements.

The sequence organization of reverse gyrases suggested an early model for the mechanism of positive supercoiling (23). This model supposed that positive supercoiling was produced by the concerted action of a translocating helicase, producing both positive and negative supercoils, the later being specifically relaxed by the topoisomerase IA domain. The most recent results on reverse gyrases (32) indicate that this model is probably not correct (see Section I.C.3).

2. Origin and Evolution of Reverse Gyrases

The conserved organization of the reverse gyrase sequence suggests possible scenarios to explain the formation of this enzyme in the course of evolution (33). In particular, it is conceivable that the reverse gyrase was generated by the fusion of ancestral helicase and topoisomerase IA domains. Whether this event was early in the evolution of prokaryotes is still a matter of speculation. Remarkably, when the topoisomerase domain alone is used in the analysis of Fig. 1, reverse gyrases again form a compact group, suggesting that the event (possibly domain fusion) that created reverse gyrase is ancient. In any case, the exclusive existence of reverse gyrase in hyperthermophiles suggests that it is essential for life at high temperature. If true, early organisms before fusion were not hyperthermophiles, putting into question the idea that the last universal common ancestor (LUCA) was a hyperthermophile (34).

The presence of reverse gyrases in eubacteria seems again to be strictly related to hyperthermophily (21, 35). Moderate thermophiles (i.e., living below 65 °C) do not contain a reverse gyrase (36). This is also true in the archaeal kingdom since *T. acidophilum*, *T. volcanium*, and *M. thermoautotrophicum* do not possess a reverse gyrase (Table I). It seems that the maximal rather than optimal growth temperature is an important parameter for the presence of reverse gyrase activity in thermophiles. Organisms that can support growth at or above 65 °C would therefore be expected to possess at least one reverse gyrase gene.

The presence of reverse gyrases in all thermophilic eubacteria is not well documented, mostly because the detection of reverse gyrase activity in crude extracts can be tedious. For instance, early reports stated that there was no reverse gyrase activity in *T. maritima* (36, 37). This was latter contradicted both by biochemical data and genome sequence analysis (4, 21). Therefore the absence of reverse gyrase activity in eubacteria living at temperatures above 65° (36) should be examined with great care and should await genome sequence data analysis.

Examination of the genes surrounding reverse gyrase in *T. maritima* and *A. aeolicus* shows an accumulation of several genes of archaeal origin, suggesting a horizontal gene transfer from archaea to eubacteria in response to thermoadaptation (38). This again supports evidence indicating that reverse gyrase is a marker of hyperthermophily. The presence of two reverse gyrase genes in the genomes of the *Sulfolobales* so far sequenced is also puzzling. Interestingly, there are two reverse gyrase versions of these genes. For each version, the sequence is closer between two species than within the same species. One corresponds to the version found in *S. acidocaldarius*, and the other to that found in *S. shibatae*, but we do not know yet if these strains possess both genes.[2] The gene environment of the two reverse gyrase genes is different. However, this environment (about 50 kbp in both cases) is conserved in the two *Sulfolobales* genomes, suggesting that if the two versions of reverse gyrase result from gene duplication, this was an early event in the evolution of crenarchaea.

3. Mechanism and Clues about the Structure

A number of mechanistic studies have been performed on reverse gyrases to try to determine how they produce positive supercoiling.

The first puzzling result for an enzyme performing "gyration" was its type IA topoisomerase activity, recently confirmed by the expression of the recombinant carboxy-terminal half of the protein in *E. coli* (28). However, this strand passage activity is strictly ATP dependent and occurs exclusively in the whole reverse gyrase to increase the Lk between the two DNA strands. As for the other topoisomerase IA, reverse gyrase cleaves DNA at specific sequences, as shown by incubating stoichiometric amounts of the enzyme with DNA at high temperature in the absence of ATP or in the presence of a nonhydrolyzable analog. The same strong preference as eubacterial topoisomerases I for a cytosine in position −4 of the cleavage point was observed (29, 39). These sites are nevertheless specific, since their export into a foreign sequence creates a new site (15).

The second feature of reverse gyrase is its ATPase activity. The enzyme exclusively hydrolyzes ATP to ADP, but not the other triphosphates, and the reaction is strictly DNA dependent, with a preference for single-stranded

---

[2] The sequence of the *S. acidocaldarius* genome should be released soon and will provide more information on the duplication of the reverse gyrase genes.

DNA (*40*). It is remarkable that this activity is located in the amino-terminal half of reverse gyrase, whose sequence exhibits putative motifs of ATP binding and hydrolysis reminiscent of helicases. However, no helicase activity has been found so far for reverse gyrase nor for the recombinant N-terminal domain (*28*).

Reconstitution of positive supercoiling activity has been successfully obtained by mixing the two halves of the enzyme separately produced as recombinant proteins in *E. coli* (*28*). This association is very specific, since reconstitution of positive supercoiling was not possible by replacing either the topoisomerase IA domain by an active topoisomerase, or the helicase-like domain by an active helicase. In the case of *M. kandleri*, reconstitution was obtained by incubating the two separate subunits of the enzyme (*31*).

Another feature of reverse gyrase is its capacity to change the DNA structure upon stoichiometric binding: closure of a DNA form II by ligase after reverse gyrase stoichiometric binding results in a decrease of the DNA Lk compared to the control without reverse gyrase. This occurs in the absence of the topoisomerase activity of the C-terminal part (Tyr to Phe mutant), and was interpreted as stoichiometric DNA unwinding, or as left-handed DNA wrapping around the enzyme (*24*). Neither the helicase-like nor topoisomerase IA domain is able to produce this change on its own, but mixing of these domains reconstitutes this stoichiometric effect on DNA structure (*28*).

These mechanistic properties led to the proposal of a new model for positive supercoiling (*32*). In this model, binding of reverse gyrase to DNA induces local DNA unwinding in a topologically closed domain, resulting in overwinding (positive supercoiling) of the rest of the DNA molecule. Then, ATP binding allows strand passage in the unwound domain by the topoisomerase IA region of reverse gyrase, which increases the DNA Lk. Finally, ATP hydrolysis would recycle reverse gyrase in a conformation competent for DNA binding/unwinding.

Recently, a crystal structure of reverse gyrase of *A. fulgidus* was obtained (*41, 42*). The conclusions of this work support the above model. As expected, the topoisomerase part has a ring structure very similar to that of other type IA enzymes, whereas the helicase-like part contains two RecA-like domains, as found in helicases and presumably responsible for local unwinding. It seems to lack the residues usually involved in translocation along DNA, confirming the absence of processive helicase activity. To explain the ATP-dependent activity of reverse gyrase, it was postulated that a small domain in the helicase part might work as an ATP-dependent latch controlling the strand passage by the topoisomerase I domain.

4. Hypotheses on the Functions of Reverse Gyrases

Despite *in vitro* investigations of the activity of reverse gyrase, little is known about its *in vivo* functions. The first idea is that reverse gyrases play an essential role in the regulation of DNA supercoiling in hyperthermophiles.

Indeed, Lk increase by positive supercoiling would stabilize the duplex at high temperatures. Supporting this is the fact that viral episomes or plasmids extracted from archaeal hyperthermophiles are relaxed or positively supercoiled (43, 44). Moreover, thermal stress in hyperthermophilic archaea transiently increases the DNA Lk (45), as was observed for mesophilic eubacteria. It was proposed that in archaeal hyperthermophiles, the global level of supercoiling is tightly controled by the antagonist activity of reverse gyrase and the unique[3] type II topoisomerase present, topoisomerase VI (46). This view was complicated by several additional observations. The first is the situation in hyperthermophilic eubacteria in which episomes or plasmids are found negatively supercoiled as in mesophiles (47). In this case, a gyrase is present in addition to reverse gyrase and topoisomerase I, and may account for this global negative supercoiling. It was proposed that reverse gyrase acts more locally to increase the DNA Lk in regions where the stability of the duplex is reduced, for instance after the passage of a transcription complex. The second complication in estimating the role of reverse gyrase is the presence of chromatin-like structures in hyperthermophilic archaea. In euryarchaea, histones are present and seem to be able to stabilize both negative (left-handed) and positive (right-handed) toroidal supercoils in nucleosome-like structures (48). Thus, it is conceivable that independently of the global level of supercoiling, free or constrained supercoils (either positive or negative) locally coexist and are the targets of topoisomerases. For instance, reverse gyrases may rapidly transform negative to positive supercoiling in a precise DNA region. Kikuchi (49) suggested that reverse gyrases may suppress a variety of "abnormal" structures into DNA, such as cruciforms, triple helices, or unpaired regions and he proposed the term "renaturase" for these enzymes. More generally, reverse gyrases might be required in all of the processes that possibly generate single-stranded DNA at high temperature, to restore the double helix. These include replication and recombination/repair intermediates, as well as transcription complexes (50).

D. Topoisomerase V (Type IB, Archaea)

1. Phylogeny

Topoisomerase V (Topo V) has so far only been found in the hyperthermophilic methanogen *M. kandleri* (51). The gene was cloned and sequenced (52), and sequence analysis revealed that several regions of Topo V had similarities to different protein families (Fig. 4). The N-terminus of Topo V has a significant similarity with the RecA motif VI, a region that is part of the monomer–monomer interface in a RecA filament. The signature of the helix-turn-helix (HTH) motif of the asnC proteins is also found within the N-terminal region

---

[3] Except for *A. fulgidus*, which contains both gyrase and Topo VI (Table I).

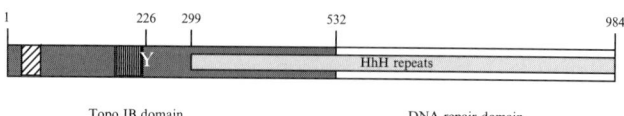

FIG. 4. Domain organization of *Methanopyrus kandleri* topoisomerase V. The Topo IB domain is shaded in dark gray. The DNA repair domain is in white. The diagonally hatched box represents the region of homology with the RecA motif VI. The black box represents the region of homology with the HTH motif of asnC proteins. The white Y (position 226) localizes the active site tyrosine. The long light gray box represents the region where the 12 repeats consisting of two similar helix-hairpin-helix (HhH) DNA binding motifs are found.

of Topo V. Within this region are also found the conserved catalytic signature of eukaryotic Topo IB and of tyrosine recombinases (52). From these alignments, the position of the catalytic tyrosine of Topo V was inferred to be Tyr-226, and mutagenesis of this residue confirmed its role in DNA cleavage and relaxation (52). Within the 684 C-terminal amino acid region, 12 repeats of about 50 amino acid long consisting of two similar but distinct helix-hairpin-helix (HhH) DNA binding motif are found. The HhH motif provides nonsequence-specific protein–DNA interactions and has been identified in 14 homologous families of DNA binding proteins (53). The 12 motifs from Topo V exhibited similarities with *E. coli* RuvA protein, human DNA polymerase β (β-pol), and several other proteins (52). The arrangement by repeats of two contiguous HhH motifs is also found in RuvA and human β-pol. Further biochemical analysis revealed that Topo V shares DNA repair activity with the latter protein (see the next section).

Topo V is (at least) a bifunctional protein, with two catalytic domains fused (Fig. 4). The N-terminal domain is responsible for the topoisomerase activity and the C-terminal domain harbors the DNA repair activity, a unique property for a topoisomerase (52). The presence of a signature motif of leucine-responsive regulatory proteins makes it also possible that Topo V might act as a transcriptional regulator. Several hypotheses have been suggested to take into account the dual catalytic activity of Topo V. Among these, an attractive hypothesis is that fusion of a DNA repair activity highly specific for apurinic/apyrimidic (AP) sites with Topo IB activity in a single protein might prevent Topo V from beeing toxic for *M. kandleri*. At temperatures up to 110 °C, which support *M. kandleri* growth, the probability of occurrence of DNA lesions (particularly apurination) is high. On the other hand, it is known that DNA lesions greatly enhance Topo IB-mediated cleavage and formation of protein–DNA adducts that are poisonous for the cell in or around AP sites (54). In eukaryotes this kind of lesions can be repaired by a Tyr-DNA phosphodiesterase, which is absent in prokaryotes (55). It is thus tempting to assume that the DNA repair activity module of Topo V would lead the enzyme to bind specifically to AP sites and initiate base-excision repair, thereby preventing Topo V from making lethal lesions.

2. ENZYMATIC PROPERTIES

Until Topo V was purified from *M. kandleri* (*1*, *56*), type IB topoisomerases were confined to eukaryotes and their viruses. Very recently, other prokaryotic type IB topoisomerases from mesophilic eubacteria were identified (*2*).

Biochemical analysis revealed that Topo V relaxes both negatively and positively supercoiled DNA in the absence of divalent cations (*1*), a property specific for type IB topoisomerases. Topo V shows a preference for positive over negative supercoiling (*56*). Further analysis revealed that Topo V forms a 3′-end covalent complex with DNA allowing it to be classified as a type IB enzyme (*1*). Topo V has a molecular mass of 110 kDa, which is in the size range of type IB enzymes purified from different eukaryotic sources and is recognized by antibodies raised against human Topo I (*1*, *56*).

The temperature dependence of Topo V activity revealed that relaxation of supercoiled DNA is possible at 50 °C, although at a low level, and reaches its maximum between 70 and 90 °C (*1*). Above 90 °C, Topo V produces unwound DNA, a phenomenon known as unlinking, which was previously described for the hyperthermophilic topoisomerase, Topo III from *D. amylolyticus* (*18*).

More recently the DNA repair activity of Topo V was biochemically characterized, HhH motifs probably being involved in catalysis (*52*). Two of the HhH motifs of Topo V are arranged very similarly to the HhH motifs of human β-pol containing residues involved in the dRP (2-deoxyribose 5-phosphate) lyase activity of this enzyme. It was shown that Topo V possesses the AP lyase and dRP lyase activities associated with the base-excision repair mechanism. Further analysis revealed that the AP/dRP lyase activity is located within the C-terminal region of Topo V and is independent of its topoisomerase activity (*52*, *57*).

E. Gyrase (Type IIa, Eubacteria and Some Archaea)

Gyrase is a specialized type II topoisomerase able to introduce negative supercoils in a circular DNA through an ATP-dependent reaction.

1. PHYLOGENIC POSITION AND SEQUENCE ORGANIZATION

Gyrase was initially discovered in *E. coli* (*58*), and it was thought for a long time to be restrained to mesophilic eubacteria, where it could participate in the regulation of DNA supercoiling and in various cellular processes such as replication, transcription, or recombination/repair. From a phylogenetic point of view, gyrases belong to the large family of type IIa topoisomerases that also comprises the eukaryotic topoisomerases II, as well as eubacterial Topo IV and some phage-encoded topoisomerases, all mesophilic.

Biochemical experiments (47) and genomic sequencing indicate that gyrase is also present in thermophilic and hyperthermophilic eubacteria and in some thermophilic archaeal strains (Table I), but is "replaced" by Topo VI in the large majority of the archaeal species. Thermophilic gyrases have the same organization in two subunits, GyrA and GyrB, as the mesophilic enzymes. Indeed, in a phylogenetic analysis, thermophilic gyrases group with the mesophilic GyrA and GyrB subunits, independently of their belonging to eubacterial or even archaeal species (Fig. 5). This suggests either that they were present in the common ancestor of eubacteria and archaea, or that they appear in a few archeal species presumably by horizontal gene transfer from eubacteria.

Gyrases are homologous and share a common sequence organization with the other type IIa topoisomerases, although they are divided into two subunits (GyrB, GyrA) as in Topo IV (ParE, ParC). By contrast, the sequence of the eukaryotic Topo II is concentrated in a single polypeptide (Fig. 6). Bioinformatics (59) and structural data (60, 61) allow several domains in the gyrase sequence to be distinguished. The GyrB subunit contains at least three domains: (1) an atypical ATP-binding domain formed by a new fold (found in Topo IIb, MutL, HSP 90, etc.) named the "Bergerat fold" (46) followed by an antiparallel β-strand region; (2) a putative nucleic acid-binding domain found in ribosomal protein S5; and (3) a domain involved in the topoisomerization reaction and shared by type IA and type IIa topoisomerases, as well as by DnaG-type primases and some nucleases (59, 62), and named the "Toprim" domain.

The GyrA subunit contains a region mainly involved in the cleavage/religation activity that is similar to the CAP DNA binding domain also found in type IA and type IIb topoisomerases. Finally, the carboxy-terminus of GyrA contains residues essential for the wrapping of DNA around the gyrase structure and is not conserved within the IIa family.

2. Enzymatic Properties

A considerable amount of biochemical and structural data has been accumulated on the *E. coli* gyrase, producing a relatively clear picture of its mechanism. The enzyme is a heterotetramer that captures a duplex DNA. Upon capture the DNA segment wraps around the enzyme in a right-handed fashion, forming a positive crossing. This is converted to a negative crossing by double-strand cleavage, passage, and religation. The mechanism is driven by ATP binding and hydrolysis, which induce conformational changes in the enzyme (63). The result is the net production of two negative supercoils in a circular DNA. The chemistry of the reaction is similar to that of Topo IA, with a transient phosphotyrosine covalent bond formed between each GyrA subunit and each 5' DNA end. Indeed, Topo IA and gyrase share the Toprim and CAP-like structural domains.

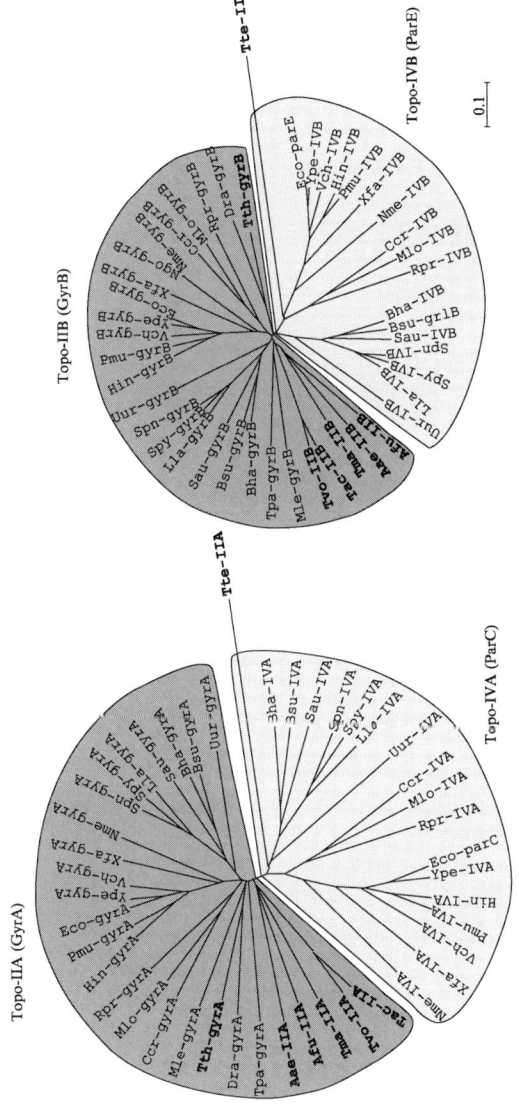

FIG. 5. Phylogenetic trees of prokaryotic topoisomerases IIa. The phylogenetic trees were deduced from Topo IIa sequence alignments produced by ClustalX. Each tree shows the two subfamilies GyrA/ParC and GyrB/ParE corresponding to the divergence between Topo II and Topo IV. (Left) A subunits from prokaryotic Topo IIa. (Right) B subunits. Organism abbreviations are as in Fig. 1, with the following eubacteria in addition: Ccr, *Caulobacter crescentus*; Sau, *Staphylococcus aureus*; Spn, *Staphylococcus pneumoniae*; Spy, *Staphylococcus pyogenes*; Tth, *Thermus thermophilus*; Ype, *Yersinia pestis*. Thermophilic organisms are indicated by bold letters. Note the very long branches of Tte-IIA and Tte-IIB.

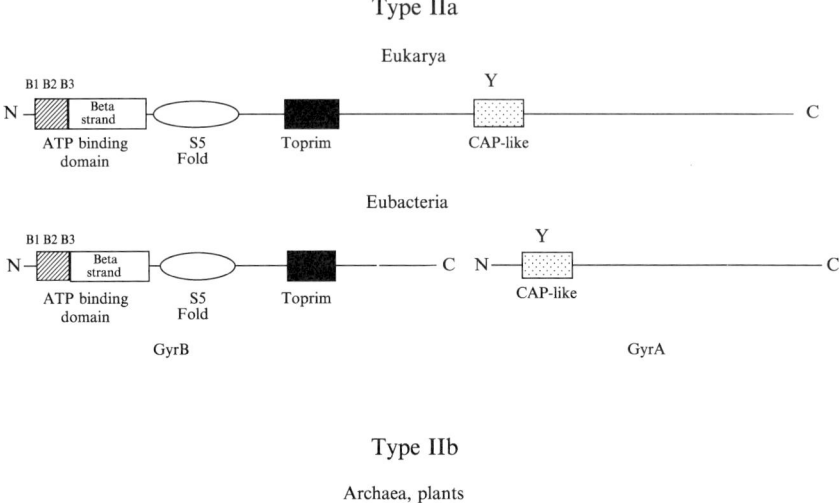

FIG. 6. Domain organization of topoisomerases II. The ATP binding domain present in all type II topoisomerases is composed of the three motifs (B1, B2, B3) comprising the Bergerat fold (hatched box) followed by a β-strand region (white box). The S5 fold is restricted to the type IIa family. The Toprim domain is located in the GyrB region of type IIa enzymes and in the A subunit of type IIb enzymes. The active site tyrosines are indicated by a Y and located within the CAP-like domain of all type II topoisomerases.

However, very little is known about the properties of thermophilic gyrases. The only study available was done on the enzyme from *T. maritima* (47) and shows that the purified protein, as the gyrase from *E. coli*, was able to perform a reaction of negative supercoiling with an optimum activity at 82–86 °C. For an unknown reason, *T. maritima* gyrase is poorly thermostable *in vitro* when incubated alone. The enzyme also differs from that of *E. coli* by the high amount of charged residues in its GyrB subunit, possibly in connection with thermophily (64), and by the tight association between its two subunits, resistant to 5 *M* urea (47).

3. Hypotheses on the Physiological Role of Gyrase in Thermophiles

Two different roles are clearly devoted to gyrase in *E. coli*. The first is to maintain a sufficient degree of negative supercoiling in order to facilitate DNA transactions in the bacterium. The other is to eliminate topological constraints, either positive supercoils or precatenanes, at the replication fork to prepare segregation of newly replicated chromosomes, in concert with Topo IV.

In hyperthermophilic organisms with a growing temperature close to the DNA melting point, the presence of gyrase in some species is puzzling, since negative supercoiling produced by gyrase would facilitate melting. Indeed, hyperthermophiles that contain a gyrase also contain a reverse gyrase that normally produces positively supercoiled DNA (Table I). Therefore, the problem was to determine which of these antagonist enzymes was predominent. In other words, what was the supercoiling state of DNA in these species. This has been investigated by measuring the degree of supercoiling of a plasmid present in *T. maritima*, assuming that the eubacterial DNA has the same topology. The result shows that, surprisingly, the plasmid extracted from the bacterium was negatively supercoiled, as a plasmid from *E. coli* (47). Moreover, when gyrase was inhibited *in vivo* by the addition of novobiocin to the culture of *T. maritima*, the plasmid became positively supercoiled (47). These results suggested that gyrase is the main determinant in the regulation of global DNA supercoiling in hyperthermophilic eubacteria. However, the proportion of free DNA supercoiling in these organisms is not known, and it is likely that a large proportion of supercoiling is constrained in nucleoprotein structures. In this case, reverse gyrase would locally take in charge the regions of naked DNA to prevent their melting, or to reform the DNA duplex after passage of transcription or replication complexes.

The second series of problems concerns the role of gyrase at the replication fork and for the decatenation of newly replicated chromosomes. At the fork, in *E. coli*, precatenanes that accumulate behind the fork are mainly removed by Topo IV, while positive supercoiling that is produced in front of the fork is mainly removed by gyrase. Again, for chromosome decatenation, the main actors seem to be Topo IV and Xer (see below): gyrase would increase its decatenation efficiency only by means of negative supercoiling. However, Topo IV is not present in bacterial thermophiles, and in that case gyrase is supposed to play the role of both topoisomerases for fork movement and decatenation. It would be interesting to test whether thermophilic gyrase is more efficient in decatenation than *E. coli* gyrase. In the case of archaeal thermophiles, decatenation is presumably performed by the unique type II topoisomerase present, Topo VI. A more puzzling case is that of *A. fulgidus* in which gyrase, reverse gyrase, and Topo VI are presumably present: the role devoted to each of these three topoisomerases is not clear.

F. Topoisomerase VI (Type IIb, Archaea)

This new family of type II topoisomerases was discovered unexpectedly at a time when it was thought that all types of topoisomerases had already been described. The first member was purified a few years ago from the archaeon

S. shibatae (65), based on its ATP-dependent decatenation activity. It immediately appeared atypical, formed by two relatively small subunits of 47 (subunit A) and 60 kDa (subunit B). Subsequent identification of the genes by polymerase chain reaction (PCR) using degenerated oligonucleotides and DNA sequencing revealed a totally new topoisomerase, which was named topoisomerase VI (46).

1. SEQUENCE ORGANIZATION AND PHYLOGENIC POSITION

Examination of the sequence from S. shibatae shows almost no similarity to the type IIa family, with the exception of three short sequences in the amino-terminus of the B subunit, which form the Bergerat fold involved in ATP binding and also found in MutL and HSP 90 (see Fig. 6 and the previous section). The A subunit, which drives the trans-esterification reactions, contains a CAP-like domain and a Toprim domain, as Topo IA and IIa, but presents no sequence similarity with these enzymes. Instead, it was shown to be similar to Spo11, a protein involved in the first step of meiotic recombination (66). In particular, the similarity covers the region of the putative active site tyrosine of the A subunit. Remarkably, replacement of this tyrosine by a phenylalanine in S. cerevisiae Spo11 impairs the formation of the double-strand breaks that initiate meiotic recombination (46). Topoisomerase VI was further found in the large majority of archaeal genomes, both thermophiles and mesophiles, but not in bacterial genomes. Recently, both subunit A (Spo11-like) and B were found in the sequence of the *Arabidopsis thaliana* genome, suggesting that a functional TopoVI may exist in plants. The presence of this enzyme besides the "normal" eukaryotic type II topoisomerase addresses the question of their respective functions.

2. BIOCHEMICAL PROPERTIES AND STRUCTURAL DATA

Both A and B subunits of Topo VI from S. shibatae were expressed as recombinant proteins in E. coli and purified. The activity of the holoenzyme (heterotetramer) was reconstituted *in vitro* by subunit mixing, and exhibited a specific relaxation activity comparable to that of the Topo VI purified from *shibatae* (67). As with the type IIa topoisomerases, it is able to relax negative or positive supercoiled DNA and to decatenate intertwined duplexes in ATP-dependent reactions. Topo VI is also able to cleave a DNA duplex, but contrary to type IIa, cleavage is highly dependent on the presence of ATP or ADPNP. The enzyme remains covalently bound to the 5' DNA ends through phosphodiester links involving the two A subunits of Topo VI. Interestingly, the double-strand breaks produced by the enzyme generate two-nucleotide overhangs, instead of the four nucleotide stagger found in type IIa (68). This feature seems consistent with the mode of DNA cleavage found for initiation of meiotic recombination by Spo11 in yeast (69). Spo11 forms a complex with several other proteins that are required for *in vivo* cleavage (66). Similarly, the isolated subunit A of Topo VI is unable to cleave DNA in the absence of the B subunit. Another original feature of Topo VI is that, despite a lack of sequence specificity for cleavage, the 5′ extensions are almost exclusively composed of A or T (68).

Recently, the 3D structure of a large fragment of the A subunit from *M. jannashii* has been obtained (70). The structure confirms the presence of a CAP-like domain and suggests that Topo VI acts as molecular tweezers whose opening and closure are dependent on the dimerization of the B subunit triggered by ATP binding and hydrolysis (68).

In the majority of thermophilic archaea where Topo VI is the sole type II topoisomerase, it is likely that the enzyme performs the essential functions devoted to this class of proteins, in particular at the replication fork and for chromosome decatenation. This does not exclude the possibility that Topo VI is involved in other not yet discovered functions.

II. Recombinases

Specialized recombinases from thermophilic organisms are still poorly studied. We will focus here on two classes of recombinases that are currently under investigation, Holliday junction resolvases and tyrosine recombinases.

Holliday junction resolvases are structure-specific endonucleases that catalyze the resolution of Holliday junctions produced by homologous recombination and DNA repair (Fig. 7). They are widespread enzymes found in all

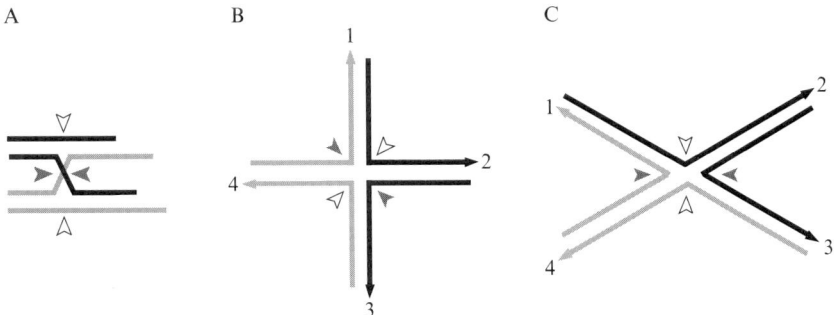

FIG. 7. Holliday junction structures. (A) Schematic diagram of a Holliday junction formed by homologous recombination. Resolution of the junction occurs by dual strand incision in either orientation (indicated by white and dark gray arrows, respectively, for cleavage of the continuous or exchange strands). (B, C) Representation of synthetic Holliday junctions in two configurations, extented square (B) and stacked X structure (C). The 3′ ends of the strands are labeled by an arrowhead. Cleavage positions for resolution are indicated as in A.

TABLE II
INVENTORY OF THERMOPHILIC HOLLIDAY JUNCTION RESOLVASES AND TYROSINE RECOMBINASES[a]

	Holliday junction resolvases	Tyrosine recombinases	
		Xer like	Integrase like
Complete archaeal genomes			
Crenarchaea			
Aeropyrum pernix	APE0461	APE0805	
Pyrobaculum aerophilum	PAE2162	—	
Sulfolobus solfataricus	SSO0575 (Hjc)	SSO0375	
	SSO1176 (Hje)		
Sulfolobus tokodaii	ST1444	ST1393	
Euryarchaea			
Archaeoglobus fulgidus	AF1965	—	
Methanobacterium thermoautotrophicum	MTH1270	MTH893	
Methanococcus jannaschii	MJ0497	MJ0367	
Methanopyrus kandleri	MK0102	—	
Pyrococcus abyssi	PAB0552	PAB0255	
Pyrococcus horikoshii	PH1328	PH1826	
Pyrococcus furiosus	PF1503	PF1868	
Thermoplasma acidophilum	Ta0069	Ta1314	
Thermoplasma volcanium	TVN0025	TVN0263	
Complete bacterial genomes			
Aquifex aeolicus	aq_1953	aq_1137	
Thermoanaerobacter tengcongensis	TTE1178	TTE1363	
Thermotoga maritima	TM0575	TM0967	
Viral genes			
SIRV1	CAC37359		
SIRV2	CAC37360		
SSV1			P20214

[a]For sequenced prokaryotic genomes, the gene numbers correspond to the notations available at http://www.ncbi.nlm.nih.gov/PMGifs/Genomes/micr.html. For viral genes the accession number is indicated.

three kingdoms. Thermophilic representatives have been identified both in eubacteria and archaea (Table II). Despite their lack of sequence homologies all Holliday junction resolvases most probably follow the same general mechanism where the targeted phosphodiester bond is hydrolyzed by using a metal-ion-activated water molecule.

Two large families of site-specific recombinases have been defined, the resolvase family and the tyrosine recombinases family (71). These two families are evolutionary unrelated and use a different chemistry to catalyze site-specific recombination. Members of the tyrosine recombinases family are found in eukaryotes, and prokaryotes and their phages. Their functions include integration and excision of viral and plasmid DNA in and out of the host chromosome, resolution of catenated DNAs, regulation of plasmid copy number, conjugative transposition, and regulation of gene expression by DNA inversion or excision. Members of the tyrosine recombinases family catalyze site-specific recombination between two DNA sites by using a Topo IB-like mechanism to cut and religate DNA strands (72, 73). Unlike topoisomerases, tyrosine recombinases perform the ligation step after strand exchange between the two DNA partners. After the first round of strand cleavage–strand exchange, a Holliday junction is formed. This recombination intermediate is resolved by a second round of strand cleavage–strand exchange, thus completing the recombination reaction. However, tyrosine recombinases are able to resolve only Holliday junctions that they generated, as opposed to Holliday junction resolvases.

A. Holliday Junction Resolvases

1. PHYLOGENY

Holliday junction resolvases (HJR) share identical biochemical properties (see the next section) but are remarkably diverse at the level of protein sequences and structure homologies supporting the idea that several independent evolutionary events led to this widespread class of enzymes. Six classes of HJR have been identified: RuvC in eubacteria, RusA in bacteriophages and eubacteria, T4 endonuclease VII and T7 endonuclease I in bacteriophages, Cce1/Ydc2 in yeasts, and Hjc in archaea (74). Thermophilic eubacteria encode HJRs homologous to mesophilic eubacterial HJRs: *Thermotoga maritima* and *Thermoanaerobacter tengcongensis* encode a RuvC homologue (75) (Table II). The *T. tengcongensis* RuvC have been identified by database searches, and the corresponding gene located next to the *ruv*A and *ruv*B genes (7) while *Aquifex aeolicus* has a RusA homologue (76) and lacks the RuvABC system, a case unique in eubacteria (77).

The situation is more complex in archaea where no orthologues of RuvC or RusA can be found. HJRs encoding genes were identified after purifying the enzymatic activity, N-terminal sequencing of the protein, and genome search for similitudes. The first archaeal HJR activity (referred as to Hjc, *H*olliday *j*unction *c*leavage) was detected in *Pyrococcus furiosus* and the corresponding gene was identified (78). Hjc is conserved in all archaea whose complete genome sequences are available (Table II) and may therefore be the cellular archaeal HJR, as is RuvC in eubacteria. It should be noted that the putative Hjc of the two *thermoplasmales* (Table II) have an N-terminal extention that make them divergent to all other Hjc that are highly conserved in size and N-terminal sequences. Assignment of these genes to the Hjc family therefore has to be taken with care.

In *S. solfataricus*, a second and distinct HJR activity (Hje, *H*olliday *j*unction *e*ndonuclease) has been found (79), and the corresponding gene identified in the genome (80) (Table II). This second HJR is similar to enzymes encoded by two viruses SIRV1 and SIRV2 infecting *Sulfolobus islandicus* (81, 82). Sequence alignments of all the archaeal HJRs and construction of a phylogenetic tree (Fig. 8) strongly suggest that Hje could have a viral origin like RusA in *E. coli*.

More recently another different HJR activity (Hjr, *H*olliday *j*unction *r*esolving) was detected in *P. furiosus* (83). However, the Hjr-encoding gene remains unidentified suggesting that the enzyme has very little or no similarities with the other HJRs. This could be consistent with a viral or plasmidic origin for Hjr, either with a strongly divergent gene or a new unrelated subclass of HJR.

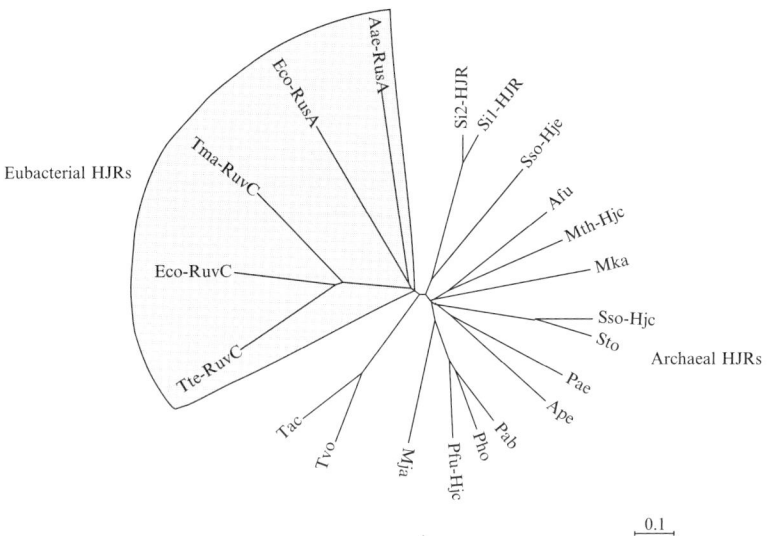

FIG. 8. Phylogenetic tree of eubacterial and archaeal Holliday junctions resolvases. The phylogenetic tree was deduced from HJRs sequence alignments produced by ClustalX. Eubacterial HJRs are within the gray shaded box. Organism abbreviations are as in Fig. 1, with Si1 and Si2 SIRV1 and SIRV2 viruses, respectively.

Sequence and structural analysis revealed that all known HJRs can be classified within three superfamilies defined by specific structural folds (25) (Table III). RuvC and its homologs belong to the Hsp70/Rnase H fold. All archaeal HJRs (84) independently of their viral or cellular origin harbor the three conserved motifs of the endonuclease superfamily (25) (Fig. 9). This superfamily also includes several other nucleases, like the recB-like nucleases, the λ exonuclease family, and the Vsr-like nuclease family (25). The last fold (Endo VII) regroups the bacteriophages HJRs and related nucleases.

2. Enzymatic Properties

All the identified junction-resolving enzymes isolated from mesophilic organisms share some biochemical properties. They are homodimeric in solution, require Mg^{2+} for cleavage activity, and dissociate from Holliday junctions with a K_d in the low nanomolar range.

Two HJRs from thermophilic eubacteria have been biochemically characterized. *In vitro* the RusA protein from *A. aeolicus* is able to resolve Holliday junctions, but its sequence preference for cleavage (3' to TG) differs from that of *E. coli* RusA (5' to CC) (76) (Table III). The *T. maritima* RuvC has also been purified and tested for activity *in vitro* and remarkably shows the same preferred cleavage site as *E. coli* RuvC, that is cleavage between the third and fourth positions of a tetranucleotide sequence with the consensus 5'-(A/T)TT (G/C)-3'(75).

TABLE III
STRUCTURAL AND CLEAVAGE PROPERTIES OF THE DIFFERENT CLASSES OF HOLLIDAY JUNCTION RESOLVASES

Enzyme	Origin	Structural fold[a]	Sequence specificity
RuvC	Eubacteria	Hsp70/Rnase H	Yes: 5'-(A/T)TT↓(G/C)
RusA	Lambdoid phage	None identified	Yes: 5'-↓CC
RusA	*A. aeolicus*	None identified	Yes: 5'-TG↓
Cce1	*S. cerivisiae* mitochondria	Hsp70/Rnase H	Yes: 5'-ACT↓A
T4 endo VII	Bacteriophage T4	Endo VII	Weak: cleaves 2 nt 3' of the junction center
Hjc	Archaea	Endonuclease	Weak: cleaves 3 nt 3' of the junction center
Hje	*S. solfataricus*	Endonuclease	Weak: cleaves 2 nt 3' of the junction center
SIRV1/2	Archaeal viruses	Endonuclease	Weak: cleaves 2 nt 3' of the junction center
Hjr	*P. furiosus*	?	Weak: cleaves 1 nt 3' of the junction center

[a]The superfamilies of the different HJRs were defined on the bases of structural folds conservation (25). No structural cognate was identified for RusA. No sequence data are available for Hjr, thus obliterating its classification.

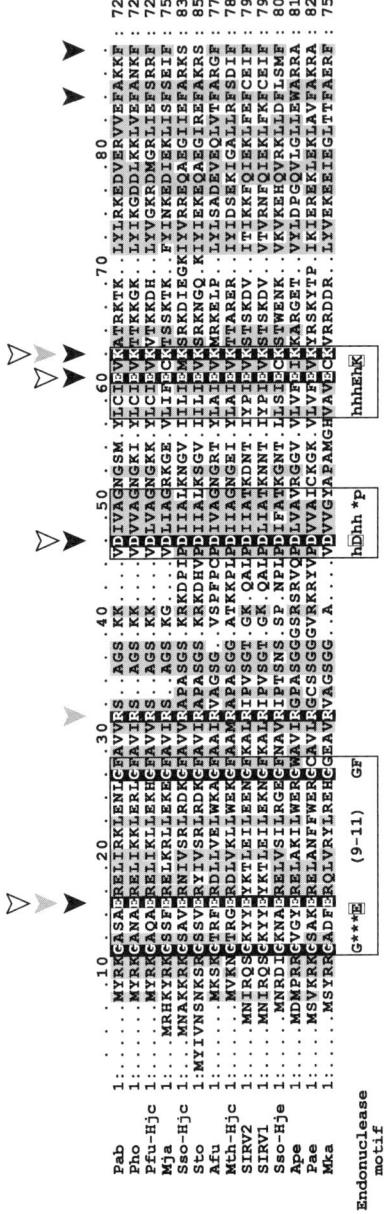

FIG. 9. Alignment of archaeal Holliday junction resolvases. Comparison of archaeal HJRs (single-letter amino acid code) identified by the abbreviations defined in Fig. 1. The sequence alignments were produced by ClustalX. Black boxed residues indicate 100% identity, gray boxed residues ≥ 60% conservation. Substitutions were considered as conservative when following the Dayhoff exchange groups: S, P, A, G, and T; I, L, M, and V; D, E, Q, and N; F, W, and Y; R, K, and H; C. The three conserved regions defining the endonuclease motif are boxed, and the consensus is indicated below the alignment with the convention defined in Aravind et al. (25): the E, D, and K boxed residues are invariant within the superfamily; upper case letters indicate the conserved residues that are present in >25% of the cases; h, hydrophobic residues (YFWLIVMA); p, polar residues (STQNEDRKH); *, any. Arrows above the alignment indicate the positions mutagenized in P. furiosus (black arrows), M. thermoautotrophicum (gray arrows), and S. solfataricus (white arrows) Hjc.

The main archaeal HJR activity from three thermophilic species, Hjc, has been characterized in vitro (78, 85–87). Recombinant Hjc from *P. furiosus*, *S. solfataricus*, and *M. thermoautotrophicum* have the same biochemical properties as mesophilic HJRs. They are homodimeric in solution and bind specifically to four-way DNA junctions in vitro (85–87). Their apparent K_d is in the same range as that observed for the eukaryotic Cce1 and eubacterial RuvC HJR. It was further demonstrated that the *P. furiosus* Hjc binds its substrate in a dimeric form (86), a property shared by all the junction-resolving enzymes.

All three enzymes cleave mobile and fixed holliday junctions and, like *E. coli* RuvC, require Mg^{2+} ions for full activity (85–87). Cleavage of fixed junctions by the three Hjc introduces symmetrical nicks located three nucleotides in 3' of the junction center, suggesting a particular fold of the junction upon Hjc binding. Experiments on *S. solfataricus* Hjc suggested that the four-way DNA junction is bound in a 2-fold symmetric conformation with estimated angles of 60°/120° between the arms of the junction (stacked-X form, Fig. 7), and that base pairing of the DNA duplex near the cleavage site was weakened or disrupted by Hjc binding (88). The three archaeal Hjc enzymes show a cleavage preference for the two continuous strands of the junction (85–87). No consensus sequence can be found around the cleavage positions, a property that differentiates the archaeal enzymes from the eukaryal and eubacterial enzymes Cce1, RuvC, and RusA (Table III). Mobile junctions are cleaved like fixed junctions, a behavior reminiscent of mammalian HJR and not of eubacterial ones. However, there is only one cleavage site on each arm of the mobile junction, suggesting the existence of a sequence preference when the enzyme is allowed to select its cleavage site.

In RuvC four acidic residues are known to be essential for the resolvase activity. Sequence alignments of all Hjc revealed that they have three conserved acidic residues and a Lys residue (Fig. 9) that belong to the endonuclease motif (85, 88, 89). These residues were suggested to be part of the catalytic pocket and therefore targeted for mutagenesis. In vitro testing of the *P. furiosus* and *S. solfataricus* mutants of Hjc revealed that binding of the enzymes to their substrates was not affected but that cleavage was fully abolished by these mutations (88, 90). A catalytic pocket composed of Glu-9, Asp-33, Glu-46, and Lys-48 for *P. furiosus* has therefore been defined. Furthermore two residues of the *P. furiosus* Hjc (Phe-68 and Phe-72) have been shown to be involved in dimer formation as revealed by sedimentation equilibrium experiments (90). These two mutants show a defect in substrate binding correlated by a loss of cleavage activity. These results are in agreement with the hypothesis that the Hjc dimer may consist of two nucleases domains that would allow DNA cleavage on either side of the branch point of the four-way junction (88).

Mutants of the Hjc from *M. thermoautotrophicum* were tested *in vivo* in *E. coli*. Indeed, the wild-type Hjc was shown to complement *E. coli ruv* mutants (85). This complementation is RecG dependent in the absence of RuvABC, and probably reflects a requirement for RecG branch migration activity to form junctions that can be resolved by Hjc. Mutants at positions corresponding to Glu-9, Arg-25, and Lys-48 in *P. furiosus* were unable to promote cell survival upon irradiation of a ΔruvAC ΔrusA strain most probably because of a catalytic defect in junction resolution (85).

All these results taken together strongly suggest that Hjc is the functional homolog of RuvC in archaea, and that formation and resolution of recombination intermediates in archaea follow mechanisms similar to those in eubacteria and probably eukarya. This is further emphasized by the observation that the *P. furiosus* Hjc is able to catalyze the cleavage of a Holliday junction formed by the RecA-mediated strand exchange reaction between plasmid DNAs *in vitro* (78). Furthermore, the enzyme interacts with RadB, a homolog of Rad51/RecA (91). RadB was shown to modulate the Hjc activity: cleavage of a Holliday junction by Hjc is inhibited by RadB in the absence but not in the presence of ATP. It was then proposed that the RadB–Hjc complex would be required for branch migration and that upon arrival at a cleavage site RadB would dissociate thus liberating the Hjc activity (91).

The second group of archaeal HJR has also been biochemically characterized. The Hje enzyme was purified from *S. solfataricus* cells (79) whereas the SIRV1 and SIRV2 HJR were expressed and purified from *E. coli* (81). All three enzymes share the Hjc biochemical properties. They bind to four-way DNA junctions and have an apparent K_d in the low nanomolar range, their cleavage activity is dependent on Mg^{2+} ions, and they are able to cleave both fixed and mobile Holliday junctions. Both Hje and SIRV HJR show a strong preference for cleavage of the continuous strands on fixed junctions, where they introduce symmetrical paired nicks two nucleotides in 3′ of the junction center (Table III), thus cutting two strands out of four in the junction (79, 81). On mobile junctions, they introduce cleavage in pairs on opposing strands, i.e., the four arms of the junction are cut symmetrically. Not all the available positions are cut, probably reflecting a specificity for cleavage of the continuous strand. None of these enzymes has a sequence specificity for cleavage. The cleavage properties of the three enzymes are identical but different from the cellular Hjc (Table III), thus supporting the hypothesis that Hje would be of viral origin, as is suggested by the phylogenetic tree of archaeal HJR (Fig. 8).

The last archaeal HJR activity, Hjr, was recently isolated from *P. furiosus* cells (83). Hjr is able to resolve fixed and mobile DNA junctions but differs in its properties from both Hjc and Hje. Cleavage of the mobile junction occurs in all four strands with preferred sites but no sequence selectivity. Hjr also cleaves the four strands of a fixed junction, and cleavage occurs one nucleotide in 3' of the junction center (83). The Hjr enzyme is clearly distinct in specificity from Hjc and Hje (Table III). By analogy with eubacteria and eukarya, it is thus tempting to assume that Hjr, like Hje, has a viral origin that remains to be identified.

3. STRUCTURE OF ARCHAEAL HJR

The structures of Hjc enzymes from *P. furiosus* and *S. solfataricus* have been solved (92, 93). Both enzymes are homodimers with a topology different from other HJRs, the prokaryotic RuvC, the bacteriophages T_7 endo I and T_4 endoVII enzymes, and the eukaryotic Ydc2 enzyme (94–97). By contrast, the dimeric architecture and active site conformation of Hjc are highly similar to type II restriction endonucleases. The topology of the Hjc monomer is a subset of that of the type II restriction enzymes suggesting that the Hjc subunit represents a minimal nuclease domain (92). Superposition of the Hjc structure onto the *Fok*I, *Eco*RV, and *Bgl*I structures revealed the spatial coincidence of Glu-9, Asp-33, Glu-46, and Lys-48 in Hjc with three acidic and one Lys residue of the restriction enzymes (93). All these residues constitute the canonical motif of type II endonucleases (Fig. 9) with the three acidic residues forming the metal binding pocket for the catalytic Mg^{2+}. To take all the biochemical data into account, modelizations of the Hjc–Holliday junction DNA complex require that the junction adopt a stacked-X form. Only in this case the two cleaved phosphodiester bonds of the two continuous strands are located exactly near the active site of each Hjc subunit (93).

The dimer formations in RuvC and Hjr involve completely different parts of the molecules. The dimer assembly could be the key control of their differing DNA-cutting patterns, Hjc cutting 3 nucleotides 3' of the point of strand exchange whereas RuvC cuts exactly at this point (92).

HJRs possess similar biochemical properties such as dimerization, binding to four-way junction DNAs, and requirement of Mg^{2+} ions for cleavage activity. However, at least six classes of HJRs have been defined on sequence homologies. Members of five of these six classes were crystallized, and each HJR appeared to differ in its fold. This strongly suggests that each protein evolved convergently to recognize the Holliday junction. For archaeal HJR, the catalytic module that evolved is a nuclease domain that was also retained in evolution to give the catalytic domain of type II restriction endonucleases.

B. Viral Tyrosine Recombinases: The Fuselloviridae Integrases

1. PHYLOGENY

Viruses from thermophilic organisms have been poorly studied, although some effort has been made to collect and classify viral particles sampled from geographically independent hot springs (98–100). Of 12 viral morphotypes described in archaea, four new virus families have been isolated from the thermophile *Sulfolobus*: Fuselloviridae, Rudiviridae, Lipothrixviridae, and Guttaviridae (101).

Members of the Fuselloviridae family have been isolated from several acidic hot springs located in Japan [SSV1 (102)], Iceland [SSV2 and SSV3 (99)], the United states [SSVY1, SSVY2, and SSVY3 (98)], and Russia (SSV-K, K. Stedman, unpublished observations). Analysis of the SSVs genomic sequences available indicates that these viruses are genetically distinct yet distinctly related (98) (K. Stedman, personal communication). All are temperate viruses of the genera *Sulfolobus* and have a lemon-shaped virion containing a 15-kbp circular double-stranded DNA genome. Another attribute of SSV particles is their potential capability to integrate their genome into the host chromosome. It was shown for SSV1 that integration is site specific and mediated by the viraly encoded integrase (103, 104).

Alignment of the integrases from SSV1, SSV2, and SSV-K shows that they are distinct but very similar to each other, suggesting that their specificity of insertion could be different (K. Stedman and M. C. Serre, unpublished observations).

2. ENZYMATIC PROPERTIES

The only viral tyrosine recombinase from a thermophile to be studied is the SSV1 integrase. It belongs to the large family of tyrosine recombinases (105–107) and is the most distantly related member of the family (105). Further the SSV1 integrase harbors substitutions at several conserved catalytic positions (106). These observations raised the possibility that the recombination mechanism might be different in archaea. The integration and excision reactions have been shown to work *in vitro* in the presence of DNA fragments containing the viral and chromosomal *att* sites where the recombination takes place (103, 108). The integrase has been expressed and purified from *E. coli*, and its cleavage properties determined (109). Like other tyrosine recombinases, the cleavage reaction leads to the formation of a 3′-phosphotyrosine intermediate between Tyr-314 and the substrate DNA. Furthermore, mutations at each of the conserved residues involved in catalysis within the tyrosine recombinases family abolish or strongly alter the enzymatic properties of the SSV1 integrase (C. Letzelter, unpublished observations). However, the charge-relay system within the active site seems to be different from the mesophilic species. Determination of the 3D structure is underway and may identify the residues involved in thermal adaptation of the catalytic pocket as well as thermostabilization of the enzyme by comparison with the structures of tyrosine recombinases from mesophiles (110). Like other tyrosine recombinases the SSV1 integrase has a relaxation activity on supercoiled DNA in the absence of divalent cations (M. C. Serre, unpublished observations). Taken together, these observations reveal that the catalytic properties of the thermophilic archaeal integrase are similar to those observed for tyrosine recombinases and type IB topoisomerases from mesophilic sources (72) suggesting that speciation of a "Tyr-recombinase/Topo IB" catalytic domain is an ancient process that probably took place in LUCA.

In vivo, SSV1 specifically integrates into a tRNA^{Arg} gene in S. shibatae or S. solfataricus (104, 111). In vitro all the sequence information for binding and cleavage is located within a minimal 19-bp sequence corresponding to the anticodon stem-loop sequence of the tRNA gene where SSV1 integration takes place and the cleavage sites target the border of the anticodon loop sequence (109). It was proposed that tRNA genes may have been the recognition sequences for the ancestral site-specific recombinase (104) and that mobile elements integrating in tRNA may use the symmetry of the anticodon stem-loop sequence as the recombination site (22). The observations made for the SSV1 integrase strongly support these hypotheses.

C. Cellular Tyrosine Recombinases: The Thermophilic Xer Proteins

The role of Xer recombination in chromosomal metabolism is to convert chromosomal multimers produded by homologous recombination to monomers. This allows the correct inheritance of the chromosomes into the daughter cells (112). The Xer system is also involved in the stable inheritance of several multicopy plasmids. In E. coli, the Xer site-specific recombination system is composed of two related tyrosine recombinases, XerC and XerD, and a specific DNA recombination site *dif*. Both XerC and XerD proteins are required to mediate site-specific recombination at the *dif* site (113), each protein performing only the cleavage of one strand (114).

Eubacteria containing both circular chromosomes and homologous recombination systems were hypothesized to share mechanisms for resolving replicons. Indeed, in most mesophilic eubacteria whose genomes have been sequenced, XerC and XerD homologues are found. The absence of Xer genes in *Borrelia burgdorferi*, *Mycoplasma genitalium*, and *Mycoplasma pneumoniae* is also consistent with the above statement since the former has a linear genome and the two latter show deficiencies in homologous recombination genes (115).

The mechanisms that ensure chromosome and plasmid segregational stability in archaea are not as well documented as in *E. coli*. Interestingly, whereas most proteins involved in DNA replication and recombination conserved in all archaea are eukaryotic like (*116, 117*), a unique eubacterial-like Xer gene can be found in most archaeal genomes sequenced (Table II). Whether the Xer-like gene is absent from *A. fulgidus, P. aerophilum,* and *M. kandleri* or has drastically diverged is not known.

The presence of a single Xer gene is not specific for the archaeal kingdom. All three genomes of the hyperthermophilic eubacteria so far sequenced also contain a single Xer-like gene (Table II) suggesting at first glance that a unique Xer-like gene would be a marker of thermophily. Alignment of the thermophilic Xer-like proteins with *E. coli* XerC and XerD reveals that similarities extend from the N- to the C-terminus. All the previously defined catalytic boxes (*105*) are conserved and all consensus residues within these boxes are present in the Xer-like proteins (Fig. 10). Although the thermophilic Xer-like proteins form a subgroup distinct from the XerC and XerD subgroups (Fig. 11), no residue conservation that might be a specific signature for thermophily can be found. The amino acid identity between the different thermophilic archaeal Xer-like genes is in the range of 19–39% (33–91% similarity) and the identity between thermophilic Xer-like proteins and mesophilic XerC/D proteins is in the range of 12–33% (31–52% similarity), compared with the 35% identity between XerC and XerD from *E. coli*. Furthermore, in the Protein Data Base all the Xer-like proteins match the XerD structure. These findings suggest that the unique Xer present in eubacterial and archaeal thermophiles is presumably involved in chromosome segregation. This Xer-like protein would be less specialized than the XerC and XerD proteins and likely be able to cleave both strands of the recombination site.

The increasing amount of prokaryotic genome data revealed, however, a more complex situation (*115*). Several mesophilic archaeal and eubacterial species also have a unique Xer-like gene (Fig. 11). Even more puzzling is the observation that *Mesorhizobium loti, Helicobacter pylori*,[4] and *Ureaplasma urealyticum* have three copies of the Xer-like gene. Sequence alignments and construction of a phylogenetic tree reveal that the Xer family can be divided in three groups, the well-defined XerC and XerD groups and the new Xer-like group (Fig. 11). Within this group are found all the unique Xer-like proteins from archaea and eubacteria, the three Xers from *H. pylori* and *U. urealyticum*, the two Xers from *Lactococcus lactis*, but only one of the three Xers from *M. loti*. The two other Xers from this organism are found in the XerC and

---

[4] The third Xer-like gene of *H. pylori* 26695 is disrupted by an *IS*.

TOPOISOMERASES AND RECOMBINASES FROM THERMOPHILES 71

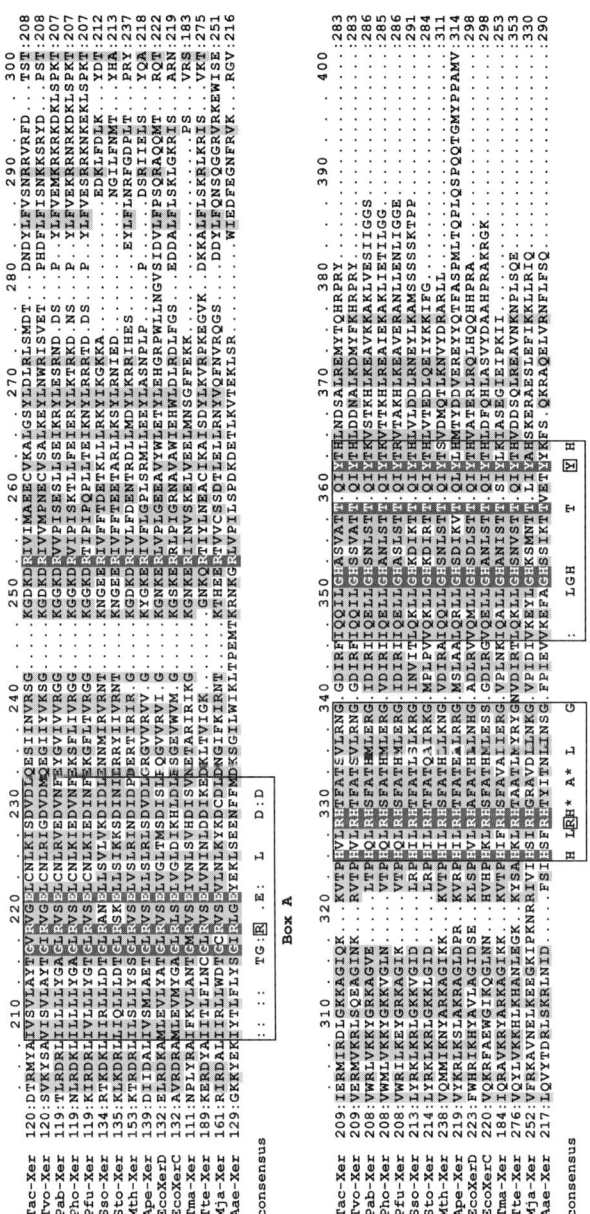

FIG. 10. Alignment of thermophilic Xer-like proteins. Comparison of the thermophilic Xer-like proteins (single-letter amino acid code) identified by the abbreviations defined in Fig. 1. The sequence alignments were produced by ClustalX. Black boxed residues indicate 100% identity, gray boxed residues ≥ 60% conservation. Substitutions were considered as conservative when following the Dayhoff exchange groups: S, P, A, G, and T; I, L, M, and V; D, E, Q, and N; F, W, and Y; R, K, and H; C. The three conserved regions defining the tyrosine recombinase active site motif are boxed. The consensus sequence was determined by aligning the motifs of 75 prokaryotic tyrosine recombinases (105). Box A, the invariant arginine is boxed. The other defined residues are identical in >50% of sequences. Colons indicate the presence of hydrophobic residues (AVILM) at >75% of the positions, and the single dot indicates residues with a small side chain (GSA) at >75% of the positions. Box B, the invariant arginine is boxed. An asterisk indicates the presence of similar residues (STA) at >75% of the positions. Box C, the catalytic tyrosine is boxed.

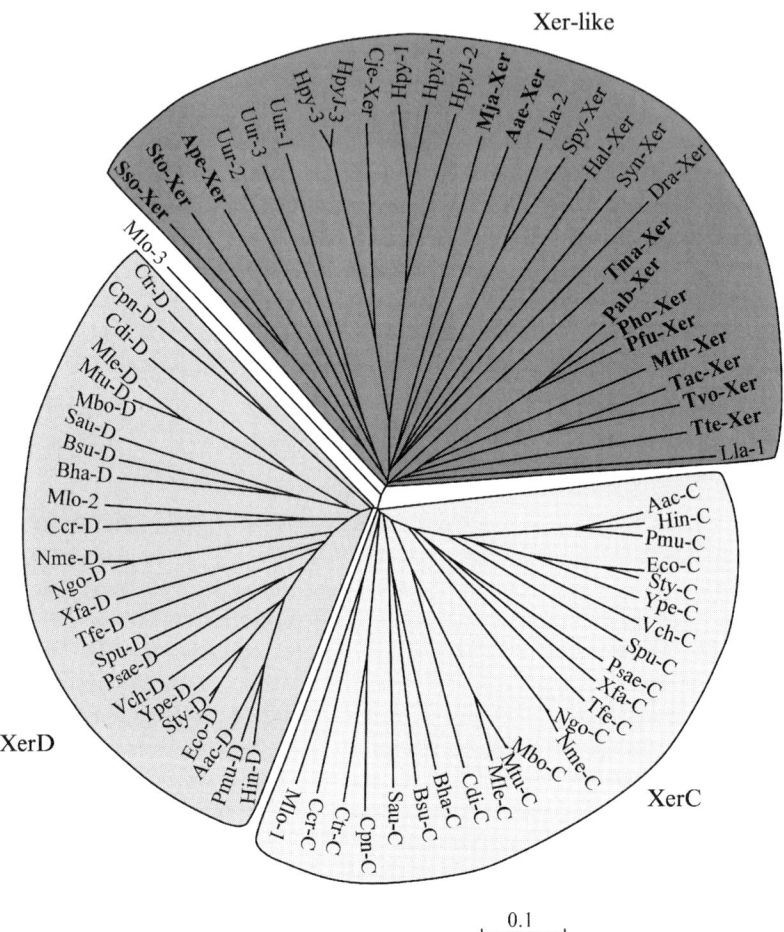

FIG. 11. Phylogenetic tree of Xers proteins. The phylogenetic tree was deduced from Xer sequence alignments produced by ClustalX. The three Xer families are boxed: XerC, light gray shading; XerD, medium gray shading; Xer-like, dark gray shading. Thermophilic organisms are indicated by bold lettering. Organism abbreviations are as in Fig. 1, with the following species additions: Archaea: Hal, *Halobacterium* sp. NRC-1. Eubacteria: Aac, *Actinobacillus actinomycetemcomitans*; Ccr, *Caulobacter crescentus*; Cdi, *Clostridium difficile*; Cpn, *Chlamydophila pneumoniae*; HpyJ, *Helicobacter pylori* J99; Mbo, *Mycobacterium bovis*; Ngo, *Neisseria gonorrhoeae*; Psae, *Pseudomonas aeruginosa*; Sau, *Staphylococcus aureus*; Spu, *Shewanella putrefaciens*; Spy, *Streptococcus pyogenes*; Sty, *Salmonella typhimurium*; Tfe, *Thiobacillus ferrooxidans*; Ype, *Yersinia pestis*.

XerD subgroups. It is thus tempting to speculate that the huge Xer family is still evolving. From a unique gene ancestor, gene duplication (or triplication in some cases) occurred, followed by speciation into the XerC and XerD groups. Because no mesophilic or thermophilic archaea have more than one Xer-like gene in their genome, the gene duplication event may have taken place after the separation of the two prokaryotic phyla. Another possibility that cannot be ruled out is that gene duplication occurred in the last prokaryotic ancestor followed by the loss of one of the alleles in several species. The hypothesis that a unique Xer-like protein would be more efficient than the XerC/D system in a high-temperature environment should also be considered as a selection pressure for the thermophilic eubacteria. The case of the radioresistant bacterium *D. radiodurans*, which although mesophilic has a unique Xer-like protein, could be interpretated in this scheme since previous phylogenetic studies have suggested that the *Deinococci* are closely related to the *Thermus* genus (*118*) and that the common ancestor of the *Deinococcus–Thermus* group was thermophilic (*119*).

Anyhow, the universality of Xer genes in prokaryotes suggests that the mechanisms involved to resolve multimers of circular replicons are conserved in most members of the two prokaryotic domains. Study of the biochemical properties of thermophilic Xer-like proteins may therefore be of great help in analyzing the replication termination mechanism in archaea.

III. Concluding Remarks

Our present picture of thermophilic topoisomerases and recombinases is deeply influenced by genomic data, since a very small fraction of the proteins described *in silico* has been expressed and biochemically characterized. From a phylogenetic or a functional point of view, there does not seem to be a clear logic in the presence or absence of a given gene in a particular genome. For instance, the exclusive presence of Topo IB in eukaryotes and Topo IA in prokaryotes is no longer valid. Other typical examples are the lack of explanation for the existence of a gyrase together with a Topo VI and a reverse gyrase in *A. fulgidus* or the puzzling distribution of Xer genes in the living world. One can always argue about putative horizontal transfers, but there is little evidence for a general phenomenon. Thus, multiplication of sequenced genomes has, for the first time, complicated our understanding of how thermophilic topoisomerases and recombinases evolved and the various functions they sustain. It is possible that the outcome of a number of additional genome sequences would provide better answers to these fundamental questions. At the same time, one can imagine that the biochemical characterization and 3D structure determination of an increasing number of these proteins would be performed. Moreover, it is noteworthy that *in silico* analysis of genomes will permit proteins to be identified whose sequence has diverged only a little. Therefore, there is a wide field of research for enzymes of unrelated or largely divergent sequences that kept similar biochemical functions. A recent example is Topo VI, which was impossible to detect on the basis of its sequence, but which was identified by its topoisomerase activity. Finally, the reason for the observed redundancy of topoisomerases and recombinases in a given genome and the question of their respective biological functions in thermophilic organisms remain to be clarified.

References

1. Slesarev, A. I., Stetter, K. O., Lake, J. A., Gellert, M., Krah, R., and Kozyavkin, S. A. (1993). DNA topoisomerase V is a relative of eukaryotic topoisomerase I from a hyperthermophilic prokaryote. *Nature* **364**, 735–737.
2. Krogh, B. O., and Shuman, S. (2002). A poxvirus-like type IB topoisomerase family in bacteria. *Proc. Natl. Acad. Sci. USA* **99**, 1853–1858.
3. Bouthier de la Tour, C., Kaltoum, H., Portemer, C., Confalonieri, F., Huber, R., and Duguet, M. (1995). Cloning and sequencing of the gene coding for topoisomerase I from the extremely thermophilic eubacterium, *Thermotoga maritima*. *Biochim. Biophys. Acta* **1264**, 279–283.
4. Nelson, K. E., Clayton, R. A., Gill, S. R., Gwinn, M. L., Dodson, R. J., Haft, D. H., Hickey, E. K., Peterson, J. D., Nelson, W. C., Ketchum, K. A., McDonald, L., Utterback, T. R., Malek, J. A., Linher, K. D., Garrett, M. M., Stewart, A. M., Cotton, M. D., Pratt, M. S., Phillips, C. A., Richardson, D., Heidelberg, J., Sutton, G. G., Fleischmann, R. D., Eisen, J. A., Fraser, C. M., *et al.* (1999). Evidence for lateral gene transfer between archaea and bacteria from genome sequence of *Thermotoga maritima*. *Nature* **399**, 323–329.
5. Kaltoum, H., Portemer, C., Confalonieri, F., Duguet, M., and Bouthier de la Tour, C. (1997). DNA topoisomerases I from thermophilic bacteria. *System. Appl. Microbiol.* **20**, 505–512.
6. Deckert, G., Warren, P. V., Gaasterland, T., Young, W. G., Lenox, A. L., Graham, D. E., Overbeek, R., Snead, M. A., Keller, M., Aujay, M., Huber, R., Feldman, R. A., Short, J. M., Olsen, G. J., and Swanson, R. V. (1998). The complete genome of the hyperthermophilic bacterium *Aquifex aeolicus*. *Nature* **392**, 353–358.
7. Bao, Q., Tian, Y., Li, W., Xu, Z., Xuan, Z., Hu, S., Dong, W., Yang, J., Chen, Y., Xue, Y., Xu, Y., Lai, X., Huang, L., Dong, X., Ma, Y., Ling, L., Tan, H., Chen, R., Wang, J., Yu, J., and Yang, H. (2002). A complete sequence of the *T. tengcongensis* genome. *Genome Res.* **12**, 689–700.
8. Lima, C. D., Wang, J. C., and Mondragon, A. (1994). Three-dimensional structure of the 67K N-terminal fragment of *E. coli* DNA topoisomerase I. *Nature* **367**, 138–146.
9. Berger, J. M., Fass, D., Wang, J. C., and Harrison, S. C. (1998). Structural similarities between topoisomerases that cleave one or both DNA strands. *Proc. Natl. Acad. Sci. USA* **95**, 7876–7881.
10. Bouthier de la Tour, C., Portemer, C., Forterre, P., Huber, R., and Duguet, M. (1993). ATP-independent DNA topoisomerase from *Fervidobacterium islandicum*. *Biochim. Biophys. Acta* **1216**, 213–220.
11. Viard, T., Lamour, V., Duguet, M., and Bouthier de la Tour, C. (2001). Hyperthermophilic topoisomerase I from *Thermotoga maritima*. A very efficient enzyme that functions independently of zinc binding. *J. Biol. Chem.* **276**, 46495–46503.

12. Zechiedrich, E. L., Khodursky, A. B., Bachellier, S., Schneider, R., Chen, D., Lilley, D. M. J., and Cozzarelli, N. R. (2000). Roles of topoisomerases in maintaining steady-state DNA supercoiling in *Escherichia coli*. *J. Biol. Chem.* **275**, 8103–8113.
13. Stewart, L., Redinbo, M. R., Qiu, X., Hol, W. G., and Champoux, J. J. (1998). A model for the mechanism of human topoisomerase I. *Science* **279**, 1534–1541.
14. Dean, F. B., and Cozzarelli, N. R. (1985). Mechanism of strand passage by *Escherichia coli* topoisomerase I. The role of the required nick in catenation and knotting of duplex DNA. *J. Biol. Chem.* **260**, 4984–4994.
15. Jaxel, C., Duguet, M., and Nadal, M. (1999). Analysis of DNA cleavage by reverse gyrase from *Sulfolobus shibatae* B12. *Eur. J. Biochem.* **260**, 103–111.
16. Zhu, C. X., Qi, H. Y., and Tse-Dinh, Y. C. (1995). Mutation in Cys662 of *Escherichia coli* DNA topoisomerase I confers temperature sensitivity and change in DNA cleavage selectivity. *J. Mol. Biol.* **250**, 609–616.
17. Dekker, N., Rybenkov, V., Duguet, M., Cozzarelli, N. R., Bensimon, D., and Croquette, V. (2002). The mechanism of type IA topoisomerases. *Proc. Natl. Acad. Sci. USA* **99**, 12126–12131.
17a. Dekker, N., Viard, T., Bouthier de la Tour, C., Duguet, M., Benisom, D., and Croquette, V. (2003). Thermophilic Topoisomerase I on a single DNA molecule. *J. Mol. Biol.* **329**, 271–282.
18. Slesarev, A. I., Zaitzev, D. A., Kopylov, V. M., Stetter, K. O., and Kozyavkin, S. A. (1991). DNA topoisomerase III from extremely thermophilic archaebacteria. ATP-independent type I topoisomerase from *Desulfurococcus amylolyticus* drives extensive unwinding of closed circular DNA at high temperature. *J. Biol. Chem.* **266**, 12321–12328.
19. Duguet, M. (1995). Reverse gyrase. In "Nucleic Acids and Molecular Biology" (F. Eckstein and D. M. J. Lilley, Eds.), Vol. 9, pp. 84–114. Springer, Berlin.
20. Kikuchi, A., and Asai, K. (1984). Reverse gyrase a topoisomerase which introduces positive superhelical turns into DNA. *Nature* **309**, 677–681.
21. Bouthier de la Tour, C., Portemer, C., Huber, R., Forterre, P., and Duguet, M. (1991). Reverse gyrase in thermophilic eubacteria. *J. Bacteriol.* **173**, 3921–3923.
22. Campbell, A. M. (1992). Chromosomal insertion sites for phages and plasmids. *J. Bacteriol.* **174**, 7495–7499.
23. Confalonieri, F., Elie, C., Nadal, M., Bouthier, de la Tour, C., Forterre, P., and Duguet, M. (1993). Reverse gyrase: A helicase-like domain and a type I topoisomerase in the same polypeptide. *Proc. Natl. Acad. Sci. USA* **90**, 4753–4757.
24. Jaxel, C., Nadal, M., Mirambeau, G., Forterre, P., Takahashi, M., and Duguet, M. (1989). Reverse gyrase binding to DNA alters the double helix structure and produces single-strand cleavage in the absence of ATP. *EMBO J.* **8**, 3135–3139.
25. Aravind, L., Makarova, K. S., and Koonin, E. V. (2000). Holliday junction resolvases and related nucleases: Identification of new families, phyletic distribution and evolutionary trajectories. *Nucleic Acids Res.* **28**, 3417–3432.
26. Jaxel, C., Bouthier de la Tour, C., Duguet, M., and Nadal, M. (1996). Reverse gyrase gene from *Sulfolobus shibatae* B12: Gene structure, transcription unit and comparative sequence analysis of the two domains. *Nucleic Acids Res.* **24**, 4668–4675.
27. Omichinski, J. G., Clore, G. M., Schaad, O., Felsenfeld, G., Trainor, C., Appella, E., Stahl, S. J., and Gronenborn, A. M. (1993). NMR structure of a specific DNA complex of Zn-containing DNA binding domain of GATA-1. *Science* **261**, 438–446.
28. Declais, A. C., Marsault, J., Confalonieri, F., Bouthier de la Tour, C., and Duguet, M. (2000). Reverse gyrase, the two domains intimately cooperate to promote positive supercoiling. *J. Biol. Chem.* **275**, 19498–19504.
29. Kozyavkin, S. A., Krah, R., Gellert, M., Stetter, K. O., Lake, J. A., and Slesarev, A. I. (1994). A reverse gyrase with an unusual structure. A type I DNA topoisomerase from the hyperthermophile *Methanopyrus kandleri* is a two-subunit protein. *J. Biol. Chem.* **269**, 11081–11089.

30. Krah, R., Kozyavkin, S. A., Slesarev, A. I., and Gellert, M. (1996). A two-subunit type I DNA topoisomerase (reverse gyrase) from an extreme hyperthermophile. *Proc. Natl. Acad. Sci. USA* **93**, 106–110.
31. Krah, R., O'Dea, M. H., and Gellert, M. (1997). Reverse gyrase from *Methanopyrus kandleri*. Reconstitution of an active extremozyme from its two recombinant subunits. *J. Biol. Chem.* **272**, 13986–13990.
32. Declais, A. C., Bouthier de la Tour, C., and Duguet, M. (2001). Reverse gyrases from bacteria and archaea. *Methods Enzymol.* **334**, 146–162.
33. Forterre, P., Confalonieri, F., Charbonnier, F., and Duguet, M. (1995). Speculations on the origin of life and thermophily: Review of available information on reverse gyrase suggests that hyperthermophilic procaryotes are not so primitive. *Orig. Life Evol. Biosph.* **25**, 235–249.
34. Forterre, P. (1996). A hot topic: The origin of hyperthermophiles. *Cell* **85**, 789–792.
35. Andera, L., Mikulik, K., and Savelyeva, N. D. (1993). Characterization of a reverse gyrase from the extremely thermophilic hydrogen-oxidizing eubacterium *Calderobacterium hydrogenophilum*. *FEMS Microbiol. Lett.* **110**, 107–112.
36. Collin, R. G., Morgan, H. W., Musgrave, D., and Daniel, R. M. (1988). Distribution of reverse gyrase in representative species of eubacteria and archaebacteria. *FEMS Microbiol. Lett.* **55**, 235–240.
37. Bouthier de la Tour, C., Portemer, C., Nadal, M., Stetter, K. O., Forterre, P., and Duguet, M. (1990). Reverse gyrase, a hallmark of the hyperthermophilic archaebacteria. *J. Bacteriol.* **172**, 6803–6818.
38. Forterre, P., Bouthier de la Tour, C., Philippe, H., and Duguet, M. (2000). Reverse gyrase from hyperthermophiles: Probable transfer of a thermoadaptation trait from archaea to bacteria. *Trends Genet.* **16**, 152–154.
39. Kovalsky, O. I., Kozyavkin, S. A., and Slesarev, A. I. (1990). Archaebacterial reverse gyrase cleavage-site specificity is similar to that of eubacterial DNA topoisomerases I. *Nucleic Acids Res.* **18**, 2801–2805.
40. Shibata, T., Nakasu, S., Yasui, K., and Kikuchi, A. (1987). Intrinsic DNA-dependent ATPase activity of reverse gyrase. *J. Biol. Chem.* **262**, 10419–10421.
41. Rodriguez, A. C. (2002). Studies of a positive supercoiling machine. Nucleotide hydrolysis and a multifunctional "latch" in the mechanism of reverse gyrase. *J. Biol. Chem.* **277**, 29865–29873.
42. Rodriguez, A. C., and Stock, D. (2002). Crystal structure of reverse gyrase: Insights into the positive supercoiling of DNA. *EMBO J.* **21**, 418–426.
43. Forterre, P., Bergerat, A., and Lopez-Garcia, P. (1996). The unique DNA topology and DNA topoisomerases of hyperthermophilic archaea. *FEMS Microbiol. Rev.* **18**, 237–248.
44. Nadal, M., Mirambeau, G., Forterre, P., Reiter, W. D., and Duguet, M. (1986). Positively supercoiled DNA in a virus-like particle of an archaebacterium. *Nature* **321**, 256–258.
45. Lopez-Garcia, P., and Forterre, P. (1999). Control of DNA topology during thermal stress in hyperthermophilic archaea: DNA topoisomerase levels, activities and induced thermotolerance during heat and cold shock in *Sulfolobus*. *Mol. Microbiol.* **33**, 766–777.
46. Bergerat, A., de Massy, B., Gadelle, D., Varoutas, P. C., Nicolas, A., and Forterre, P. (1997). An atypical topoisomerase II from Archaea with implications for meiotic recombination. *Nature* **386**, 414–417.
47. Guipaud, O., Marguet, E., Noll, K. M., Bouthier de la Tour, C., and Forterre, P. (1997). Both DNA gyrase and reverse gyrase are present in the hyperthermophilic bacterium *Thermotoga maritima*. *Proc. Natl. Acad. Sci. USA* **94**, 10606–10611.
48. Musgrave, D., Forterre, P., and Slesarev, A. (2000). Negative constrained DNA supercoiling in archaeal nucleosomes. *Mol. Microbiol.* **35**, 341–349.

49. Kikuchi, A. (1990). Reverse gyrase and other archaebacterial topoisomerases. In "DNA Topology and Its Biological Effects" (N. R. Cozzarelli and J. Wang, Eds.), pp. 285–298. Cold Spring Harbor Laboratory Press, Cold Spring Harbor, NY.
50. Duguet, M. (1997). When helicase and topoisomerase meet! *J. Cell Sci.* **110**, 1345–1350.
51. Slesarev, A. I., Mezhevaya, K. V., Makarova, K. S., Polushin, N. N., Shcherbinina, O. V., Shakhova, V. V., Belova, G. I., Aravind, L., Natale, D. A., Rogozin, I. B., Tatusov, R. L., Wolf, Y. I., Stetter, K. O., Malykh, A. G., Koonin, E. V., and Kozyavkin, S. A. (2002). The complete genome of hyperthermophile *Methanopyrus kandleri* AV19 and monophyly of archaeal methanogens. *Proc. Natl. Acad. Sci. USA* **99**, 4644–4649.
52. Belova, G. I., Prasad, R., Kozyavkin, S. A., Lake, J. A., Wilson, S. H., and Slesarev, A. I. (2001). A type IB topoisomerase with DNA repair activities. *Proc. Natl. Acad. Sci. USA* **98**, 6015–6020.
53. Doherty, A. J., Serpell, L. C., and Ponting, C. P. (1996). The helix-hairpin-helix DNA-binding motif: A structural basis for non-sequence-specific recognition of DNA. *Nucleic Acids Res.* **24**, 2488–2497.
54. Pourquier, P., Ueng, L. M., Kohlhagen, G., Mazumder, A., Gupta, M., Kohn, K. W., and Pommier, Y. (1997). Effects of uracil incorporation, DNA mismatches, and abasic sites on cleavage and religation activities of mammalian topoisomerase I. *J. Biol. Chem.* **272**, 7792–7796.
55. Pouliot, J. J., Yao, K. C., Robertson, C. A., and Nash, H. A. (1999). Yeast gene for a Tyr-DNA phosphodiesterase that repairs topoisomerase I complexes. *Science* **286**, 552–555.
56. Slesarev, A. I., Lake, J. A., Stetter, K. O., Gellert, M., and Kozyavkin, S. A. (1994). Purification and characterization of DNA topoisomerase V. An enzyme from the hyperthermophilic prokaryote *Methanopyrus kandleri* that resembles eukaryotic topoisomerase I. *J. Biol. Chem.* **269**, 3295–3303.
57. Belova, G. I., Prasad, R., Nazimov, I. V., Wilson, S. H., and Slesarev, A. I. (2002). The domain organization and properties of individual domains of DNA topoisomerase V, a type 1B topoisomerase with DNA repair activities. *J. Biol. Chem.* **277**, 4959–4965.
58. Gellert, M., Mizuuchi, K., O'Dea, M. H., and Nash, H. A. (1976). DNA gyrase: An enzyme that introduces superhelical turns into DNA. *Proc. Natl. Acad. Sci. USA* **73**, 3872–3876.
59. Aravind, L., Leipe, D. D., and Koonin, E. V. (1998). Toprim–a conserved catalytic domain in type IA and II topoisomerases, DnaG-type primases, OLD family nucleases and RecR proteins. *Nucleic Acids Res.* **26**, 4205–4213.
60. Morais, Cabral, J. H., Jackson, A. P., Smith, C. V., Shikotra, N., Maxwell, A., and Liddington, R. C. (1997). Crystal structure of the breakage-reunion domain of DNA gyrase. *Nature* **388**, 903–906.
61. Wigley, D. B., Davies, G. J., Dodson, E. J., Maxwell, A., and Dodson, G. (1991). Crystal structure of an N-terminal fragment of the DNA gyrase B protein. *Nature* **351**, 624–629.
62. Podobnik, M., McInerney, P., O'Donnell, M., and Kuriyan, J. (2000). A TOPRIM domain in the crystal structure of the catalytic core of *Escherichia coli* primase confirms a structural link to DNA topoisomerases. *J. Mol. Biol.* **300**, 353–362.
63. Orphanides, G., and Maxwell, A. (1994). Evidence for a conformational change in the DNA gyrase-DNA complex from hydroxyl radical footprinting. *Nucleic Acids Res.* **22**, 1567–1575.
64. Guipaud, O., Labedan, B., and Forterre, P. (1996). A gyrB-like gene from the hyperthermophilic bacterion *Thermotoga maritima*. *Gene* **174**, 121–128.
65. Bergerat, A., Gadelle, D., and Forterre, P. (1994). Purification of a DNA topoisomerase II from the hyperthermophiic archaeon *Sulfolobus shibatae*. A thermostable enzyme with both bacterial and eucaryal features. *J. Biol. Chem.* **269**, 27663–27669.

66. Keeney, S., and Kleckner, N. (1995). Covalent protein-DNA complexes at the 5' strand termini of meiosis-specific double-strand breaks in yeast. *Proc. Natl. Acad. Sci. USA* **92**, 11274–11278.
67. Buhler, C., Gadelle, D., Forterre, P., Wang, J. C., and Bergerat, A. (1998). Reconstitution of DNA topoisomerase VI of the thermophilic archaeon *Sulfolobus shibatae* from subunits separately overexpressed in *Escherichia coli*. *Nucleic Acids Res.* **26**, 5157–5162.
68. Buhler, C., Lebbink, J. H., Bocs, C., Ladenstein, R., and Forterre, P. (2001). DNA topoisomerase VI generates ATP-dependent double-strand breaks with two-nucleotide overhangs. *J. Biol. Chem.* **276**, 37215–37222.
69. Liu, J., Wu, T. C., and Lichten, M. (1995). The location and structure of double-strand DNA breaks induced during yeast meiosis: Evidence for a covalently linked DNA-protein intermediate. *EMBO J.* **14**, 4599–4608.
70. Nichols, M. D., DeAngelis, K., Keck, J. L., and Berger, J. M. (1999). Structure and function of an archaeal topoisomerase VI subunit with homology to the meiotic recombination factor Spo11. *EMBO J.* **18**, 6177–6188.
71. Stark, W. M., Boocock, M. R., and Sherratt, D. J. (1992). Catalysis by site-specific recombinases. *Trends Genet.* **8**, 432–439.
72. Cheng, C., Kussie, P., Pavletich, N., and Shuman, S. (1998). Conservation of structure and mechanism between eukaryotic topoisomerase I and site-specific recombinases. *Cell* **92**, 841–850.
73. Sherratt, D. J., and Wigley, D. B. (1998). Conserved themes but novel activities in recombinases and topoisomerases. *Cell* **93**, 149–152.
74. Sharples, G. J. (2001). The X philes: Structure-specific endonucleases that resolve Holliday junctions. *Mol. Microbiol.* **39**, 823–834.
75. Gonzalez, S., Rosenfeld, A., Szeto, D., and Wetmur, J. G. (2000). The Ruv proteins of *Thermotoga maritima*: Branch migration and resolution of Holliday junctions. *Biochim. Biophys. Acta* **1494**, 217–225.
76. Sharples, G. J., Bolt, E. L., and Lloyd, R. G. (2002). RusA proteins from the extreme thermophile *Aquifex aeolicus* and lactococcal phage rlt resolve Holliday junctions. *Mol. Microbiol.* **44**, 549–559.
77. Sharples, G. J., Ingleston, S. M., and Lloyd, R. G. (1999). Holliday junction processing in bacteria: Insights from the evolutionary conservation of RuvABC, RecG and RusA. *J. Bacteriol.* **181**, 5543–5550.
78. Komori, K., Sakae, S., Shinagawa, H., Morikawa, K., and Ishino, Y. (1999). A holliday junction resolvase from *Pyrococcus furiosus*: Functional similarity to *Escherichia coli* RuvC provides evidence for conserved mechanism of homologous recombination in bacteria, eukarya, and archaea. *Proc. Natl. Acad. Sci. USA* **96**, 8873–8878.
79. Kvaratskhelia, M., and White, M. F. (2000). An archaeal holliday junction resolving enzyme from *Sulfolobus solfataricus* exhibits unique properties. *J. Mol. Biol.* **295**, 193–202.
80. She, Q., Singh, R. K., Confalonieri, F., Zivanovic, Y., Allard, G., Awayez, M. J., Chan-Weiher, C. C., Clausen, I. G., Curtis, B. A., De Moors, A., Erauso, G., Fletcher, C., Gordon, P. M., Heikamp-de Jong, I., Jeffries, A. C., Kozera, C. J., Medina, N., Peng, X., Thi-Ngoc, H. P., Redder, P., Schenk, M. E., Theriault, C., Tolstrup, N., Charlebois, R. L., Doolittle, W. F., Duguet, M., Gaasterland, T., Garrett, R. A., Ragan, M. A., Sensen, C. W., and Van der Oost, J. (2001). The complete genome of the crenarchaeon *Sulfolobus solfataricus* P2. *Proc. Natl. Acad. Sci. USA* **98**, 7835–7840.
81. Birkenbihl, R. P., Neef, K., Prangishvili, D., and Kemper, B. (2001). Holliday junction resolving enzymes of archaeal viruses SIRV1 and SIRV2. *J. Mol. Biol.* **309**, 1067–1076.
82. Peng, X., Blum, H., She, Q., Mallok, S., Brügger, K., Garrett, R. A., Zillig, W., and Prangishvili, D. (2001). Sequences and replication of genomes of the archaeal rudiviruses SIRV1 and

SIRV2: Relationships to the archaeal lipothrixvirus SIFV and some eukaryal viruses. *Virology* **291,** 226–234.
83. Kvaratskhelia, M., Wardleworth, B. N., and White, M. F. (2001). Multiple Holliday junction resolving enzyme activities in the crenarchaeota and euryarchaeota. *FEBS Lett.* **491,** 243–246.
84. Daiyasu, H., Komori, K., Sakae, S., Ishino, Y., and Toh, H. (2000). Hjc resolvase is a distantly related member of the type II restriction endonuclease family. *Nucleic Acids Res.* **28,** 4540–4543.
85. Bolt, E. L., Lloyd, R. G., and Sharples, G. J. (2001). Genetic analysis of an archaeal Holliday junction resolvase in *Escherichia coli*. *J. Mol. Biol.* **310,** 577–589.
86. Komori, K., Sakae, S., Fujikane, R., Morikawa, K., Shinagawa, H., and Ishino, Y. (2000). Biochemical characterization of the Hjc holliday junction resolvase of *Pyrococcus furiosus*. *Nucleic Acids Res.* **28,** 4544–4551.
87. Kvaratskhelia, M., and White, M. F. (2000). Two holliday junction resolving enzymes in *Sulfolobus solfataricus*. *J. Mol. Biol.* **297,** 923–932.
88. Kvaratskhelia, M., Wardleworth, B. N., Norman, D. G., and White, M. F. (2000). A conserved nuclease domain in the archaeal holliday junction resolving enzyme Hjc. *J. Biol. Chem.* **275,** 25540–25546.
89. Nishino, T., Komori, K., Ishino, Y., and Morikawa, K. (2001). Dissection of the regional roles of the archaeal Holliday junction resolvase Hjc by structural and mutational analysis. *J. Biol. Chem.* **276,** 35735–35740.
90. Komori, K., Sakae, S., Daiyasu, H., Toh, H., Morikawa, K., Shinagawa, H., and Ishino, Y. (2000). Mutational analysis of the *Pyrococcus furiosus* Holliday junction resolvase Hjc revealed functionally important residues for dimer formation, junction DNA binding, and cleavage activities. *J. Biol. Chem.* **275,** 40385–40391.
91. Komori, K., Miyata, T., DiRuggiero, J., Holley-Shanks, R., Hayashi, I., Cann, I. K. O., Mayanagi, K., Shinagawa, H., and Ishino, Y. (2000). Both RadA and RadB are involved in homologous recombination in *Pyrococcus furiosus*. *J. Biol. Chem.* **275,** 33782–33790.
92. Bond, C. S., Kvaratskhelia, M., Richard, D., White, M. F., and Hunter, W. N. (2001). Structure of Hjc, a holliday junction resolvase, from *Sulfolobus solfataricus*. *Proc. Natl. Acad. Sci. USA* **98,** 5509–5514.
93. Nishino, T., Komori, K., Tsuchiya, D., Ishino, Y., and Morikawa, K. (2001). Crystal structure of the archaeal holliday junction resolvase Hjc and implications for DNA recognition. *Structure* **9,** 197–204.
94. Ariyoshi, M., Vassylyev, D. G., Iwasaki, H., Nakamura, H., Shinagawa, H., and Morikawa, K. (1994). Atomic structure of the RuvC resolvase: A holliday junction-specific endonuclease from *E. coli*. *Cell* **78,** 1063–1072.
95. Ceschini, S., Keeley, A., McAlister, M. S. B., Oram, M., Phelan, J., Pearl, L. H., Tsaneva, I. R., and Barrett, T. E. (2001). Crystal structure of the fission yeast mitochondrial Holliday junction resolvase Ydc2. *EMBO J.* **20,** 6601–6611.
96. Hadden, J. M., Convery, M. A., Declais, A. C., Lilley, D. M., and Phillips, S. E. (2001). Crystal structure of the Holliday junction resolving enzyme T7 endonuclease I. *Nat. Struct. Biol.* **8,** 62–67.
97. Raaijmakers, H., Vix, O., Toro, I., Golz, S., Kemper, B., and Suck, D. (1999). X-ray structure of T4 endonuclease VII: A DNA junction resolvase with a novel fold and unusual domain-swapped dimer architecture. *EMBO J.* **18,** 1447–1458.
98. Rice, G., Stedman, K., Snyder, J., Wiedenheft, B., Willits, D., Brumfield, S., McDermott, T., and Young, M. J. (2001). Viruses from extreme thermal environments. *Proc. Natl. Acad. Sci. USA* **98,** 13341–13345.

99. Zillig, W., Arnold, H. P., Holz, I., Prangishvili, D., Schweier, A., Stedman, K., She, Q., Phan, H., Garrett, R., and Kristjansson, J. K. (1998). Genetic elements in the extremely thermophilic archaeon *Sulfolobus*. *Extremophiles* **2**, 131–140.
100. Zillig, W., Kletzin, A., Schleper, C., Holz, I., Janekovic, D., Hain, J., Lanzendorfer, M., and Kristjansson, J. K. (1994). Screening for *Sulfolobales*, their plasmids and their viruses in Icelandic solfataras. *Syst. Appl. Microbiol.* **16**, 609–628.
101. Prangishvili, D., Stedman, K., and Zillig, W. (2001). Viruses of the extremely thermophilic archaeon *Sulfolobus*. *Trends Microbiol.* **9**, 39–43.
102. Martin, A., Yeats, S., Janekovic, D., Reiter, W.-D., Aicher, W., and Zillig, W. (1984). SAV1, a temperate U.V.-inducible DNA virus-like particle from the archaebacterium *Sulfolobus acidocaldarius* isolate B12. *EMBO J.* **3**, 2165–2168.
103. Muskhelishvili, G., Palm, P., and Zillig, W. (1993). SSV1-encoded site-specific recombination system in *Sulfolobus shibatae*. *Mol. Gen. Genet.* **237**, 334–342.
104. Reiter, W.-D., Palm, P., and Yeats, S. (1989). Transfer RNA genes frequently serve as integration sites for prokaryotic genetic elements. *Nucleic Acids Res.* **17**, 1907–1914.
105. Esposito, D., and Scocca, J. J. (1997). The integrase family of tyrosine recombinases: Evolution of a conserved active site domain. *Nucleic Acids Res.* **25**, 3605–3614.
106. Nunes-Düby, S. E., Kwon, H. J., Tirumalai, R. S., Ellenberger, T., and Landy, A. (1998). Similarities and differences among 105 members of the Int family of site-specific recombinases. *Nucleic Acids Res.* **26**, 391–406.
107. Palm, P., Schleper, C., Grampp, B., Yeats, S., McWilliam, P., Reiter, W.-D., and Zillig, W. (1991). Complete nucleotide sequence of the virus SSV1 of the archaebacterium *Sulfolobus shibatae*. *Virology* **185**, 242–250.
108. Muskhelishvili, G. (1994). The archaeal SSV integrase promotes intermolecular excisive recombination *in vitro*. *Syst. Appl. Microbiol.* **16**, 605–608.
109. Serre, M. C., Letzelter, C., Garel, J. R., and Duguet, M. (2002). Cleavage properties of an archaeal site-specific recombinase, the SSV1 integrase. *J. Biol. Chem.* **277**, 16758–16767.
110. Gopaul, D. N., and Duyne, G. D. (1999). Structure and mechanism in site-specific recombination. *Curr. Opin. Struct. Biol.* **9**, 14–20.
111. Schleper, C., Kubo, K., and Zilig, W. (1992). The particle SSV1 from the extremely thermophilic archaeon *Sulfolobus* is a virus: Demonstration of infectivity and of transfection with viral DNA. *Proc. Natl. Acad. Sci. USA* **89**, 7645–7649.
112. Sherratt, D. J., Arciszewska, L. K., Blakely, G., Colloms, S., Grant, K., Leslie, N., and McCulloch, R. (1995). Site-specific recombination and circular chromosome segregation. *Philos. Trans. R. Soc. Lond. B Biol. Sci.* **347**, 37–42.
113. Blakely, G., May, G., McCulloch, R., Arciszewska, L. K., Burke, M., Lovett, S. T., and Sherratt, D. J. (1993). Two related recombinases are required for site-specific recombination at *dif* and *cer* in *E. coli* K12. *Cell* **75**, 351–361.
114. Arciszewska, L. K., and Sherratt, D. J. (1995). Xer site-specific recombination *in vitro*. *EMBO J.* **14**, 2112–2120.
115. Recchia, G. D., and Sherratt, D. J. (1999). conservation of xer site-specific recombination in bacteria. *Mol. Microbiol.* **34**, 1146–1148.
116. Forterre, P. (1999). Non-orthologous gene displacement in the evolution of DNA informational proteins. *Mol. Microbiol.* **33**, 457–465.
117. Leipe, D. D., Aravind, L., and Koonin, E. V. (1999). Did DNA replication evolve twice independently? *Nucleic Acids Res.* **27**, 3389–3401.

118. Weisburg, W. G., Giovannoni, S. J., and Woese, C. R. (1989). The *Deinococcus-Thermus* phylum and the effect of rRNA composition on phylogenetic tree construction. *Syst. Appl. Microbiol.* **11,** 128–134.
119. White, O., Eisen, J. A., Heidelberg, J. F., Hickey, E. K., Peterson, J. D., Dodson, R. J., Haft, D. H., Gwinn, M. L., Nelson, W. C., Richardson, D. L., Moffat, K. S., Qin, H., Jiang, L., Pamphile, W., Crosby, M., Shen, M., Vamathevan, J. J., Lam, P., McDonald, L., Utterback, T., Zalewski, C., Makarova, K. S., Aravind, L., Daly, M. J., Fraser, C. M., *et al.* (1999). Genome sequence of the radioresistant bacterium *Deinococcus radiodurans* R1. *Science* **286,** 1571–1577.

Molecular Mimicry in the Decoding of Translational Stop Signals

Elizabeth S. Poole, Marjan E. Askarian-Amiri, Louise L. Major,[1] Kim K. McCaughan, Debbie-Jane G. Scarlett,[2] Daniel N. Wilson,[3] and Warren P. Tate

Department of Biochemistry, University of Otago, Dunedin, New Zealand

I. Introduction	84
II. Atomic Level Structures of the Ribosome and Its Subunits	85
III. The Active Center of the Ribosome	87
IV. The Three-Dimensional Space of the Active Center	90
V. The Mimicry Complex in Termination	94
VI. The RF Structure: Does It Look Like a tRNA?	99
VII. The Paradox between the Structural and Functional Evidence for the Bacterial RF "Anticodon"	103
VIII. Specific Interactions within the Bacterial Termination Complex	106
IX. The Termination Mechanism	111
X. Summary	114
References	115

The structures of the ribosome and its subunits are now available at atomic detail, as well as those of several factors that bind to its active center. Of particular interest are the protein release factors that decode stop signals. In contrast to the codons specifying the different amino acids, the stop signals are not decoded by RNA molecules, the tRNAs. The tRNA analogue hypothesis (1994) for the decoding of stop signals was proposed to explain how the release factors might mimic a tRNA to span the decoding site of the small subunit and the enzyme center of the large subunit of the ribosome. The specific term "molecular mimicry" was applied soon after to include proteins or their

---

[1] Present address: Centre for Biomolecular Sciences, North Haugh, The University, St Andrews, Fife KY16 9ST, Scotland.

[2] Present address: Malaghan Institute of Medical Research, Newtown, Wellington, New Zealand.

[3] Present address: Max-Planck Institut für Molekulare Genetik, D-14195, Berlin, Germany.

domains that enter the tRNA binding sites on the ribosome. The solution crystal structures of the two release factors already solved (one eubacterial and one eukaryotic), although quite distinct in their folds, each resembles the shape of a tRNA. The eukaryotic factor, like a tRNA, seems to have specific motifs at the tips of two of its domains that interact with the decoding site and the enzyme center as predicted in the tRNA analogue model. Biochemical and genetic studies had identified two analogous motifs in the bacterial factors, but these are quite close together in the solution structure, suggesting a major conformational change may take place when the factor binds to the ribosome. Indeed, reconstructed cryoelectron microscopic images support an unfolding of the structure. A second class of release factor functions as a translational G-protein in the same manner as the two elongation factors and forms part of the termination mimicry complex. Undoubtedly, the molecular mimicry concept will be refined as the conformational changes that take place in the active center of the ribosome and in the proteins that bind to it are better understood.

I. Introduction

Major advances in thinking as well as understanding of fundamental mechanisms in biology often occur subsequent to a significant technological advance. Such an advance occurred in protein synthesis in the late 1960s with the solving of the genetic code and predicted a new era of major discovery. Novel protein factors that interact with the ribosome were discovered and the details of the steps of how a protein was put together have gradually unfolded over the past 30 years. The first phase of this discovery was limited by our understanding of the cell organelle where these molecular events occur. Indeed, initially ribosomes were added to an experiment as a dark stage on which the players could weave their magic so that the key steps in making a protein could be studied individually. The lights on this stage gradually intensified with technologies such as electron microscopy revealing first an image of a ribosome and then coupled with immunology revealing where the various ribosomal components were found.

Herculean attempts through the 1970s to understand the structures of the RNA components of the ribosome took on new importance as a result of the chance discovery by Cech and colleagues in 1982 (1) that an intron could splice itself out of the pre-RNA primary transcript without the need for protein enzymes. This finding refocused thinking on how protein synthesis might be catalyzed. The RNA components of the ribosome went from being regarded merely as the scaffold to position "functional" ribosomal proteins, to being regarded as the likely functional centers of the ribosome. This created a new era of thought, characterized by the concept of the "RNA World" that had as its genesis the discovery of Cech and his colleagues. Others, such as Crick in 1968 (2), had speculated before that RNA could be the critical ribosomal molecule for protein synthesis but the idea was difficult to study experimentally and attracted few devotees at the time. Cech's discovery provided a framework to understand how a prototype of the modern ribosome could have been an RNA molecule that evolved an ability to link amino acids together. Could the modern ribosomal RNA have as its origin a primitive RNA replicase? Perhaps it could have developed via an aminoacyl-tRNA synthetase to an RNA that now provided a template for other RNAs and provided an environment whereby amino acids could be linked together. The many proteins of the modern ribosome also took on a different significance within this theoretical framework. Now, they could be regarded as molecules to hold the structure of the ribosomal RNA in a stable and active conformation, adding to the efficiency of the process but not critically required for the catalytic events at the active center.

The true revelation of the significance of the ribosomal RNAs awaited the next burst in technology; a second phase in a wonderful 30 years of discovery. Cryoelectron microscopy developed rapidly during the 1990s and promised to provide our first near atomic-level detail of the ribosome. In the event, despite the very valuable images and insights that this technology has provided, the dramatic new images of the ribosome came from an unexpected quarter with the dawning of the new millennium. Apparently intractable problems in applying X-ray crystallographic methods to such a complex particle as the ribosome were solved, and there has been a rush of structures supplying amazing images at atomic detail to illuminate brightly the previously darkened stage where the magic of protein synthesis occurs. Now, the new structures could be tested with the hindsight of the many years of accumulated biochemical data, and interpretation of how protein synthesis works examined in a new light.

II. Atomic Level Structures of the Ribosome and Its Subunits

The first significant problem in resolving structures of such complex particles as the ribosome was to obtain three-dimensional crystals. The initial focus was to solve the individual subunits. The first 50 S subunit crystals were obtained from *Bacillus stearothermophilis* (3) and then later from *Haloarcula marismortui* (4). The impetus for resolving the 30 S subunit, and indeed the full 70 S ribosome came from reports of successful attempts to crystallize 30 S particles from *Thermus thermophilis* (5, 6). However, to analyze such complex crystals there were still major problems to overcome such as X-ray data collection and the hardware and software of computation. Significant improvements in these areas allowed progression inexorably closer to what still seemed unattainable, namely, a successfully resolved crystal structure. Toward the end of the 1990s, crystals were being manipulated with cryomethodology but there was still no solution to the classic problem of resolving the phases of diffraction patterns. The first structure, although at low resolution, of the archaean *H. marismortui* 50 S subunit appeared in 1998 (7), and then just a year later tantalizing higher resolution structures of both 30 S (8) and 50 S (9) subunits and the 70 S ribosome (10). For those that had been collecting biochemical data on the ribosome and protein synthesis for many years these structures were exquisite. To the crystallographers themselves they were still fuzzy images. Remarkably, it took only another year for truly high-resolution structures with atomic detail for both subunits to be published (11–13). Ramakrishnan and Moore (14) reflect their own excitement with the following statement: "A complete atomic structure of the 30 S subunit that, like its 50 S counterpart will serve as a reference for years to come." The structure of a 50 S subunit of a mesophilic bacterium has now been completed (15) and it is very similar to that of the archaean. These high-resolution structures have assisted construction of a model for the RNA and protein backbone of the 70 S ribosome (16) even though the collected data do not give a structure at a resolution that would normally allow an accurate model to be obtained.

What have these models told us about the ribosome that is relevant to protein synthesis and, specifically for this article, the decoding of stop signals in mRNA on the ribosome by polypeptide chain release factors? The first striking feature of the subunit structures was that the intersubunit region in each case, highly relevant to the decoding and enzyme functions of the complete ribosome, was almost completely free of ribosomal proteins. This means that it is predominantly rRNA that forms the active center and lines the space between the subunits and allows substrates and protein factors to bind. This is clearly seen if the interface planes of the subunits are compared with the solvent planes. The former view has proteins dispersed around the edges and the latter view has proteins dispersed equally over the surface of each subunit. This observation alone provides enormous impetus for rRNA having the key functional roles of the ribosome, mRNA decoding and enzyme synthesis, as the rRNA-rich regions were where these functions were mediated. What the functional studies had implied strongly but could not prove definitively was now made clear as structures with tRNAs spanning the decoding and active centers were resolved (13). Cech triumphantly stated "The ribosome is a ribozyme" in his commentary in 2000 (17) that accompanied the *Science* publications of the atomic resolution of the 50 S subunit and a proposal for the structural basis of ribosomal activity for peptide synthesis.

In the 50 S subunit, the ribosomal exit tunnel for the growing polypeptide spans the peptidyltransferase center of the inner face through to the outer solvent face such that the protein is threaded through the subunit during synthesis. Also, the subunit structure shows that the tunnel is almost entirely lined with rRNA involving four of the six domains of the 23 S rRNA. Interestingly, at one point, two ribosomal proteins intrude to constrict the tunnel. Many of the ribosomal proteins not only have globular domains but also have extended regions that interleave with rRNA. For example, the proteins known functionally to be important for maintaining the integrity of the ribosomal peptidyltransferase center (18, 19) have extended tails snaking toward the rRNA at the center. These protein tails clearly play a role in rRNA folding but are not directly involved in enzyme function. It is possible that they are in evolutionary transit to extend further and eventually will exert a functional role in the enzymatic reaction of peptide bond formation.

The three-dimensional resolution of such a large RNA structure permitted a detailed examination of the motifs that comprise secondary and tertiary interactions. Surprisingly, the same motifs that had been found in small RNA secondary structures were rediscovered, indicating a limited range of motifs. Tertiary interaction stabilization of rRNA helices and loops involves both minor groove to minor groove packing and the insertion of phosphate ridges into the minor grooves. Highly conserved adenines are used as "plugs" inserted into adjacent helices made from a distant sequence, and unpaired purines facilitate packing of two helices perpendicular to each other. Of interest is that the overall packing of the rRNAs in the two subunits is quite different. In the 30 S subunit, the domains fold as distinct entities and are delicately interposed with a single helical thread connecting the head of the subunit to its body, giving some understanding to how each might be able to move with respect to the other. In contrast, the six domains of the 23 S rRNA of the 50 S subunit are interwoven in a complex manner to create a less flexible, apparently single, "superdomain." This suggests that the two rRNA molecules might have evolved independently and have been refined for their quite different functional interactions of decoding and enzyme activity separately.

How do the crystal structures contribute to our understanding of specific events in protein synthesis? Mostly they provide encyclopedic information on the interactions between individual atoms of rRNAs and proteins, as well as providing a global architecture of the subunits both together and in isolation. The structures where ligands have been included indicate the types of interaction that specific ligands make, or might make at various steps of protein synthesis.

III. The Active Center of the Ribosome

This center, now known to be an RNA structure that is stabilized from a distance by ribosomal proteins, is composed of elements from both subunits that line the interface region of the complete ribosome. Functionally, the active center has been divided into two parts reflecting the role of the subunits. The elements from the 30 S subunit form the decoding functions and those from the 50 S subunit form the region where the protein is assembled and removed. These parts of the active center are quite distant and allow the mRNA ligand with its template function to stay physically separated from the polypeptide assembly units and product. The tRNA spans the distance between these active "subcenters" to act as a molecular adapter.

The ribosome accommodates the tRNA in three characteristic forms. First, the tRNA acts as a substrate when it carries the incoming amino acid, and the region of the active center it occupies in this state has been defined functionally as the A-site. The tRNA also carries the growing polypeptide and in this state is the enzyme intermediate, shuttling between the A-site and a second site at the active centre, the P-site (20). This means that two tRNAs are on the ribosome at the same time and in contact with the mRNA at the decoding site and the peptidyltransferase at the enzyme center. When the A- and P-sites are occupied by these two tRNAs, one of the major states in polypeptide elongation is represented. This is defined functionally as the *pretranslocation state*. A third tRNA binding site, the E-site (20), was defined at first by function, and, although once highly controversial, the E-site is now established as an entity and was elegantly confirmed by the ribosome structures. After peptide bond formation at the enzyme center and translocation of the tRNA complexes, the A-site is free to accept a new substrate aminoacyl-tRNA. Both original tRNAs remain on the ribosome, with the "intermediate" polypeptidyl-tRNA at the P-site and the now free deacylated-tRNA at the E-site. This is a second major state of elongation, defined functionally as the *posttranslocation state*.

Biochemically, the active center of the ribosome is divided into two regions in one dimension (the decoding site and enzyme center) and into three regions in another dimension (the three tRNA sites) (Fig. 1). Examination of what that means at the structural level of the ribosome is now possible with the structures that were published in 2000 and 2001. Intersubunit bridges that had been seen first in low- and recently in high-resolution cryoelectron microscopic studies (21) were resolved in the high-resolution 5.5-Å structure of the 70 S ribosome (16). Intersubunit bridges can be seen between RNA:RNA contacts at the interface. Those arising from the 30 S subunit are immediately adjacent to tRNA contacts with the decoding center, while those on the 50 S subunit form a triangle across the surface of the interface wall where the peptidyltransferase center and E-site are separated (16). Most intersubunit bridges arise from domain IV of the 23 S rRNA.

Interactions between the substrate, intermediate, and product tRNAs can be seen with rRNAs at the active center. Not only do these provide generic stabilization of tRNA binding, but they also serve specific functional purposes

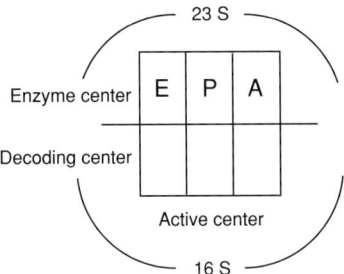

FIG. 1. A diagram of the ribosomal active center indicating interaction sites of the tRNA (A, P, and E) with the enzyme and decoding centers of the 23 S rRNA and 16 S rRNA, respectively.

such as discrimination to ensure accuracy of decoding and aminoacyl-tRNA selection, stability to maintain the correct translational reading frame, conformational changes that facilitate translocation of the tRNAs from the pretranslocation to the posttranslocation state, and to ensure interactions that are important for catalysis during peptide bond formation. These interactions confirm previously inferred contacts deduced from footprint, cross-link, and cleavage strategies (22–24). Different tRNAs bind at each of the ribosomal A-, P-, and E-sites in the active center. Therefore, it was expected that universal features of their structures, either primary sequence or sequence motifs, would be the sites of interaction, and, indeed, this is the case. The A-site tRNA seems to be bound with no change to its solution conformation, whereas when tRNAs are in the other two sites there are distortions to their structures. The P-site tRNA has its anticodon loop angled toward the A-site through a distortion at the junction of the D and anticodon stems. The E-site tRNA contains major conformational changes with the angle of the elbow in a more open state and the anticodon loop sharply twisted.

As expected, the anticodon loops of all three tRNAs are in contact with the 30 S subunit, and other major features such as the D stem, elbow, and acceptor arm make contact with the 50 S subunit (16). The A- and P-site tRNAs are closest to each other at the enzyme center, at 5 Å apart, whereas at the decoding center they are 10 Å distant to each other, which is made possible by a twist in the mRNA backbone so that two codons can be read simultaneously. The E-site tRNA is twisted 50 Å away from the P-site tRNA at the enzyme center, while retaining a relatively close contact of 6 Å at the 30 S subunit (Fig. 2A). The anticodon stem-loop is in an environment of high-density molecular interactions while distal parts of the E-site tRNA make several interactions with the 50 S subunit. For example, the acceptor arm CCA end is secured by a number of ribosomal contacts while buried in a hole separate from the peptidyltransferase cleft. It is interesting that the P-site tRNA has numerous interactions

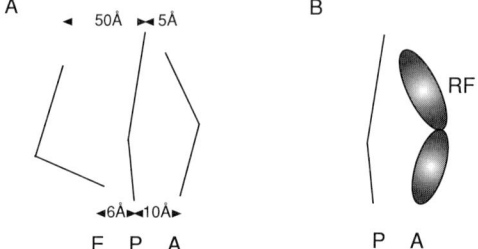

FIG. 2. Diagrams indicating tRNA and RF interactions with the ribosome. (A) The distances between the A-, P-, and E-site tRNA anticodon and CCA ends at the decoding and enzyme centers, respectively, reflect that observed in the structures (16). (B) The tRNA analogue model of the decoding RF proposing that a domain makes contact with the mRNA at the decoding site with a second domain interacting with the peptidyltransferase center.

with rRNAs, having six sets of interactions from the tRNA anticodon stem-loop with the 16 S rRNA of the 30 S subunit with the rRNA bases previously inferred from base modification or modification interference studies (25–27). Several further interactions with the 23 S rRNA complete a complex set of interactions, indicating that the P-site tRNA is very stably gripped in the active centre. The A-site tRNA anticodon loop is positioned close to three residues, G530, A1492, and A1493, invoked from previous biochemical studies to be key determinants of A-site decoding. A bulged-base (C1054) points toward the tip of the anticodon loop but similar stabilizing interactions found with the P-site tRNA are not present, although there are more complex interactions with rRNA in the distal parts of the A-site tRNA (16).

IV. The Three-Dimensional Space of the Active Center

The ribosomal active center and its rRNA lining seem to have been finely honed for the two states (pre- and posttranslocation), each of which accommodates two tRNAs simultaneously. The intimate contacts described above are between two different types of RNA nucleic acid, although many tRNA bases are highly modified. When a tRNA is occupying the ribosomal A-site, the modern ribosome seems finely crafted to decode sense signals in the mRNA for the amino acids at this center. Moreover, image reconstructions from cryoimmunoelectron microscopy suggest that the space in the inner intersubunit cavern containing the A-site spanning between the 30 S and 50 S subunits is shaped like a tRNA (28), indicating that any molecule entering this site must be able to accommodate itself within this space. Indeed, cryoelectron microscopy shows that the A-site aminoacyl-tRNA accommodates to the space by a new conformation of the anticodon arm that facilitates interaction with the codon of the mRNA (29, 29a). Therefore, a paradox is created for decoding stop signals in the mRNA.

Stop signals are not decoded by tRNAs but by proteins called the polypeptide chain release factors (RFs) (30, 31). The class I RFs must enter the ribosomal A-site in the same manner as a tRNA and somehow make the same kinds of intimate interactions with the rRNA, but now the structural considerations are quite different. How does the RF protein use the "imprint" of another species, the tRNA, for its functional activities? Biochemical evidence accumulated through the 1980s and 1990s showed that the RF influenced both the decoding and enzyme functions of the active center, and this eventually resulted in the tRNA analogue hypothesis proposed by Moffat and Tate in 1994 (32) (Fig. 2B). This hypothesis proposed that the RF spanned the two main centers on the ribosome just like a tRNA, with a domain that made intimate contact with the mRNA at the decoding site, and a domain that made sufficiently close contact with the peptidyltransferase center to alter its function. The consequence would be the conversion from a water-excluding center catalyzing peptide bond formation, to one allowing water access to catalyze the more natural peptidyl-tRNA hydrolysis.

What is the evidence supporting the tRNA analogue hypothesis? Site-directed cross-links from the first (the U of, e.g., UGAN) and fourth positions of the stop signal to the RF provided excellent evidence that the RF was in intimate contact with the signal just like a tRNA at the decoding site (33–35). The cross-link moiety was a 4-thiouridine base where the S is only slightly larger than the natural O at the N4 position. Cross-links from this base are effectively zero length, suggesting that for a successful cross-link the two molecules had to be in close contact. Significant cross-links were observed, particularly with UGA-containing mRNAs that presumably reflected a more favorable three-dimensional disposition of the S to the amino acid side chains of the RF when UGA-containing signals were decoded compared with the other two stop signals.

Evidence for interaction at the enzyme center was more indirect, with cleavage of Tyr-244 and Arg-245 in the bacterial RF2 (UGA decoding) within a then-hypothesized exposed loop abolishing peptidyl-tRNA hydrolysis when the RF was in the ribosomal A-site (32). The recently published structure of RF2 confirms this region is in an exposed loop of domain III of the protein (36). Subsequently, other residues close to the cleavage site have been shown to be important for peptidyl-tRNA hydrolysis. For example, Thr-246 is a key residue that is clearly important for the conformation of the hydrolysis domain in the RF2 protein and facilitates the loss of the peptidyl-tRNA hydrolysis function in an overexpressed RF (37) (Fig. 3). Moreover, the GGQ motif found

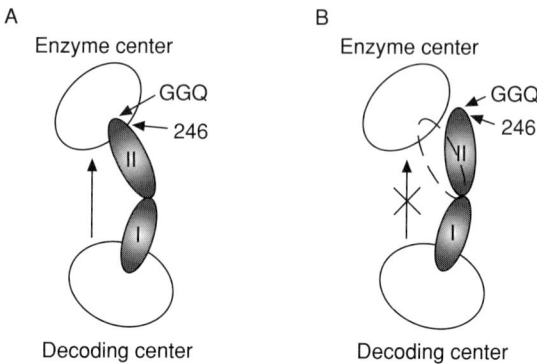

FIG. 3. A two-domain model for RF2 conformation during protein synthesis termination. Position 246 is proposed to orient the GGQ motif to provide efficient peptidyltransferase center contact of domain II of RF2. (A) With native RF2, correct stop codon recognition activates a conformational signal (vertical arrow) from domain I to domain II allowing functional peptidyl-tRNA hydrolysis. (B) When native RF2 is overexpressed (or contains certain residues at position 246), the inactive alignment of domain II is resistant to the conformational signal resulting in inefficient peptidyl-tRNA hydrolysis. In addition, any disturbed codon recognition may prevent transfer of a signal from domains I to II to align correctly the GGQ motif, and, therefore, prevent peptidyl-tRNA hydrolysis.

at positions 250–252 is conserved in RFs and is critical for the peptidyl-tRNA hydrolysis function of eRF1 (38). These indirect data showed that a region of the RF important for catalysis of the hydrolysis function was relatively close to the enzyme center.

Other functional evidence had implicated L11 as a key ribosomal binding determinant of the bacterial RF into the A-site. Initially, it was discovered that ribosomes lacking L11, produced by either construction *in vitro* by omitting L11 from a reconstitution mixture of ribosomal proteins (39), or naturally occurring mutants (40), had very low activity for RF1-mediated binding but highly elevated RF2 binding. This suggested that the position of L11 within the ribosome structure was a key interaction site for RF proteins. This finding was reinforced when antibodies against the N-terminus of L11 specifically and differentially affected the activities of RF1 and RF2 (40). A single iodination of Tyr at position 7 of L11, the most highly reactive Tyr residue within the ribosome, abolished RF activity. In relation to the peptidyltransferase center, L11 is part of the GTPase center associated with domain II of the rRNA (41) (Fig. 4). In the recently published structure of the 70 S ribosome (16) protein L11 and its associated rRNA do not interact directly with the tRNA in the A-site, but are close to the T-loop of the tRNA. L11 contact is possible with relatively modest conformational change within either the tRNA or rRNA. Moreover,

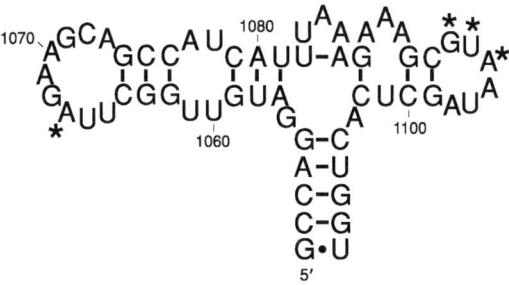

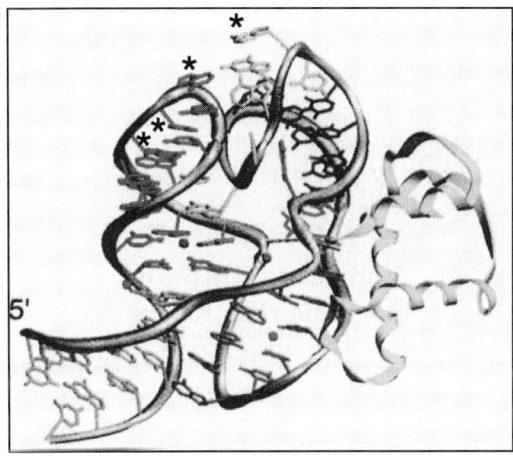

FIG. 4. L11 and the GTPase center interactions with RF2. The RF2 interactive sites (*) identified from mutational studies (*41*) on the GTPase rRNA (above and below) and the GTPase rRNA (below, left) complexed with the C-terminal domain of L11 (below, right). The structure (below) is derived from the work of Draper and colleagues (*103*).

genetic selection for stop signal suppression identified a number of bases in rRNA specific for a particular stop codon. For example, a G1093A transition was specific for UGA and strongly affected the effective association rate constant for RF2 with the ribosome and to a lesser extent the association rate of RF1 (*42*). In the GTPase center other nucleotides in the 1093–1098 loop were implicated in RF2 termination as altering them caused UGA suppression (*43*).

Collectively, the functional data place the RF at key sites in the 50 S subunit in domains II and domain V of the 23 S rRNA and suggests the RF plays a significant part in 50 S-mediated mechanisms during protein synthesis termination.

V. The Mimicry Complex in Termination

The fact that the space formed by the inner cavity of the ribosomal active center is shaped rather like a tRNA implies that the external complexes entering the site at each stage of protein synthesis must either conform to the architecture or must trigger a significant conformational change at the boundaries of the cavity. Evidence for the former case was evident with the first structures of the bacterial protein synthesis factors in the mid-1990s. The two elongation factors, EF-Tu and EF-G, were crystallized with their ligands (Fig. 5). In the case of the EF-Tu, the ligand was a tRNA that was seen to hang off the bottom of the structure (44) and this factor could place the tRNA in the ribosomal A-site while making contact with other parts of the ribosome. Remarkably, the EF-G structure was revealed (45, 46) to have a shape similar to the EF-Tu.tRNA.GTP ternary complex, but with one of the EF-G protein domains, domain IV, occupying the equivalent space as the tRNA in the EF-Tu.tRNA.GTP ternary complex. Subsequent evidence has suggested that the tip of domain IV is important for translocation (47), and that a conformational change in 16 S rRNA helix 34 occurs on EF-G binding. This places domain IV into the depths of the A-site, consistent with cryoelectron microscopy reconstructions that position domain IV at the shoulder of the 30 S ribosomal subunit prior to translocation. During translocation, these reconstructions suggest that domain IV reaches into the decoding center like a tRNA (48). The conclusions reinforce the mapping on the ribosome of domain IV

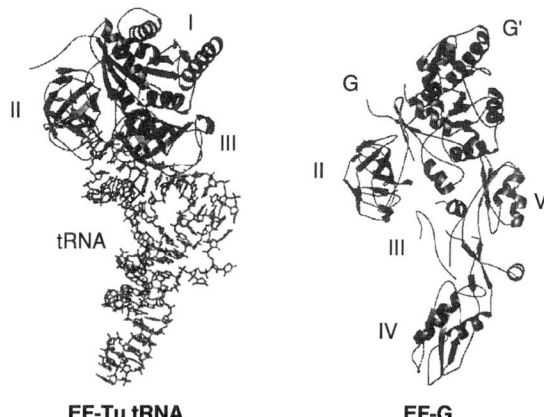

FIG. 5. Structural diagrams of the ternary complex EF-Tu.tRNA, and EF-G aligned in a similar orientation. EF-G domains III–V appear to mimic the tRNA moiety of the ternary complexed EF-Tu.

using rRNA cleavage by hydroxyl radicals generated from specific sites in EF-G. For example, hydroxyl radical generation from the tip of domain IV cleaved the rRNA in the decoding site (49). Therefore, domain IV is acting as a structural mimic of the anticodon stem-loop of the tRNA in particular, and while conformational changes are clearly happening, its tRNA-like shape allows this domain to bind initially into the three-dimensional space available.

Initiation factor 1, IF1, is a small protein synthesis factor that has been an enigma as it defies a clear definition of function despite being known and studied for many years. Some key observations emerged in the 1990s. One observation was that IF1 protects some of the same bases in 16 S rRNA as an A-site tRNA indicating that it occupies the same site as the anticodon stem-loop at the decoding site (50). This suggested that the function of IF1 might be to prevent premature binding of an aminoacyl-tRNA into the A-site during the initiation process. Indeed, the sequence of IF1 has some homology to the most N-terminal part of EF-G domain IV (51). This implies that IF1 has structural homology with a part of EF-G and, thereby, with a part of the tRNA and has extended the concept of structural mimicry at the ribosomal active center to complexes that bind at stages of protein synthesis other than at the elongation phase of the growing polypeptide chain.

Of great interest was the discovery that the C-terminal half of IF1 had a high degree of homology to about 10% of the sequence of both class I bacterial release factors, RF1 and RF2 (52) (Fig. 6). It alleviated the disappointment that there is no obvious sequence homology between the decoding RFs and EF-G using computer algorithms as might have been expected if both proteins were occupying the A-site. Nakamura and colleagues had painstakingly compared the sequence of EF-G and RF2 manually and identified possible critical residues in the C-terminal region of RF2 that might be important for structural mimicry, but the conclusions were understandably highly speculative (53). The identities and similarities of amino acids to residues in IF1 common to both bacterial decoding RFs but also unique to each provided more compelling data, although indirectly for mimicry with domain IV of EF-G. This served to reinforce the importance of the homology between the factors from these three different stages of protein synthesis. It raised the probability that this sequence region of the class I RFs homologous to IF1 was mimicking the anticodon stem-loop of the tRNA and was involved in stop signal decoding. Such a conclusion reinforced one element of the tRNA analogue model (32), namely, that a domain of the RF occupied part of the A-site during stop signal decoding, just like the anticodon stem-loop of the tRNA during decoding of each codon specifying an amino acid.

The possibility of a more complex pattern for molecular mimicry in termination had been recognized when the gene for the class II RF in bacteria, RF3, was first identified in 1994 and the characteristics of this protein deduced from

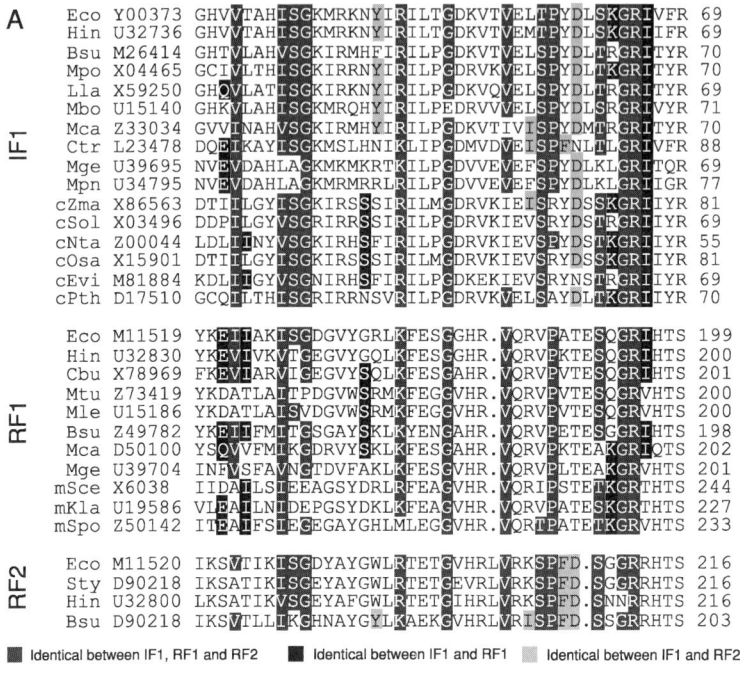

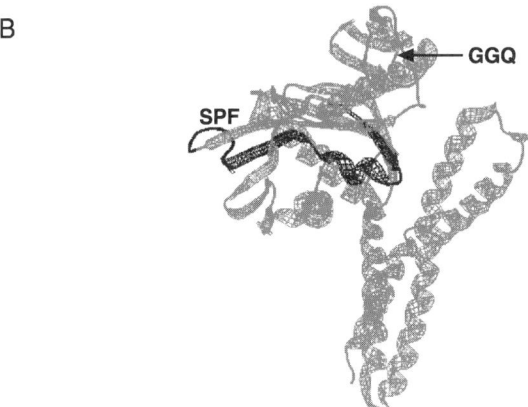

FIG. 6. Regions of similarity between prokaryotic and organellar IF1s and RF1 and RF2. (A) Alignment of distantly related members of the IF1 and RF families using the PILEUP program (GCG package, University of Wisconsin Genetics Computer Group). (B) The *E. coli* RF2 structure (36) showing the V179–R212 region (black) with similarity to IF1.

the sequence (54, 55). This protein had remained, like IF1 during initiation, an enigma for many years as being a stimulatory factor for termination, but its mode of action had been quite unclear. The sequence showed RF3 was highly homologous with the elongation factors, EF-Tu and the G and G' domains of EF-G. These signatures indicated that RF3 was also a translational G-protein. Most significantly, the new data implied that the two classes of RFs, the class I decoding RFs and the class II "stimulatory" RF, were equivalent to the EF-Tu. tRNA ternary complex, and the equivalent of the EF-G structure in two parts where the class II RF mimicked the G, G', and domain II of EF-G, and the class I RF possibly mimicked EF-G domains III, IV, and V (56). The structure of RF3 is yet to be determined and so the exact positioning of RF3 in the termination mimicry complex relies on deductions from sequence analysis. In just such an analysis from a dot plot between RF3 and EF-G we discovered a highly unexpected finding that was not found in an analysis of the class I RFs with EF-G. There was strong sequence homology between a C-terminal region of RF3 and a short stretch of EF-G domain III, and also to a lesser extent a separate stretch of domain IV (52, 57) (Fig. 7A). Both of these regions were distant from the more easily recognized homology between RF3 and the G domains of EF-G. A threading analysis of the sequences of RF3 against EF-G gave the possible structure shown in Fig. 7B. Not surprisingly, RF3 models well against the G, G', and domain II of EF-G consistent with its role as a G-protein. Strikingly, it also models with an extension of structure that protrudes from the body of the globular domain, like domains III and IV of EF-G, although with a simpler structure. These structural elements are at the C-terminus of RF3 and represent those sequences from the dot plot that have homology to domains III and IV of EF-G.

What does this mean? The class I RFs are supposed to be the mimics of these domains in the classic mimicry model and not the class II RF. The homology and modeling suggest that both the decoding RF and RF3 might occupy the decoding site with parts of their structure. The decoding RF can certainly bind to the ribosome in the absence of RF3 and catalyze the termination event *in vitro* (58) and catalyze the event in some organisms such as *Mycoplasma genitalium* that lack an RF3 (59). Therefore, RF3 is not essential for termination. RF3 is now known to be a recycling factor (60) recycling the decoding RF by destabilizing its interaction with the ribosome and, thereby, increasing the likelihood of catalytic termination efficiency *in vivo*. A structural element of RF3 at the decoding site might, when in a certain conformation, eject the decoding RF from the A-site. Moreover, the homology of the decoding RF with IF1 is only partial at the three-dimensional structural level. IF1 is an OB-fold protein containing β-strands, interconnecting loops and one α-helix (61). The decoding RF can model to this except that it is missing two β-strands and an interconnecting helical region. For this reason, it cannot be classed as a

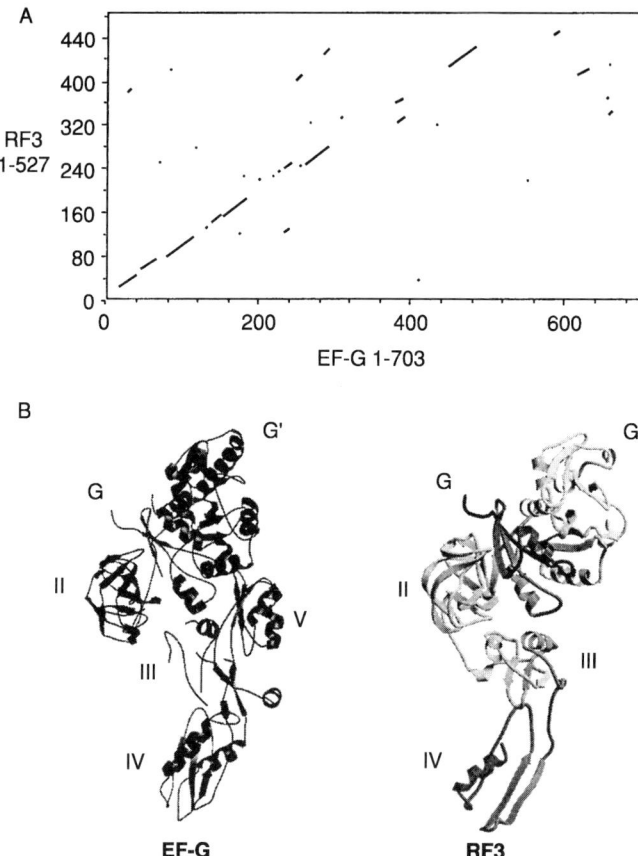

FIG. 7. A comparison of the RF3 sequence with the of EF-G. (A) A dot plot of *E. coli* RF3 against EF-G showing regions of similarity (52). The sequences of RF3 (1–527) and EF-G (1–703) were compared using the Genetics Computer Group programs Compare and Dotplot. (B) A structural diagram of RF3 based on a threading analysis with the structure of EF-G. The threading analysis was performed with SWISS-MODEL (104).

member of the OB-fold family of proteins. Intriguingly, the RF3 structural element homologous to EF-G domain IV models as two β-strands. Does this element complete the OB-fold with the decoding RF *in vivo*? This would be compatible with the RF functionally decoding stop signals when in one conformational state but disrupting it in another. These hypotheses are currently being tested with RF3 variants altered in key residues of the postulated domain IV "tip region."

VI. The RF Structure: Does It Look Like a tRNA?

The tRNA analogue model in 1994 had predicted that the RF would have at least two domains, one that interacts with the decoding center of the small ribosomal subunit and one that interacts with the enzyme center of the large subunit. It did not attempt to define a domain that interacted with the class II RF because, at that time, the function of RF3 was a mystery and prediction as to how the two classes of RF proteins might interact was premature. Functional experiments with eukaryotic RFs in the 1970s and 1980s had almost certainly studied the heterodimer, eRF1 and eRF3 (62). This provides an explanation for the then puzzling major differences between characteristics of the eukaryotic and prokaryotic RFs as class I bacterial RFs were being studied independently of RF3.

The first RF structure to emerge was that of eRFI (63), which generated enormous interest. A first impression was that the structure was indeed tRNA-like in overall shape, sustaining the prediction of an RF being a structural tRNA mimic. However, whereas a tRNA has an "L" shape, the eRF1 was shaped more like a "Y." Not surprisingly, there were three domains each with unique folds. One domain invoked mimicry of the anticodon stem and loop (domain I), a second corresponded with the aminoacyl tRNA stem (domain II), and the third was like the T stem of a tRNA that interacted with eRF3 (domain III). By the time the eRF structure was published in 2000, eRF3 was better characterized and such an interacting domain was expected.

A key feature of the eRF1 structure was the distance from the tip of domain I at one arm of the "Y" to the tip of domain II forming the stem of the "Y." This indicated that eRF1 could interact with the two major active centers of the ribosome: the decoding site and the enzyme center. An anchor point in the structure in domain I was an antiparallel helix hairpin that forms a prominent groove invoked as a possible place where codon recognition might occur. Connecting these two helices is the sequence NIKS, a motif that from functional studies clearly seems to influence codon recognition (64). Domain II has a tight turn at its tip within a motif composed of GRGGQS. The GGQ sequence is interesting because in a context of basic residues, it is the only obvious conserved motif among all RFs and suggests an essential role for this motif for structure or function. This sequence represents a turn-forming motif of the form GGX, but in both bacterial and eukaryotic RFs, modification of these residues affects peptidyl-tRNA hydrolysis and so places the motif in the vicinity of the ribosomal enzyme center. Combining functional data with the new structure gave excellent evidence that the NIKS motif at the tip of domain 1 marks where the factor contacts the decoding site, and where the GGQ motif at the tip of domain II interacts with the enzyme center.

Although the spans of eRF1 domains I and II are sufficient for it to be a decoding molecule that not only recognizes the stop signal but also influences polypeptide release, can the ribosomal active center accommodate the three-dimensional coordinates of eRF1? Normally, a tRNA is accommodated in the A-site within a confined three-dimensional space. The width of eRF1 at ~70 Å is very close to that of phenylalanine tRNA, although eRF1 is slightly "fatter" with a thickness of 27 Å compared with 22 Å for the tRNA. These dimensions suggest that the factor can be accommodated in the ribosomal active center without major conformational changes within the subunits to create more space or to change significantly the proportion of space available. However, reexamination of the structure (36, K. K. McCaughan and W. P. Tate, unpublished observations) has suggested that the NIKS and GGQ motifs are almost 100 Å apart and this suggests that it might be difficult to accommodate the factor into the available space of the active center. Nevertheless, the conformational flexibility of eRF1 domain II relative to domain I would allow some structural accommodation and this may be facilitated by a small number of interactions between the domains of the eRF1 itself, or between eRF1 and ribosomal components lining the active center.

There was a suspicion that the bacterial RF might be conformationally flexible because it had resisted attempts to produce crystals that would diffract satisfactorily in spite of significant effort from several groups. The fact that there was virtually no sequence identity or similarity within the primary sequence meant that the class I RFs could not be modeled easily against the eRF1 structure. The conserved GGQ motif was at residues 183–185 in the human eRF1 primary sequence, and at residues 250–252 in *Escherichia coli* RF2. In contrast, the eRF1 decoding motif, NIKS, was near the N-terminus (residues 61–64) but its apparent equivalent in the bacterial sequence, SPF, was nearer the C-terminus at residues 205–207 in RF2. This meant that it was unlikely the protein architectures would be the similar between the two proteins. Song and colleagues (63) note that secondary structure predictions based on 34 multiply-aligned prokaryotic and mitochondrial class I RF sequences indicate a pattern of structures that is inconsistent with the organization of eRF1.

This can be illustrated when the bacterial RF sequences are threaded against the eRF1 structure. This is shown for RF2 in Fig. 8. Without fixing any motifs as anchor points, the modeled structure has the GGQ away from the tip of domain II and the SPF is distant from the tip of domain I. Instead, the residues at the tip were 96–99 near to the N-terminus of the protein. This suggested that this structure was not a real representation of what the bacterial RF2 might be. Fixing the GGQ motif at the tip still allowed threading, but now the proposed decoding motif was sited between domains I and II. Fixing both the GGQ at one tip, and fixing residues 148, 201, and 209 proposed to be at the decoding site both from our own studies with *E. coli* RF2 (65) and those of

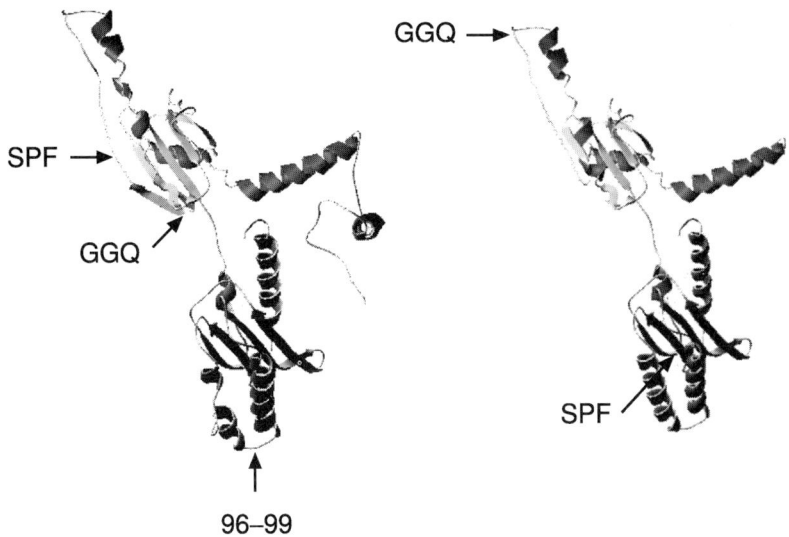

FIG. 8. Structural diagrams of *E. coli* RF2 derived from sequences threaded against the eRF1 structure (63). Without fixing any amino acids, the residues (96–99) at the tip of domain I were from the N-terminus of the protein (left). The structure (right) resulted when the GGQ motif was fixed at one tip and RF2 residues 148, 201, and 209 were fixed with eRF1 residues 71, 123, and 132 proposed to be in the decoding site (66). The threading analyses were performed using Deep View and SWISS-MODEL (104–106).

Bertram and colleagues with eRF1 (66), a credible RF2 structure was possible. *E. coli* RF1 could be modeled in a similar way.

The need to continue these kinds of studies was finally superseded when the first structure of a bacterial RF was published in December 2001 (36). The title summed up an important conclusion that "bacterial polypeptide release factor 2 is structurally distinct from eukaryotic eRF1." Depending on perspective it is possible to conclude that this structure looks more or less like a tRNA rather than eRF1. Although the structure is L-shaped like a tRNA, the top of the "L" is very bulky. It is composed of four domains as compared with three for eRF1 (Fig. 9). Domains II, III, and IV form a very compact structure, and are interconnected to resemble a single superdomain with domain I protruding from the superdomain body. It is the disposition of key sequence motifs that caused the most surprise. Domain I, expected to be equivalent to the anticodon loop and stem, shown as four helices, is from the N-terminal sequence where there is most variation among the bacterial RFs. No previous functional data had identified this region as important for stop signal decoding. Both the SPF motif and the GGQ motif of *E. coli* RF2 were in exposed loops on a large curved surface on one side of the structure with the large superdomain

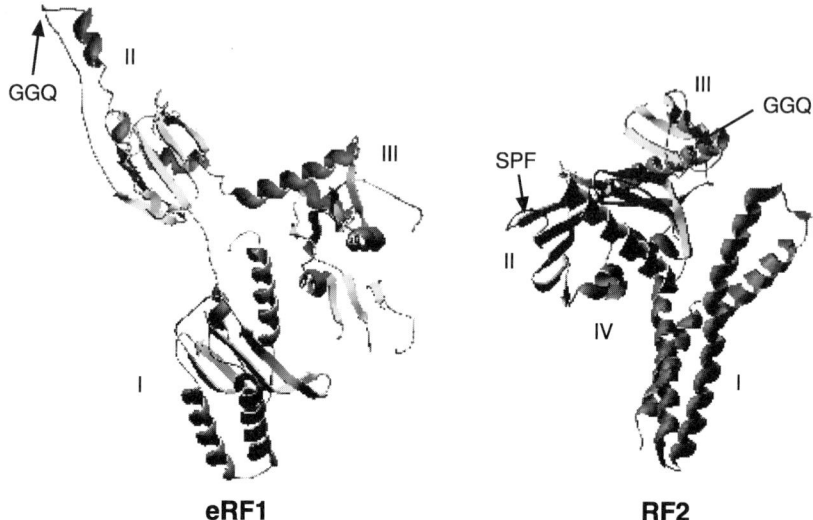

FIG. 9. A comparison of the structures of eRF1 (63) and RF2 (36).

only ~20 Å apart. The SPF motif, proposed to be a recognition sequence for the second and third nucleotides of the stop codon and responsible for codon specificity (67), would be expected to be distant from the GGQ motif and in separate domains of the structure. Moreover, the region of RF2 homologous to IF1 was in the bulky superdomain (see Fig. 6B).

The major paradox between the structure and the accumulated diverse functional data was that the biochemical studies sited the SPF region in the decoding site and associated with stop signal decoding. Vestergaard and colleagues (36) attempted to model the RF2 structure against a tRNA and then attempted to dock it into the ribosomal active center using the 70 S structure (16). The SPF domain could not fit into the A-site without major conformational change in the factor, the ribosome or both, to create the necessary space. In contrast, domain I could be modeled into the decoding site avoiding spatial clashes with the ribosome.

There has recently become available a reconstruction of cryoelectron microscopic images of a ribosome containing RF2 in a posttermination complex that, indeed, strongly supports the fact that RF2 undergoes a conformational change from that seen in the solution structure (68, 68a). It was deduced from these studies that domain I of RF2 contacts the N-terminal domain of L11 previously implicated by functional experiments to be involved in RF binding (69). The cryoelectron microscopic studies also show that the SPF loop is near the decoding site and that the GGQ motif is near the

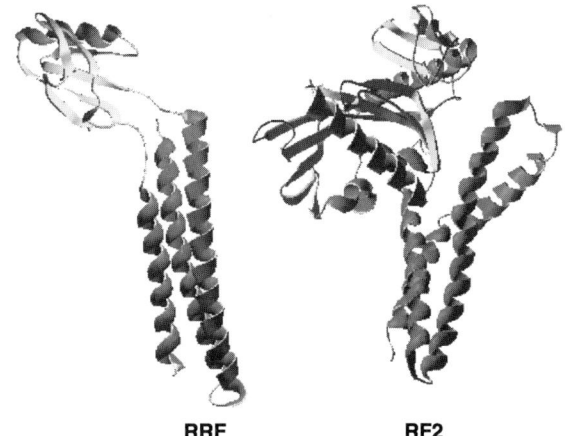

FIG. 10. A comparison of the structures of RRF (*107*) and RF2 (*36*). The three-stranded coiled-coil structures are similar for both factors.

peptidyltransferase center. Moreover, the region of RF2 homologous to IF1 is now near the decoding site as expected. This establishes concordance between the way eRF1 and the bacterial RF2 function despite their different structural folds.

Is there any precedent for another protein with a domain fold like that of RF2 domain I? The ribosome recycling factor (RRF) acts after the polypeptide has been released to recycle the ribosome after a round of protein synthesis. The RRF structure showed it had an unusual long three-helix bundle that resembles the barrel of a gun mimicking the anticodon stem and loop of a tRNA (*70*). The RRF also has flexibility in the hinge region between the two domains critical for function (*71*). The three-stranded coiled-coil in domain I of the RF2 structure is the same as that found in the RRF structure and so there is a precedent for the fold perhaps being a factor anchor site on the ribosome (*36*). Very recent localization of RRF on the ribosome positions the three-stranded coil-coil more toward the enzyme center rather than the decoding site (*71a*). The RF2 and RRF structures are compared in Fig. 10.

VII. The Paradox between the Structural and Functional Evidence for the Bacterial RF "Anticodon"

Collectively, the body of biochemical data for motif function in eRF1 is completely consistent with its structure. The evidence that the GGQ motif at the tip of domain II is in the vicinity of the enzyme center is undisputed, although the role this motif plays in peptidyl-tRNA hydrolysis is still not resolved and remains the subject of several theories. The evidence that the NIKS motif at the tip of domain I is at the decoding site is also convincing and has been strengthened recently with the finding that the first U of the stop codon contacts the K residue of this motif (72), although how the mRNA stop signal is recognized is also the subject of several hypotheses and interesting ideas. In contrast, this is not the case for bacterial RF2. The tip of domain I in the RF2 structure is rich in acidic residues whereas the tip of domain I in eRF1 is rich in basic residues. As yet, there are no functional data implicating residues around a highly conserved aspartate (D92) at the tip of domain I in the RF2 structure in decoding events, notwithstanding the fact that this part of the sequence has not been studied in detail. This would be more supportive of the cryoelectron microscopic location of domain I being at the GTPase center.

The previously proposed tripeptide "anticodon" motif (67) is located within a region of sequence that is both RF-type specific and has a high degree of homology to IF1 (52). There is a prevalence of basic residues making it more like the tip of domain I of eRF1 than that shown in the structure of RF2. Nakamura and colleagues (67) discovered that tripeptide motifs in the bacterial RF1 and RF2 seemed to control which codons they recognized, resulting in their proposal that they act as discriminators. This implied a defined part of the sequence and, thereby, a small part of the structure was the likely equivalent of the anticodon of the tRNA. Previous studies had shown that zero-length cross-links could be obtained from the first base of the stop signal in the mRNA directed to the bacterial RF. This seemed to be a clear indication that there certainly was a motif in the factor that "read" the stop signals (33, 34, 73). Moreover, a zero-length cross-link could still be obtained when it was directed from the three nucleotides beyond the stop codon indicating recognition might be different from that of a tRNA in that more than just three nucleotides were being scanned (35). So far, the sites on the bacterial factor involved in these cross-links have defied identification.

Despite the compelling data, key residues in the proposed "anticodon" motif (SPF in RF2) can be substituted with another residue with retention of codon recognition, function, and specificity (65). Only the first of the three residues (S205 in RF2) seemed particularly sensitive to such change. This is puzzling if these residues are so important in an "anticodon" domain. Two residues flanking this key motif have been targeted to place hydroxyl radical-generating centers that can be used to map ribosomal neighbors when the termination complex forms. Although a radical that is generated can travel quite a distance from the site of origin and result in rRNA cleavages that provide confusing information, in reality there are multiple cleavage sites produced from each particular source. These cleavage sites result from radicals scattering in a plethora of directions and so the chances of neighboring structures being avoided is low. We substituted a conserved leucine residue with cysteine four residues upstream from SPF in RF2, and in a separate RF2 variant a less highly conserved serine was substituted for cysteine two residues downstream from SPF. This was after first removing the natural cysteine residues in RF2 by replacing them with residues that did not result in loss of activity. Therefore, each variant had a single cysteine with an attached moiety from which radicals were generated.

What were the results of these studies? Cleavages were detected only in the 16 S rRNA. They targeted a binding pocket on the molecular surface of the small ribosomal subunit. Significant cleavages were detected in the highly conserved 530 loop of 16 S rRNA when radicals were generated from the downstream residue (C209S in RF2). This is a particularly interesting result as residues in this loop cross-link 3′ from the A-site codon in the mRNA (74) and are protected by the A-site tRNA (75). Residue 209 is, therefore, close to an important element of the ribosomal decoding center. Cleavages from radicals generated from both chosen positions spanning the SPF motif of RF2 were also detected at other positions in the vicinity of the decoding site in the ribosome structure. Cleavages were also detected in helix 30 surrounding nucleotide 1230. The 3Å crystal structure of the small ribosomal subunits showed that the subunits were packed so that a protruding spur of one was inserted into the P-site of the neighboring subunit (76). Several regions of the 16 S rRNA including nucleotides 1229–1230 were shown to be in contact with this spur, suggesting that residues 201 and 209 of the A-site bound RF2 also interact with elements of the ribosomal P-site. The 790 loop located in the platform of the small ribosomal subunit at the subunit interface was also a cleavage target (Fig. 11). Identical cleavages were detected in this loop from the tip of domain IV of EF-G (the tRNA anticodon mimicry domain) (49).

These data seemed to provide compelling evidence for the exposed SPF loop of RF2 being at the decoding site and highlights further the paradox between the solution structure of RF2 and the functional and cryoelectron microscopic studies. Further support for a major conformational change comes from hydroxyl radical footprint studies with the other bacterial class I decoding RF, although, as yet, no structure of RF1 is available. Two sites on RF1 close to the proposed "anticodon" equivalent motif also showed cleavages in the 1230 region (73). This seems to suggest that there might be a major conformational change in the protein ligand when the RF interacts at the A-site of the bacterial ribosome.

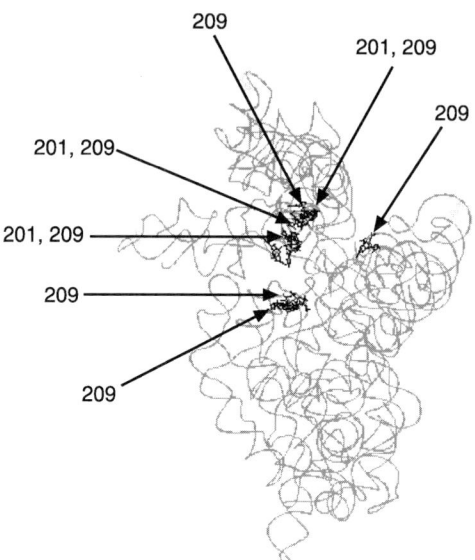

Fig. 11. Sites of 16 S rRNA cleavages generated from residues spanning the putative "anticodon" in *E. coli* RF2. The cleavage sites induced by hydroxyl radicals generated from residues 201 and 209 are shown in black (arrows) on *T. thermophilus* 16 S rRNA (*108*). The 16 S rRNA interface is oriented toward the foreground.

VIII. Specific Interactions within the Bacterial Termination Complex

As described above, site-directed cross-links from the mRNA have shown the bacterial decoding RF makes close contact in the decoding site with the mRNA stop signal including the stop codon and the three bases beyond (*35*). Functional studies support this conclusion. Not only do decoding RFs respond to each of the codons with specificity (RF1, UAA, and UAG; RF2, UAA, and UGA) but they also respond to the base following the stop codon in particular (*77, 78*). There can be dramatic differences in termination efficiency depending on whether C/A or U/G immediately follows the stop codon (A-site), and less marked but measurable differences depending on the identity of the bases in the remaining two positions. Bases within individual signals downstream from the stop codon seem to act independently of each other to strengthen or weaken a signal. This implies there may be a secondary or tertiary orientation of the mRNA that influences any interactions a particular signal has with the decoding RF. At the interface between the two subunits, X-ray crystallography has shown eight nucleotides of the mRNA are exposed (−1, the two codons, +7) with about 30 nucleotides wrapped in a groove around the neck of the small subunit (79). The mRNA has up to nine nucleotides beyond the A-site codon within the ribosome during translation, potentially allowing for interactions between RFs and the nucleotides in the mRNA.

Upstream sequences spanning two codons prior to the stop signal also affect the efficiency of decoding by an RF (80, 81). This is at first puzzling, because these sequences are involved in codon:anticodon interactions with the P-site and E-site tRNAs. However, we have discovered the preferences for particular codons reflect the preferences for a small subset of tRNAs especially at the P-site (78). Most of the preferred codons have only one isoacceptor tRNA species, and a modification near the anticodon is highly overrepresented among the preferred tRNAs. This implies that the P-site tRNA might prepare part of the A-site active centre for the decoding RF in the termination complex. It implies that the factor makes close contact with the P-site tRNA, and, perhaps, makes interactions with specific nucleotides in the tRNA. Although all tRNAs have similar shapes, the particular nucleotide sequence and the post-translational modifications frequently found at specific sites throughout the tRNA may be important for such interactions. Indeed, we have found zero-length site-directed crosslinks to RF2 in a termination complex from two positions in the tRNA located in the P-site. These were from nucleotide 32 contiguous with the anticodon at the decoding site, and from the elbow of the tRNA (nucleotide 8) positioned towards the enzyme centre (E. S. Poole and W. P. Tate, unpublished observations). This implies the decoding RF fits tightly into the A-site perhaps more tightly than the tRNA

What contacts does the decoding RF have with the rRNA and ribosomal proteins? We have taken a global approach using a modification of SELEX developed to define interactions of ribosomal proteins with rRNAs (82). The principle of the approach for our studies was to fragment the rRNA genes and clone the fragments into a transcription vector so that a random array of rRNA fragments could be generated as a "bait" for the decoding RFs. The rRNA was passed over immobilized RF and then iterative cycles (SERF) carried out for *in vitro* selection of random rRNA fragments that had bound specifically to the RF. After several cycles the selected rRNA was cloned, sequenced, identified, and mapped to the ribosome structures. Fragments of rRNA from 16 S rRNA, 23 S rRNA, and 5 S rRNA structures were selected, with some fragments identified multiple times (for example, the last 30 nucleotides of 5 S rRNA). Common fragments were associated with each of three decoding RFs, the bacterial RF1 and RF2, and mitochondrial RF1 that can also bind to the bacterial ribosome. In addition, there were fragments that were unique to each factor (M. E. Askarian-Amiri and W. P. Tate, unpublished observations). The 16 S rRNA fragments were derived from the platform region below the decoding site and those from 23 S rRNA were clustered in several domains surrounding the peptidyltransferase region and at the RF entry site to the active center. The 5 S rRNA fragment was the part of the rRNA bridging 23 S rRNA domains II and V and again at the RF entry site to the active center. These analyses have provided a good overview of the span of the decoding RF in the termination complex at the A-site. Sites common to the two bacterial factors and the mitochondrial factors are shown in Fig. 12.

Some evidence for interaction of the decoding RF with the peptidyltransferase center of the 50 S subunit has come from two strategies. One strategy used a classic SELEX procedure to select RNA aptamers that bound to a thermophilic bacterial RF1 (83). Two groups of sequences were found that contained short invariant single-stranded motifs that mapped to the peptidyltransferase center.

We used a second strategy that used specific sites in the biochemically defined putative peptidyl-tRNA hydrolysis domain of RF2 (residues 238–274) to generate hydroxyl radicals from positions (243 and 246) close to the $^{250}GGQ^{252}$ motif as well as from a naturally occurring cysteine at 274. Perhaps the most interesting finding of our experiments is that when radicals are generated from the residues close to the GGQ motif, many of the cleaved bases in the 23 S rRNA are located within the peptidyltransferase center. For example, hydroxyl radicals originating from positions 246 and 274 of RF2 specifically cleave at A2602, a universally conserved bulged nucleotide that is protected by the aminoacyl moiety of the A-site-bound tRNA (22), and a base also cleaved by

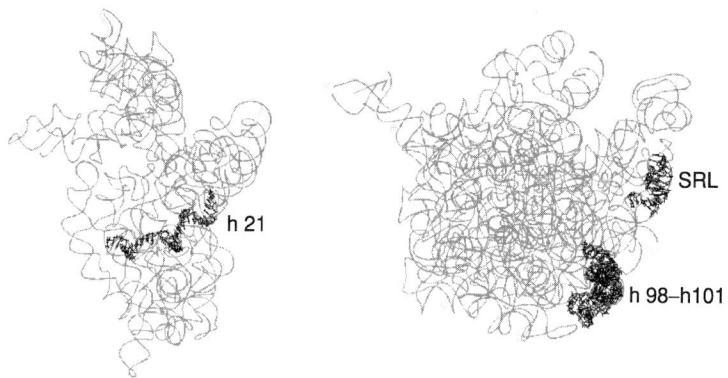

FIG. 12. rRNA fragment sites that have interactions common to RF1, RF2, and mitochondrial RF1. The site of the helix 21 fragment (left, black) is shown on *T. thermophilus* 16 S rRNA (108). The position of fragments encompassing helices 98–101 and the sarcin–ricin loop (SRL) (right, black) are shown on *D. radiodurans* 23 S rRNA (15). The rRNA interfaces are oriented toward the foreground.

radicals generated from the 5′-CCA terminus of the deacylated tRNA (*84*). Mutations at A2602 lead to ribosomes incapable of catalyzing efficient peptide release but not peptide bond formation (*85*). In addition, sites within the central peptidyltransferase ring of 23 S rRNA domain V were cleaved by hydroxyl radicals originating from RF2 positions 243 or 246. Mutations in some of these cleavage sites have been shown to dramatically reduce peptidyltransferase activity (*86*). The universally conserved nucleotide A2451 was cleaved by radicals generated from position 246. Interestingly, A2451 was hypothesized to be the "magic bullet" general base catalyst for peptide bond formation (*87*), although there is now uncertainty as to its role as activity is not significantly affected by mutations at this site (*88, 89*). Hydroxyl radicals originating from position 246 of RF2 also specifically cleaved nucleotides within the A-loop of domain V of the 23 S rRNA. These nucleotides are protected by the 3′ terminus of the A-site-bound tRNA (*22, 89, 90*). Radicals from both RF2 positions 246 and 274 cleaved nucleotides within the P-loop of 23 S rRNA domain V, a region protected by the 3′ terminus of the P-site tRNA (*22, 90, 91*).

Radicals generated from positions 243 and 246 of the proposed peptidyl-tRNA hydrolysis domain of RF2 also cleaved sites in the GTPase center, supporting the previously discussed proposed interaction of the factor with this region of the 23 S rRNA. In the crystal structure of a GTPase center rRNA fragment bound to ribosomal protein L11 (*92, 93*), two loops (1093 to 1098 and 1065 to 1073) are situated very close to each other. A 23 S rRNA mutation located within the GTPase center, G1093A, was previously shown *in vitro* to decrease the effective association rate of RF2 with the ribosome (*42, 94*). Although cleavages were detected only around 1093 and not detected around nucleotide 1067 from the selected positions on RF2, RF1 was shown to form a "cleavage-print" on both nucleotides 1067 and 1093 of the GTPase domain (*73*). This suggests that both loops of the GTPase center may interact with a compact region of the decoding RFs. These data are consistent with the importance of ribosomal protein L11 in RF binding to the ribosome. L11 has been shown both *in vitro* (*39, 40, 69*), and *in vivo* to affect differentially the two bacterial decoding RFs (*41, 95*). L11 is essential for RF1-dependent ribosome binding and UAG termination, but dampens RF2-dependent binding and UGA termination. A comparison of cleavage patterns in 23 S rRNA generated from specific sites in RF1 and RF2 is shown in Fig. 13.

The second class of RF important in bacterial termination of protein synthesis, RF3, has not as yet been subjected to the same detailed analysis as the decoding RF. As described above, RF3 has significant homology to EF-Tu and EF-G and in particular the domains of these proteins that mark all three as translational G-proteins (*54, 55*). Presumably, the main site of RF3 interaction with the ribosome is at the outer binding region of the ribosomal active center (*30*) despite evidence that strands of the structure may penetrate the inner

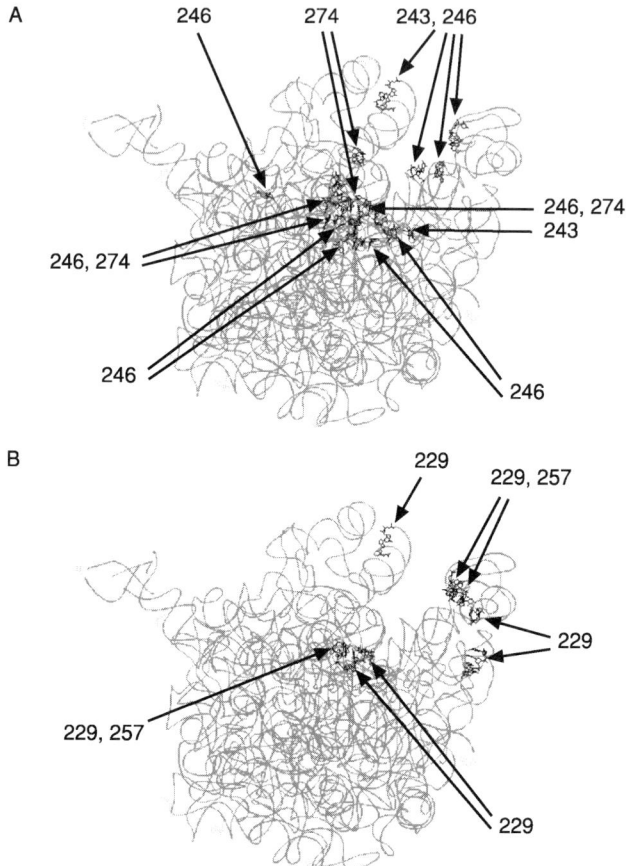

FIG. 13. Sites of 23 S rRNA cleavages generated from residues spanning the putative peptidyl-tRNA hydrolysis domains of *E. coli* RF1 and RF2. (A) The cleavage sites induced by hydroxyl radicals generated from residues 243, 246, and 274 of RF2 (black, arrows). (B) The cleavage sites induced by hydroxyl radicals generated from residues 229 and 257 of RF1 (black, arrows) (73). The cleavage sites are shown on *D. radiodurans* 23 S rRNA (15) with the interface oriented toward the foreground.

active center (52). Early attempts using reconstitution of ribosomal subunits to identify ribosomal proteins important for RF activity had highlighted L11 and several other proteins important for peptidyltransferase integrity for the decoding RFs (39), but did not identify specifically any additional proteins important for RF3 function (W. P. Tate, unpublished observations). Nevertheless, RF3 can bind to the ribosome independent of the decoding RF (96), although binding is stimulated by the decoding RF. This suggests that RF3

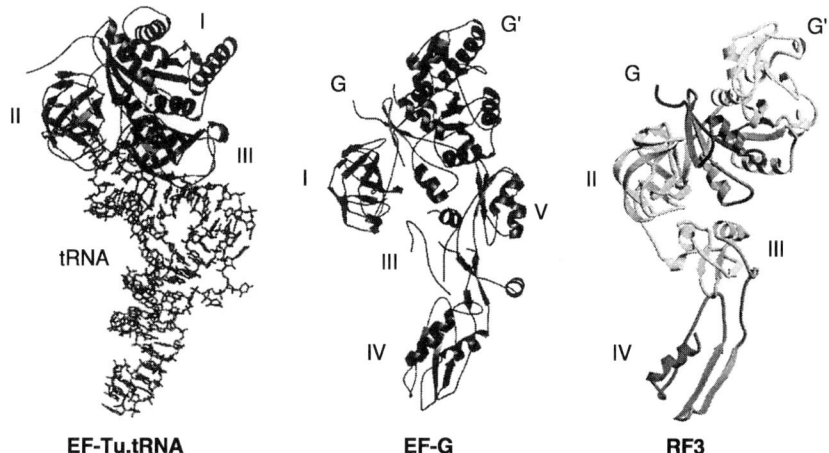

FIG. 14. A comparison of the structures of the ternary complex EF-Tu.tRNA, EF-G, and RF3 threaded to the structure of EF-G.

binding is stabilized by further sites of interaction between the two factors. Footprinting and mapping experiments with RF3 alone on the ribosome or together with other ligands will reveal the nature of these interactions. Now that there is a much better understanding of the mechanism of RF3 function through the recent studies of Zavialov and colleagues (97) sensible strategies can be developed.

As yet, no structure of a bacterial RF3 has been resolved but given its high degree of homology with EF-Tu and EF-G, RF3 can be threaded against these structures with some confidence. These two structures and the threaded RF3 are shown together in Fig. 14. If the actual structure of decoding RF2 and the threaded structure of RF3 are compared with the structure of EF-G, as together the two classes of RFs mimic the multidomain EF-G molecule, then some idea of an actual mimicry complex in termination can be visualized. However, if the decoding RF undergoes significant conformational unfolding in order to bind to the ribosome, then these structures might be quite different at the ribosomal active center.

IX. The Termination Mechanism

The decoding RF enters the ribosomal A-site when a stop codon is in the decoding position. For both eubacterial and eukaryotic decoding RFs, there is evidence that the factor is in close contact with the stop signal (33–35, 72, 98) Recognition most likely involves at least six and maybe a greater number of nucleotides (78), although for this to occur the mRNA may have to form a loop in some way. There are several intriguing hypotheses as to how these signals are recognized. Nakamura and Ito (99) have identified a tripeptide in the decoding RF that they invoke as the functional mimic of the anticodon and as a functional discriminator between the UAG and UGA stop codons. Indeed, if the RF2 were to dock into the decoding site with domain I as proposed by Vestergaard and colleagues (36), then this model would be incompatible with the structure. On the other hand, if, as seems likely from cryoelectron microscopic (68, 68a) and functional studies, there is extensive conformational unfolding of the RF at the ribosome, then the putative tripeptide anticodon could indeed occupy the decoding site. These two possibilities are currently being distinguished by mapping experiments. Results suggest that the tripeptide anticodon is in the vicinity of the decoding site (73, K. K. McCaughan and W. P. Tate, unpublished observations). Changes in the charge surrounding the tripeptide caused a loss of codon specificity allowing both noncognate stop codons and sense codons to be decoded (99). Therefore, this part of the RF is certainly a key region for binding the factor correctly into the ribosomal active center and/or in the codon recognition mechanism.

Bertram and colleagues have suggested that pockets in the eRF1 structure are capable of binding trinucleotide codons (66). The eukaryotic factor is somewhat different from each of the bacterial equivalents in that one factor recognizes all three stop codons, and, therefore, must only discriminate between stop and sense codons and not between the second and third positions of the stop codons themselves as do RF1 and RF2. Therefore, different recognition principles may apply in each case. Moreover, given the lack of sequence homology between the eubacterial and eukaryotic RFs, they may have arisen through separate evolutionary events to develop somewhat different mechanisms of recognition.

There is even less certainty about how the RFs participate in the peptidyl-tRNA hydrolysis event partly because the question of their direct involvement or indirect perturbation of the center has not been established. After solving the structure of the eRF1, Song and colleagues (63) proposed a model where the Q of the GGQ motif might be involved in coordinating and orienting a water molecule. However, Seit-Nebi and colleagues (100) showed that this residue could be replaced with other residues to give unaltered or reduced eRF1 binding to the ribosome lowering termination activity. This seems to exclude this residue playing a highly specific role such as water presentation at the catalytic site of peptidyl-tRNA hydrolysis.

There is growing evidence for rRNA involvement ensuring accurate and efficient protein synthesis termination (101). Many observed effects of decreased activity resulting from a change of an amino acid at a specific site in the RF might be simply due to an altered interaction with the rRNA. In that way, the presentation of the decoding RF to the functional centers at the decoding site and enzyme center might be altered. The major questions for both codon recognition and peptidyl-tRNA hydrolysis will be answered once the quest for a structure of a decoding RF on the ribosome is resolved. Such a structure will provide an impetus for more specific studies on the mechanisms of these two key functions of the decoding RFs.

Zavialov and colleagues (97) have shown that RF3 is stably bound to GDP and in that form RF3 binds to a ribosome resulting in the release of GDP. Both of these steps happen most likely but not necessarily after the polypeptide has been released. Once GDP has been dissociated, GTP can bind and the decoding RF is released. At the active center of the ribosome, the GTP form of RF3 and the decoding RF are somehow mutually exclusive. The process of GTP hydrolysis is then simply a mechanism to reject RF3 from the ribosome.

This multistep mechanism provides an excellent framework for attempting to understand how rejection of the decoding RF might occur. The fact that RF3 has a subdomain mimicking the domains of EF-G at the decoding site (52) means that the function of this region of EF-G is relevant to understanding how RF3 might function in rejecting the decoding RF. The tip of EF-G (domain IV) has been shown to be important for translocation with substitution of a single residue decreasing the translocation rate 100-fold, and small deletions decreasing the rate even further (47). It has been suggested that this region might couple the conformational change in EF-G as a result of GTP hydrolysis, to structural rearrangements of the ribosome leading to tRNA translocation. The equivalent region of RF3 has a unique sequence motif of KRKFEEFKK within a domain of two β-strands and a loop (57). Perhaps, these are already interleaved with the strands of the RF at the decoding site when bound in the GDP or nucleotide-free form of RF3. It is intriguing to speculate that this may also, as suggested for EF-G, be involved in a conformational coupling when GTP binds to RF3 disrupting RF interaction at the decoding site. Whether this alone would be sufficient to displace other interactions the decoding RF makes with the rRNA awaits further studies.

The final stages of returning the ribosome to a state where it can accept another round of protein synthesis are mediated by the RRF. Following termination the ribosome is left with a deacylated tRNA in the P-site, perhaps still a tRNA in the E-site and the mRNA still attached. Recently, Hirokawa and colleagues (102) have provided evidence consistent with the proposal that the RRF binds into the A-site as a tRNA mimic rather like the decoding RF, but with EF-G on the ribosome at the same time. How this occurs is not clear. Following binding, the model presented has the RRF translocated into the P-site with release of the deacylated tRNAs from the P- and E-sites. Subsequent to this, EF-G, RRF, and the mRNA dissociate.

X. Summary

Molecular mimicry was a concept that was revived as we understood more about the ligands that bound to the active center of the ribosome, and the characteristics of the active center itself. It has been particularly useful for the termination phase of protein synthesis, because for many years this major process seemed not only to be out of step with the initiation and elongation phases but also there were no common features of the process between eubacteria and eukaryotes. As the facts that supported molecular mimicry emerged, it was seen that the protein factors that facilitated polypeptide chain release when the decoding of an mRNA was complete had common features with the ligands involved in the other phases. Moreover, now common features and mechanisms began to emerge between the eubacterial and eukaryotic RFs and suddenly there seemed to be remarkable synergy between the external ligands and commonality in at least some features of the mechanistic principles.

Almost 10 years after molecular mimicry took hold as a framework concept, we can now see that this idea is probably too simple. For example, structural mimicry can be apparent if there are extensive conformational changes either in the ribosome active center or in the ligand itself or, most likely, both. Early indications are that the bacterial RF may indeed undergo extensive conformational changes from its solution structure to achieve this accommodation. Thus, as important if not more important than structural and functional mimicry among the ligands, might be their accommodation of a common single active center made up of at least three parts to carry out a complex series of reactions. One part of the ribosomal active center is committed to decoding, a second is committed to the chemistry of putting the protein together and releasing it, and a third part, perhaps residing in the subdomains, is committed to binding ligands so that they can perform their respective single or multiple functions. It might be more accurate to regard the decoding RF as the cuckoo taking over the nest that was crafted and honed through evolution by another, the tRNA. A somewhat ungainly RF, perhaps bigger in dimensions than the tRNA, is able, nevertheless, like the cuckoo, to maneuvre into the nest. Perhaps it pushes the nest a little out of shape, but is still able to use the site for its own functions of stop signal decoding and for facilitating the release of the polypeptide.

The term molecular mimicry has been dominant in the literature for a period of important advances in the understanding of protein synthesis. When the first structures of the ribosome appeared, the concept survived and was seen to be valid still. Now, we are at the stage of understanding the more detailed molecular interactions between ligands and the rRNA in particular, and how subtle changes in localized spatial orientations of atoms occur within these interactions. The simplicity of the original concept of mimicry will inevitably be blurred by this more detailed analysis. Nevertheless, it has provided a significant set of principles that allowed development of experimental programs to enhance our understanding of the dynamic events at this remarkable active site at the interface between the two subunits of this fascinating cell organelle, the ribosome.

REFERENCES

1. Kruger, K., Grabowski, P. J., Zaug, A. J., Sands, J., Gottschling, D. E., and Cech, T. R. (1982). Self-splicing RNA: Autoexcision and autocyclization of the ribosomal RNA intervening sequence of Tetrahymena. *Cell* **31**, 147–157.
2. Crick, F. (1968). The origin of the genetic code. *J. Mol. Biol.* **38**, 367–379.
3. Yonath, A., Mussig, J., Tesche, B., Lorenz, S., Erdmann, V. A., and Wittmann, H. G. (1980). Crystallization of the large ribosomal subunits from *Bacillus stearothermophilus*. *Biochem. Int.* **1**, 428–435.
4. Shevack, A., Gewitz, H. S., Hennemann, B., Yonath, A., and Wittmann, H. G. (1985). Characterization and crystallization of ribosomal particles from *Halobacterium marismortui*. *FEBS Lett.* **184**, 68–71.
5. Trakhanov, S. D., Yusupov, M. M., Agalarov, S. C., Garber, M. B., Ryazantsev, S. N., Tischenko, S. V., and Shirokov, V. A. (1987). Crystallization of 70S ribosomes and 30 S ribosomal subunits from *Thermus thermophilus*. *FEBS Lett.* **220**, 319–322.
6. Glotz, C., Mussig, J., Gewitz, H. S., Makowski, I., Arad, T., Yonath, A., and Wittmann, H. G. (1987). Three-dimensional crystals of ribosomes and their subunits from eu- and archaebacteria. *Biochem. Int.* **15**, 953–960.
7. Ban, N., Freeborn, B., Nissen, P., Penczek, P., Grassucci, R. A., Sweet, R., Frank, J., Moore, P. B., and Steitz, T. A. (1998). A 9 Å resolution X-ray crystallographic map of the large ribosomal subunit. *Cell* **93**, 1105–1115.
8. Tocilj, A., Schlunzen, F., Janell, D., Gluehmann, M., Hansen, H. A., Harms, J., Bashan, A., Bartels, H., Agmon, I., Franceschi, F., and Yonath, A. (1999). The small ribosomal subunit from *Thermus thermophilus* at 4.5 Å resolution: Pattern fittings and the identification of a functional site. *Proc. Natl. Acad. Sci. USA* **96**, 14252–14257.
9. Ban, N., Nissen, P., Cappel, M., Moore, P. B., and Steitz, T. A. (1999). Placement of protein and RNA structures into a 5 Å-resolution map of the 50 S ribosomal subunit. *Nature* **400**, 841–847.
10. Cate, J. H., Yusupov, M. M., Yusupova, G. Zh., Earnest, T. N., and Noller, H. F. (1999). X-ray crystal structures of 70S ribosome functional complexes. *Science* **285**, 2095–2104.
11. Schluenzen, F., Tocilj, A., Zarivach, R., Harms, J., Gluehmann, M., Janell, D., Bashan, A., Bartels, H., Agmon, I., Franceschi, F., and Yonath, A. (2000). Structure of functionally activated small ribosomal subunit at 3.3 angstroms resolution. *Cell* **102**, 615–623.
12. Wimberly, B. T., Brodersen, D. E., Clemons, W. M., Jr., Morgan-Warren, R. J., Carter, A. P., Vonrhein, C., Hartsch, T., and Ramakrishnan, V. (2000). Structure of the 30S ribosomal subunit. *Nature* **407**, 327–339.
13. Ban, N., Nissen, P., Hansen, J., Moore, P. B., and Steitz, T. A. (2000). The complete atomic structure of the large ribosomal subunit at 2.4 Å resolution. *Science* **289**, 905–921.
14. Ramakrishnan, V., and Moore, P. B. (2001). Atomic structures at last: The ribosome in 2000. *Curr. Opin. Struct. Biol.* **11**, 144–154.

15. Harms, J., Schluenzen, F., Zarivach, R., Bashan, A., Gat, S., Agmon, I., Bartels, H., Franceschi, F., and Yonath, A. (2001). High resolution structure of the large ribosomal subunit from a mesophilic eubacterium. *Cell* **107,** 679–688.
16. Yusupov, M. M., Yusupova, G. Zh., Baucom, A., Liebermann, K., Earnest, T. N., Cate, J. H., and Noller, H. F. (2001). Crystal structure of the ribosome at 5.5 Å resolution. *Science* **292,** 883–896.
17. Cech, T. R. (2000). The ribosome is a ribozyme. *Science* **289,** 878–879.
18. Schulze, H., and Nierhaus, K. H. (1982). Minimal set of ribosomal components for reconstitution of the peptidyltransferase activity. *EMBO J.* **1,** 609–613.
19. Sumpter, V. G., Tate, W. P., Nowotny, P., and Nierhaus, K. H. (1991). Modification of histidine residues on proteins from the 50 S subunit of the *Escherichia coli* ribosome. Effects on subunit assembly and peptidyl transferase centre activity. *Eur. J. Biochem.* **196,** 255–260.
20. Nierhaus, K. H. (1993). Solution of the ribosome riddle: How the ribosome selects the correct aminoacyl-tRNA out of 41 similar contestants. *Mol. Microbiol.* **9,** 661–669.
21. Frank, J., Verschoor, A., Li, Y., Zhu, J., Lata, R. K., Radermacher, M., Penczek, P., Grassucci, R., Agrawal, R. K., and Srivastava, S. (1995). A model of the translational apparatus based on a three-dimensional reconstruction of the *Escherichia coli* ribosome. *Biochem. Cell Biol.* **73,** 757–765.
22. Moazed, D., and Noller, H. F. (1989). Interaction of tRNA with 23 S rRNA in the ribosomal A, P, and E sites. *Cell* **57,** 585–597.
23. Osswald, M., Döring, T., and Brimacombe, R. (1995). The ribosomal neighbourhood of the central fold of tRNA: Cross-links from position 47 of tRNA located at the A, P or E site. *Nucleic Acids Res.* **23,** 4635–4641.
24. Joseph, S., Weisser, B., and Noller, H. F. (1997). Mapping the inside of the ribosome with an RNA helical ruler. *Science* **278,** 1093–1098.
25. Moazed, D., and Noller, H. F. (1986). Transfer RNA shields specific nucleotides in 16 S ribosomal RNA from attack by chemical probes. *Cell* **47,** 985–994.
26. von Ashen, U., and Noller, H. F. (1995). Identification of bases in 16S rRNA essential for tRNA binding at the 30 S ribosomal P site. *Science* **267,** 234–237.
27. Moazed, D., and Noller, H. F. (1990). Binding of tRNA to the ribosomal A and P sites protects two distinct sets of nucleotides in 16S rRNA. *J. Mol. Biol.* **211,** 135–145.
28. Frank, J., and Agrawal, R. K. (1998). The movement of tRNA through the ribosome. *Biophys. J.* **74,** 589–594.
29. Valle, M., Sengupta, J., Swami, N. K., Grassucci, R. A., Burkhart, N., Nierhaus, K. H., Agrawal, R. K., and Frank, J. (2002). Cryo-EM reveals an active role for aminoacyl-tRNA in the accommodation process. *EMBO J.* **21,** 3557–3567.
29a. Stark, H., Rodnina, M. V., Weiden, H. J., Zemlin, F., Wintermeyer, W., and van Heel, M. (2002). Ribosome interactions of aminoacyl-tRNA and elongation factor Tu in the codon-recognition complex. *Nat. Struct. Biol.* **9,** 849–854.
30. Poole, E. S., and Tate, W. P. (2000). Release factors and their role as decoding proteins: Specificity and fidelity for termination of protein synthesis. *Biochim. Biophys. Acta* **1493,** 1–11.
31. Kisselev, L. L., and Frolova, L. Y. (1999). Termination of translation in eukaryotes: New results and new hypotheses. *Biochemistry (Mosc.)* **64,** 8–16.
32. Moffat, J. G., and Tate, W. P. (1994). A single proteolytic cleavage in release factor 2 stabilizes ribosome binding and abolishes peptidyl-tRNA hydrolysis activity. *J. Biol. Chem.* **269,** 18899–18903.
33. Brown, C. M., and Tate, W. P. (1994). Direct recognition of mRNA stop signals by *Escherichia coli* polypeptide chain release factor two. *J. Biol. Chem.* **269,** 33164–33170.

34. Poole, E. S., Brimacombe, R., and Tate, W. P. (1997). Decoding the translational termination signal: The polypeptide chain release factor in *Escherichia coli* crosslinks to the base following the stop codon. *RNA* **3**, 974–982.
35. Poole, E. S., Major, L. L., Mannering, S. A., and Tate, W. P. (1998). Translational termination in *Escherichia coli*: Three bases following the stop codon crosslink to release factor 2 and affect the decoding efficiency of UGA-containing signals. *Nucleic Acids Res.* **26**, 954–960.
36. Vestergaard, B., Van, L. B., Andersen, G. R., Nyborg, J., Buckingham, R. H., and Kjelgaard, M. (2001). Bacterial polypeptide release factor RF2 is structurally distinct from eukaryotic eRF1. *Mol. Cell* **8**, 1375–1382.
37. Wilson, D. N., Guévremont, D., and Tate, W. P. (2000). The ribosomal binding and peptidyl-tRNA hydrolysis functions of *Escherichia coli* release factor 2 are linked through residue 246. *RNA* **6**, 1704–1713.
38. Frolova, L. Y., Tsivkovskii, R. Y., Sivolobova, G. F., Oparina, N. Y., Serpinsky, O. I., Blinov, V. M., Tatkov, S. I., and Kisselev, L. L. (1999). Mutations in the highly conserved GGQ motif of class 1 polypeptide release factors abolish ability of human eRF1 to trigger peptidyl-tRNA hydrolysis. *RNA* **8**, 1014–1020.
39. Tate, W. P., Schulze, H., and Nierhaus, K. H. (1983). The *Escherichia coli* ribosomal protein L11 suppresses release factor 2 but promotes the release factor 1 activities in peptide chain termination. *J. Biol. Chem.* **258**, 12816–12820.
40. Tate, W. P., Dognin, M. J., Noah, M., Stöffler-Mielicke, M., and Stöffler, G. (1984). The NH2-terminal domain of *Escherichia coli* ribosomal protein L11. *J. Biol. Chem.* **259**, 7317–7324.
41. Murgola, E. J., Arkov, A. L., Chernyaeva, N. S., Hedenstierna, K. O. F., and Pagel, F. T. (2000). rRNA functional sites and structures for peptide chain termination. *In* "The Ribosome: Structure, Function, Antibiotics, and Cellular Interactions" (R. A. Garrett, S. R. Douthwaite, A. Liljas, A. T. Matheson, P. B. Moore, and H. F. Noller, Eds.), pp. 509–518. ASM Press, Washington, D.C.
42. Arkov, A. L., Freistroffer, D. V., Ehrenberg, M., and Murgola, E. J. (1998). Mutations in RNAs of both ribosomal subunits cause defects in translation termination. *EMBO J.* **17**, 1507–1514.
43. Xu, W., and Murgola, E. J. (1996). Functional effects of mutating the closing GxA base-pair of a conserved hairpin loop in 23S ribosomal RNA. *J. Mol. Biol.* **264**, 407–411.
44. Nissen, P., Kjeldgaard, M., Thirup, S., Polekhina, G., Reshetnikova, L., Clark, B. F. C., and Nyborg, J. (1995). Crystal structure of the ternary complex of Phe-tRNAPhe, EF-Tu, and a GTP analog. *Science* **270**, 1464–1472.
45. Czworkowski, J., Wang, J., Steitz, T. A., and Moore, P. B. (1994). The crystal structure of elongation factor G complexed with GDP, at 2.7 Å resolution. *EMBO J.* **13**, 3661–3668.
46. Ævarsson, A., Brazhnikov, E., Garber, M., Zheltonosova, J., Chirgadze, Yu., Al-Karadaghi, A., Svensson, L. A., and Liljas, A. (1994). Three-dimensional structure of the ribosomal translocase: Elongation factor G from *Thermus thermophilus*. *EMBO J.* **13**, 3669–3677.
47. Savelsbergh, A., Matassova, N. B., Rodnina, M. V., and Wintermeyer, W. (2000). Role of domains 4 and 5 in elongation factor G functions on the ribosome. *J. Mol. Biol.* **300**, 951–961.
48. Stark, H., Rodnina, M. V., Wieden, H. J., van Heel, M., and Wintermeyer, W. (2000). Large-scale movement of elongation factor G and extensive conformational change of the ribosome during translocation. *Cell* **100**, 301–309.
49. Wilson, K. S., and Noller, H. F. (1998). Mapping the position of translational elongation factor EF-G in the ribosome by directed hydroxyl radical probing. *Cell* **92**, 131–139.
50. Moazed, D., Samaha, R. R., Gualerzi, C., and Noller, H. F. (1995). Specific protection of 16 S rRNA by translational initiation factors. *J. Mol. Biol.* **248**, 207–210.

51. Brock, S., Szkaradkiewicz, K., and Sprinzl, M. (1998). Initiation factors of protein biosynthesis in bacteria and their structural relationship to elongation and termination factors. *Mol. Microbiol.* **29**, 409–417.
52. Wilson, D. N., Dalphin, M. E., Pel, H. J., Major, L. L., Mansell, J. B., and Tate, W. P. (2000). Factor-mediated termination of protein synthesis: A welcome return to the mainstream of translation. In "The Ribosome: Structure, Function, Antibiotics, and Cellular Interactions" (R. A. Garrett, S. R. Douthwaite, A. Liljas, A. T. Matheson, P. B. Moore, and H. F. Noller, Eds.), pp. 495–507. ASM Press, Washington, D.C.
53. Ito, K., Ibihara, K., Uno, M., and Nakamura, Y. (1996). Conserved motifs in prokaryotic and eukaryotic polypeptide release factors: tRNA-protein mimicry hypothesis. *Proc. Natl. Acad. Sci. USA* **93**, 5443–5448.
54. Mikuni, O., Ito, K., Moffat, J., Matsumura, K., McCaughan, K., Nobukuni, T., Tate, W., and Nakamura, Y. (1994). Identification of the *prfC* gene, which encodes peptide-chain-release factor 3 of *Escherichia coli*. *Proc. Natl. Acad. Sci. USA* **91**, 5798–5802.
55. Grentzmann, G., Brechemier-Baey, D., Heurgue, V., Mora, L., and Buckingham, R. H. (1994). Localisation and characterization of the gene encoding release factor RF3 in *Escherichia coli*. *Proc. Natl. Acad. Sci. USA* **91**, 5848–5852.
56. Nakamura, Y., Ito, K., and Isaksson, L. A. (1996). Emerging understanding of translation termination. *Cell* **87**, 147–150.
57. Napier, K. (2000). Construction of RF3 variants to test a model for the function of the factor at the ribosomal decoding site. BSc (Hons) Thesis, University of Otago, Dunedin, NZ.
58. Crawford, D.-J. G., Ito, K., Nakamura, Y., and Tate, W. P. (1999). Indirect regulation of translational termination efficiency at highly expressed genes and recoding sites by the factor recycling function of *Escherichia coli* release factor RF3. *EMBO J.* **18**, 727–732.
59. Fraser, C. M., Cocayne, J. D., White, O., Adams, M. D., Clayton, R. A., Fleischmann, R. D., Bult, C. J., Kerlavage, A. R., Sutton, G., Kelley, J. M. *et al.* (1995). The minimal gene complement of *Mycoplasma genitalium*. *Science* **270**, 397–403.
60. Friestroffer, D. V., Pavlov, M. Yu., MacDougall, J., Buckingham, R. H., and Ehrenberg, M. (1997). Release factor RF3 in *E. coli* accelerates the dissociation of release factors RF1 and RF2 from the ribosome in a GTP-dependent manner. *EMBO J.* **16**, 4126–4133.
61. Sette, M., van Tilborg, P., Spurio, R., Kaptein, R., Paci, M., Gualerzi, C. O., and Boelens, R. (1997). The structure of the translational initiation factor IF1 from *E. coli* contains an oligomer-binding motif. *EMBO J.* **16**, 1436–1443.
62. Konecki, D. S., Aune, K. C., Tate, W. P., and Caskey, C. T. (1977). Characterization of reticulocyte release factor. *J. Biol. Chem.* **252**, 4514–4520.
63. Song, H., Mugnier, P., Das, A. K., Webb, H. M., Evans, D. R., Tuite, M. F., Hemmings, B. A., and Barford, D. (2000). The crystal structure of human eukaryotic release factor eRF1—mechanism of stop codon recognition and peptidyl-tRNA hydrolysis. *Cell* **100**, 311–321.
64. Frolova, L., Seit-Nebi, A., and Kisselev, L. (2002). Highly conserved NIKS tetrapeptide is functionally essential in eukaryotic translation termination factor eRF1. *RNA* **8**, 129–136.
65. Scarlett, D.-J. G. (2002). Macromolecular mimicry in translation: The decoding release factor RF2 maps to the ribosomal A site of the decoding tRNA. Ph.D. Thesis, University of Otago, Dunedin, NZ.
66. Bertram, G., Bell, H. A., Ritchie, D. W., Fullerton, G., and Stansfield, I. (2000). Terminating eukaryotic translation: Domain 1 of release factor eRF 1 functions in stop codon recognition. *RNA* **6**, 1236–1247.
67. Ito, K., Uno, Y., and Nakamura, Y. (2000). A tri-peptide 'anticodon' deciphers stop codons in messenger RNA. *Nature* **403**, 680–684.

68. Klaholz, B. P., Pape, T., Zavialov, A. V., Myasnikov, A. G., Orlova, E. V., Vestergaard, B., Ehrenberg, M., and van Heel, M. (2003). Structure of the E. coli ribosomal termination complex with release factor 2. *Nature* **421**, 90–94.
68a. Rawat, U. B., Zavialov, A. V., Sengupta, J., Valle, M., Grassucci, R. A., Linde, J., Vestergaard, B., Ehrenberg, M., and Frank, J. (2002). A cryo-electron microscopic study of ribosome-bound termination factor RF2. *Nature*. **421**, 87–90.
69. Tate, W. P., McCaughan, K. K., Ward, C. D., Sumpter, V. G., Trotman, C. N. A., Stöffler-Mielicke, M., Maly, P., and Brimacombe, R. (1986). The ribosomal binding domain of the *Escherichia coli* release factors. Modification of tyrosine in the N-terminal domain of ribosomal protein L11 affects release factors 1 and 2 differentially. *J. Biol. Chem.* **261**, 2289–2293.
70. Selmer, M., Al-Karadaghi, S., Hirokawa, G., Kaji, A., and Liljas, A. (1999). Crystal structure of *Thermotoga maritima* ribosome recycling factor: A tRNA mimic. *Science* **286**, 2349–2352.
71. Toyoda, T., Tin, O. F., Ito, K., Fujiwara, T., Kumasaka, T., Yamamoto, M., Garber, M. B., and Nakamura, Y. (2000). Crystal structure combined with genetic analysis of the *Thermus thermophilus* ribosome recycling factor shows that a flexible hinge may act as a functional switch. *RNA* **6**, 1432–1444.
71a. Lancaster, L., Kiel, M., Kaji, A., and Noller, H. (2002). Orientation of the ribosome recycling factor in the ribosome from directed hydroxyl radical probing. *Cell* **111**, 129.
72. Chavatte, L., Seit-Nebi, A., Dubovaya, V., and Favre, A. (2002). The invariant uridine of stop codons contacts the conserved NIKSR loop of human eRF1 in the ribosome. *EMBO J.* **21**, 5302–5311.
73. Wilson, K. S., Ito, K., Noller, H. F., and Nakamura, Y. (2000). Functional sites of interaction between release factor RF1 and the ribosome. *Nat. Struct. Biol.* **7**, 866–870.
74. Tate, W., Greuer, B., and Brimacombe, R. (1990). Codon recognition in polypeptide chain termination: Site directed crosslinking of termination codon to *Escherichia coli* release factor 2. *Nucleic Acids Res.* **18**, 6537–6544.
75. Rinke-Appel, J., Junke, N., Brimacombe, R., Dokudovskaya, S., Dontsova, O., and Bogdanov, A. (1993). Site-directed cross-linking of mRNA analogues to 16S ribosomal RNA; a complete scan of cross-links from all positions between '+1' and '+16' on the mRNA, downstream from the decoding site. *Nucleic Acids Res.* **21**, 2853–2859.
76. Carter, A. P., Clemons, W. M., Brodersen, D. E., Morgan-Warren, R. J., Wimberley, B. T., and Ramakrishnan, V. (2000). Functional insights from the structure of the 30S ribosomal subunit and its interactions with antibiotics. *Nature* **407**, 340–348.
77. Poole, E. S., Brown, C. M., and Tate, W. P. (1995). The identity of the base following the stop codon determines the efficiency of *in vivo* translational termination in *Escherichia coli*. *EMBO J.* **14**, 151–158.
78. Major, L. L. (2001). Is the prokaryotic termination signal a simple triplet codon or an extended sequence element? Ph.D. Thesis, University of Otago, Dunedin, NZ.
79. Yusupova, G. Zh., Yusupov, M. M., Cate, J. H. D., and Noller, H. F. (2001). The path of messenger RNA through the ribosome. *Cell* **106**, 233–241.
80. Mottagui-Tabar, S., Björnsson, A., and Isaksson, L. A. (1994). The second to last amino acid in the nascent peptide as a codon context determinant. *EMBO J.* **13**, 249–257.
81. Björnsson, A., Mottagui-Tabar, S., and Isaksson, L. A. (1996). Structure of the C-terminal end of the nascent peptide influences translation termination. *EMBO J.* **15**, 1696–1704.
82. Stelzl, U., Spahn, C. M. T., and Nierhaus, K. H. (2000). Selecting rRNA binding sites for the ribosomal proteins L4 and L6 from randomly fragmented rRNA: Application of a method called SERF. *Proc. Natl. Acad. Sci. USA* **97**, 4597–4602.
83. Szkaradkiewicz, K., Nanninga, M., Nesper-Brock, M., Gerrits, M., Erdmann, V. A., and Sprinzl, V. A. M. (2002). RNA aptamers directed against release factor 1 from *Thermus thermophilus*. *FEBS Lett.* **514**, 90–95.

84. Joseph, S., and Noller, H. F. (1996). Mapping the rRNA neighborhood of the acceptor end of tRNA in the ribosome. *EMBO J.* **15**, 910–916.
85. Polacek, N., Gomez, M. J., Ito, K., Xiong, L., Nakamura, Y., and Mankin, A. (2003). The critical role of the universally conserved A2602 of the 23 S ribosomal RNA in the release of the nascent peptide during translation termination. *Mol. Cell* **11**, 103–112.
86. Porse, B. T., and Garett, R. A. (1995). Mapping important nucleotides in the peptidyl transferase centre of 23 S rRNA using a random mutagenesis approach. *J. Mol. Biol.* **249**, 1–10.
87. Muth, G. W., Ortoleva-Donnelly, L., and Strobel, S. A. (2000). A single adenosine with a neutral pK_a in the ribosomal peptidyl transferase center. *Science* **289**, 947–950.
88. Polacek, N., Gaynor, M., Yassin, A., and Mankin, A. S. (2001). Ribosomal peptidyl transferase can withstand mutations at the putative catalytic nucleotide. *Nature* **411**, 498–501.
89. Bayfield, M. A., Dahlberg, A. E., Schulmeister, U., Dorner, S., and Barta, A. (2001). A conformational change in the ribosomal peptidyl transferase center upon active/inactive transition. *Proc. Natl. Acad. Sci. USA* **98**, 10096–10101.
90. Steiner, G., Kuechler, E., and Barta, A. (1988). Photo-affinity labelling at the peptidyl transferase centre reveals two different positions from the A- and P-sites in domain V of 23S rRNA. *EMBO J.* **7**, 3949–3955.
91. Green, R., Switzer, C., and Noller, H. F. (1998). Ribosome-catalyzed peptide-bond formation with an A-site substrate covalently linked to 23S ribosomal RNA. *Science* **280**, 286–288.
92. Conn, G. L., Draper, D. E., Lattman, E. E., and Gittis, A. G. (1999). Crystal structure of a conserved ribosomal protein-RNA complex. *Science* **284**, 1171–1174.
93. Wimberley, B. T., Guymon, R., McCutcheon, J. P., White, S. W., and Ramakrishnan, V. (1999). A detailed view of a ribosomal active site: The structure of the L11-RNA complex. *Cell* **97**, 491–502.
94. Arkov, A. L., Freistroffer, D. V., Pavlov, M. Y., Ehrenberg, M., and Murgola, E. J. (2000). Mutations in conserved regions of ribosomal RNAs decrease the productive association of peptide-chain release factors with the ribosome during translation termination. *Biochimie* **82**, 671–682.
95. Van Dyke, N., Xu, W., and Murgola, E. J. (2002). Limitation of ribosomal protein L11 availability *in vivo* affects translation termination. *J. Mol. Biol.* **319**, 329–339.
96. Pel, H. J., Moffat, J. G., Ito, K., Nakamura, Y., and Tate, W. P. (1998). *Escherichia coli* release factor 3: Resolving the paradox of a typical G protein structure and atypical function with guanine nucleotides. *RNA* **4**, 47–54.
97. Zavialov, A., Buckingham, R. H., and Ehrenberg, M. (2001). A posttermination ribosomal complex is the guanine nucleotide exchange factor for peptide release factor RF3. *Cell* **107**, 115–124.
98. Chavatte, L., Frolova, L., Kisselev, L., and Favre, A. (2001). The polypeptide chain release factor eRF1 specifically contacts the s^4UGA stop codon located in the A site of eukaryotic ribosomes. *Eur. J. Biochem.* **268**, 2896–2904.
99. Nakamura, Y., and Ito, K. (2002). A tripeptide discriminator for stop codon recognition. *FEBS Lett.* **514**, 30–33.
100. Seit-Nebi, A., Frolova, L., Justesen, J., and Kisselev, L. (2001). Class-I translation termination factors: Invariant GGQ minidomain is essential for release activity and ribosome binding but not for stop codon recognition. *Nucleic Acids Res.* **29**, 3982–3987.
101. Arkov, A. L., and Murgola, E. J. (1999). Ribosomal RNAs in translation termination: Facts and hypotheses. *Biochemistry (Mosc.)* **64**, 1354–1359.
102. Hirokawa, G., Kiel, M. C., Muto, A., Selmer, M., Raj, V. S., Liljas, A., Igarashi, K., Kaji, H., and Kaji, A. (2002). Post-termination complex disassembly by ribosome recycling factor, a functional tRNA mimic. *EMBO J.* **21**, 2272–2281.

103. Draper, D. E., Conn, G. L., Gittis, A. G., GuhaThakurta, D., Lattman, E. E., and Reynaldo, L. (2000). RNA tertiary structure and protein recognition in an L11-RNA complex. In "The Ribosome: Structure, Function, Antibiotics, and Celullar Interactions" (R. A. Garrett, S. R. Douthwaite, A. Liljas, A. T. Matheson, P. B. Moore, and H. F. Noller, Eds.), pp. 105–114. ASM Press, Washington, D.C.
104. Guex, N., and Peitsch, M. C. (1997). SWISS-MODEL and the Swiss-Pdb Viewer: An environment for comparative protein modelling. *Electrophoresis* **18,** 2714–2723.
105. Peitsch, M. C. (1995). Protein modelling by email. *BioTechnology* **13,** 658–660.
106. Peitsch, M. C. (1996). ProMod and Swiss-Model: Internet based tools for automated comparative protein modelling. *Biochem. Soc. Trans.* **24,** 274–279.
107. Kim, K. K., Min, K., and Suh, S. W. (2000). Crystal structure of the ribosome recycling factor from *Escherichia coli*. *EMBO J.* **19,** 2362–2370.
108. Ogle, J. M., Brodersen, D. E., Clemons, W. M., Jr., Tarry, M. J., Carter, A. P., and Ramakrishnan, V. (2001). Recognition of cognate transfer RNA by the 30S ribosomal subunit. *Science* **292,** 897–902.

Phylogenetics and Functions of the Double-Stranded RNA-Binding Motif: A Genomic Survey

BIN TIAN[*,†] AND MICHAEL B. MATHEWS[†]

*Johnson and Johnson Pharmaceutical Research and Development, San Diego, California 92121

†Department of Biochemistry and Molecular Biology, New Jersey Medical School, UMDNJ, Newark, New Jersey 07101

I. Introduction	124
II. Survey Methods	125
III. dsRBM Occurrence, Frequency, and Conservation	126
IV. Domain Structures of dsRBM Proteins	129
V. dsRBM Protein Families in Five Type Species	131
A. RNase III and Dicer: RNA Processing Enzymes	134
B. RHA: A DNA/RNA Helicase	139
C. ADARs: RNA Editases	140
D. The Protein Kinase PKR	140
E. Staufen: mRNA Transport and Translation	141
F. TRBP and PACT: Regulators of PKR	143
G. NF90: A Multifunctional Regulator	143
H. NRF: A Transcriptional Repressor	144
I. SON: Another Transcriptional Regulator	144
J. Kanadaptin	144
K. XRN1: RNA Degradation	145
L. DRMs: Unassigned dsRBM-Containing Proteins	145
VI. dsRBMs in Other Taxa	146
A. Archaea	146
B. Plant	146
C. Viruses	147
VII. Concluding Remarks	148
A. Origin and Evolution of the dsRBM	148
B. The dsRBM and Protein Function	150
C. Scope of the dsRBM Protein Family	151
References	152

The double-stranded RNA-binding motif (dsRBM) is found in a diverse array of proteins that regulate gene expression in many organisms. We surveyed all known and predicted protein sequences from several organisms for the presence of the dsRBM. Among the major taxa sampled, only archebacteria lacked this motif. Plants and viruses appear to exploit the motif in proteins not found in animals. For five genomes (a bacterium, yeast, nematode, fly, and human), the proteins were annotated and examined in detail. The number of dsRBM-containing proteins increased along this series, as did the incidence of multiple dsRBMs. The dsRBM-containing proteins of the five type species are grouped into 21 families according to their similarities in domain and overall sequences, and the presence of other domains. These data provide a basis for understanding the evolution of the dsRBM, as well as the functions of proteins that contain it.

I. Introduction

Completion of the genome sequences for several organisms provides an unprecedented opportunity to examine the composition of each individual genome and to make comparisons among them. Proteins are built from modular units, called motifs or domains, which have distinct structures and carry out different functions. One intensively studied protein motif, the double-stranded RNA-binding motif (dsRBM; sometimes called the dsRNA-binding domain, dsRBD, DRBM, or DRBD), has been found in numerous proteins—including nucleases, helicases, deaminases, and kinases, inter alia—that are implicated in various biological processes. Although this group of proteins is functionally heterogeneous, a common feature is that most of them play parts in the control of gene expression.

The dsRBM is usually 65–70 amino acids (aa) in length and displays a distinctive protein folding pattern, $\alpha-\beta-\beta-\beta-\alpha$ (1–4). dsRBMs, by definition, bind to duplexed RNA and this binding is independent of dsRNA sequence in most cases. The atomic interactions responsible for dsRNA–dsRBM binding are manifested in the crystal structure of dsRBM2 from the *Xenopus laevis* protein Xlrbpa (a TRBP homologue) complexed with synthetic dsRNA (2). Three regions of the dsRBM make contact with A-form dsRNA, most of the interactions involving the RNA's phosphodiester backbone and its ribose 2'-OH groups. This explains why dsRBMs bind to dsRNA rather than single-stranded RNA (ssRNA) or DNA, and why the interaction is chiefly dependent upon structure rather than sequence. In addition to perfect duplex RNA, highly structured but single-stranded RNAs with bulges or mismatches can also be bound by dsRBMs (5, 6). In fact, several cellular RNAs, known for binding to dsRBM-containing proteins, contain bulges and internal loops.

Despite their conservation in both sequence and structure, dsRBMs vary markedly in their biochemical and cytological properties. In the canonical property of RNA binding, some dsRBMs evince robust binding to various RNAs whereas the dsRNA-binding activity of others is weak (7). Additional functions that have been associated with dsRBMs, such as dsRNA annealing, ribosome binding, protein dimerization, and subcellular localization, also vary among dsRBMs (8–15). It is arguable that some of these functions are secondary to the binding of dsRNA, whereas others are independent functions.

We have conducted a comprehensive search for all the dsRBMs in the genomes of a diverse set of organisms, on the premise that such a survey will illuminate the evolution of this domain and the functions of proteins that contain it. The results show that the motif is widespread in nature but apparently absent from the archaea. Although several dsRBM-containing proteins are common to the animals studied, the motif is present in a radically different spectrum of plant proteins. Broadly speaking, the more evolutionarily advanced organisms possess a greater number of proteins with this domain, and their proteins tend to contain more dsRBMs per molecule as well as more domains of other types. The survey identified dsRBMs in several novel proteins and in some proteins that were previously characterized but not known to contain this motif. Finally, we summarize the biochemical functions of human dsRBM-containing proteins, and speculate on the motif's biological roles.

II. Survey Methods

We took a systematic approach to find and annotate all the dsRBMs in several completely sequenced genomes, namely those of *Escherichia coli* (eubacteria), *Saccharomyces cerevisiae* (bakers' yeast), *Caenorhabditis elegans* (nematode worm), *Drosophila melanogaster* (fruit fly), and *Homo sapiens* (human). In addition, for increased breadth, the genomes of a plant (*Arabidopsis thaliana*) and several archaea (*Aeropyrum pernix, Archaeoglobus fulgidus, Halobacterium* sp., *Methanobacterium thermoautotrophicum, Methanococcus jannaschii, Pyrococcus horikoshii,* and *Thermoplasma acidophilum*) and viruses were also searched.

To create a database for proteins that contain dsRBMs, it is of prime importance to employ a sensitive and selective tool to mine the genome sequences. We first utilized the dsRBMs documented by the InterPro database at the European Bioinformatics Institute (EBI). This is one of the most comprehensive collections of protein domains as it is built upon several underlying signature databases, such as PROSITE, pfam, SMART, PRINTS, and ProDom, each of which has its own focus (16). One challenge in using the

InterPro database, however, is the redundancy of protein entries. Because of the presence of alternatively spliced variants and truncated proteins in the Swiss-Prot and TrEMBL databases on which InterPro is built, a single gene is often represented by more than one protein. To solve this problem, we first downloaded all the InterPro domain sequences, then carried out pair-wise BLAST comparison to discard redundant sequences. This strategy ensures that each annotated entry represents a gene rather than its multiple transcripts or protein products.

Because dsRBM domain sequences are loosely conserved in some regions and comprehensiveness was critical for our survey, we deployed two additional steps. Using protein sequences that are already known to have dsRBMs, we did BLAST searches to find orthologues across different species, and manually inspected sequences by pair-wise BLAST search for the presence of dsRBMs. We then constructed a domain profile with all the dsRBMs using ClustalX and HMM (Hidden Markov Model) tools (17, 18), and used the profile to search all genomes for more dsRBMs. To maximize sensitivity, the last two steps were further iterated until no new hits were found.

Finally, all domain sequences were extracted from protein sequences, aligned with ClustalX, and displayed by TreeView (19). We examined different combinations of parameters for the ClustalX alignment at both pair-wise and multiple alignment stages, but present only the results (alignment and cladogram) obtained with the following settings: gap open penalty, 10; gap extension penalty, 0.1; protein weight matrix, Gonnet 250 (20). Inferences from the cladogram, however, were based on several trees generated using different parameters.

III. dsRBM Occurrence, Frequency, and Conservation

Using the above method, we annotated 56 genes containing 97 dsRBMs, as well as two putative pseudogenes with three dsRBMs, in the five type species *E. coli*, *S. cerevisiae*, *C. elegans*, *D. melanogaster*, and *H. sapiens*. The genes are listed according to their common names and accession numbers, together with other information, in Table I. The number of genes recorded is higher than that in any reported dsRBM compilation, attesting to the high sensitivity of our search. Because all of the dsRBMs were manually inspected, the quality is also ensured.

The results of the survey represented statistically are summarized in Table II. *E. coli* has one dsRBM in a single dsRBM-containing gene, whereas the human genome has 45 dsRBMs in 24 genes excluding two suspected pseudogenes. The yeast, worm, and fly genomes have intermediate numbers of dsRBMs and genes: 2 in 2, 20 in 13, and 29 in 16, respectively. Thus, the

TABLE I
dsRBM-CONTAINING GENES IN FIVE SPECIES[a]

Family name	Domain structure	Gene name	Human	Fly	Worm	Yeast	Bacterium
RNase IIIA	999–1159	RNase IIIA1	BAB14234	AAF52025	CAA79619	CAA89127	CAA26692
		RNase IIIA2				AAB04172	
RNase IIIB	694–999–999–1159	RNase IIIB	AAF80558	AAF59169	CAB03006		
Dicer	1410(1650)–5034–3100–999–999–1159	Dicer1	BAA78691	AAF56056	AAA28101		
		Dicer2		BAB69959			
RHA helicase	1159–1159–1410(1650)	RHA1	CAA71668	AAC41573	CAA90409		
		RHA1B	XP_062934				
		RHA2	BAA74913				
ADAR1	607–607–1159–1159–1159–2466	ADAR1	CAA55968				
ADAR2	1159–1159–2466	ADAR2A1	CAB09392	CAA22774	CAB09530		
		ADAR2A2	AAF78094				
	1159–2466	ACAR2B1	XP_091235		AAC25097		
		ACAR2B2	BAB71416				
PKR	1159–1159–719	PKR	AAA36409				
Staufen	1159–1159–1159–1159–1159	STAU1	AAD17531	AAF57752	AAB07566		
		STAU2	BAA92111		AAB07562		
TRBP	1159–1159–1159	TRBP	AAB50581	AAF53295	CAA83012		
		TRBP2	XP_070394				
		PaCT	AAC25672				
NF90	DZF–1159–1159	NF90	AAD51099				

(Continues)

TABLE I (Continued)

Family name	Domain structure	Gene name	Human	Fly	Worm	Yeast	Bacterium
NRF	1159–1159–467–1374	SPNR	BAA92120				
NRF	467–1159	NRF	XP_010360				
SON	253–1159	SON	CAA45282	AAF54409	AAK21366		
Kanadaptin	1159	Kanadaptin	BAA91718	AAF55582	CAA80179		
XRN1		XRN1	CAB63749				
DRM1	1269–3009–1159	DRM1	BAA91143	AAF48360	CAA21662		
DRM2	1202–1159–1159	DRM2	BAB83032	AAF57175	CAB03384		
DRM3	1159	DRM3	AAF68967				
DRM4	1159–1159	DRM4		AAF50777			
DRM5	1159–1159	DRM5		CAC17710			
DRM6	1159–1159	DRM6		AAF47937			
DRM7	1159–1159	DRM7		AAF52561			

[a] Shown is a complete compilation of the dsRBM-containing proteins in the human, fly, worm, yeast, and bacterial genomes (*Homo sapiens*, *Drosophila melanogaster*, *Caenorhabditis elegans*, *Saccharomyces cerevisiae*, and *Escherichia coli*). The protein families are discussed in the text. Protein domain layouts are presented as a string of motifs (N-terminal to C-terminal), named according to their InterPro accession numbers. Because of space constraints, the InterPro accession numbers are abbreviated by omission of the letters IPR and string of zeros (for example, IPR000999 is listed as 999). All representative accession numbers are from the SwissProt, TrEMBL, and GenBank databases. The DZF motif in NF90 is named in the SMART database, and has not yet been categorized by InterPro, so no number is given.

TABLE II
DISTRIBUTIONS AND SIZES OF dsRBMs IN FIVE SPECIES[a]

	Human	Fly	Worm	Yeast	Bacterium
Number of dsRBMs	45	29	20	2	1
Number of dsRBM-containing proteins	24	16	13	2	1
Average size of dsRBMs	67	67	68	68	69

[a]Listed for each species is the number of dsRBMs and the number of dsRBM-containing proteins encoded in the genomes. Average sizes of dsRBMs are given in amino acids.

numbers of dsRBMs and of dsRBM-containing genes both increase with evolutionary complexity.

The lengths of the dsRBMs in the five type species range from 62 aa (two domains in human NRF) to 83 aa (the first dsRBM in *Drosophila* Staufen, Dm.STAU1). The average length is 67.5 and does not vary much among the five species (Table II). Sequences of the individual dsRBMs are aligned in Fig. 1. Inspection of the alignment reveals a high degree of conservation in the C-terminal one-third of the motif and relatively low conservation in the N-terminal two-thirds, as noted previously (21). The homology profile, shown at the foot of Fig. 1, emphasizes the distribution of conserved residues along the dsRBM.

IV. Domain Structures of dsRBM Proteins

Most, but not all, dsRBM-containing proteins also contain domains of other types. Notable exceptions include TRBP and Staufen, which respectively contain three and five dsRBMs without any other recognized domains. Our survey disclosed the existence of 16 other recognized domains in the dsRBM-containing proteins of the five type species (Table III). The domain structures of the proteins are listed in Table I, using an abbreviated version of their InterPro domain numbers. For example, PKR contains two tandem dsRBMs (designated 1159, short for IPR001159) followed by a kinase domain (designated 719 for IPR000719). The domain structures are depicted for representative proteins in Fig. 2. When other domains are present, the dsRBMs may occupy positions near the protein's N-terminus (as in PKR), C-terminus (as in RNase III), or center (as in ADAR1).

All the metazoans studied have proteins containing more than a single dsRBM. For proteins with multiple dsRBMs in the five type species, the dsRBMs are arranged in tandem without interspersion of other domains (Table I), although an exception to this rule occurs in the plant *Arabidopsis*

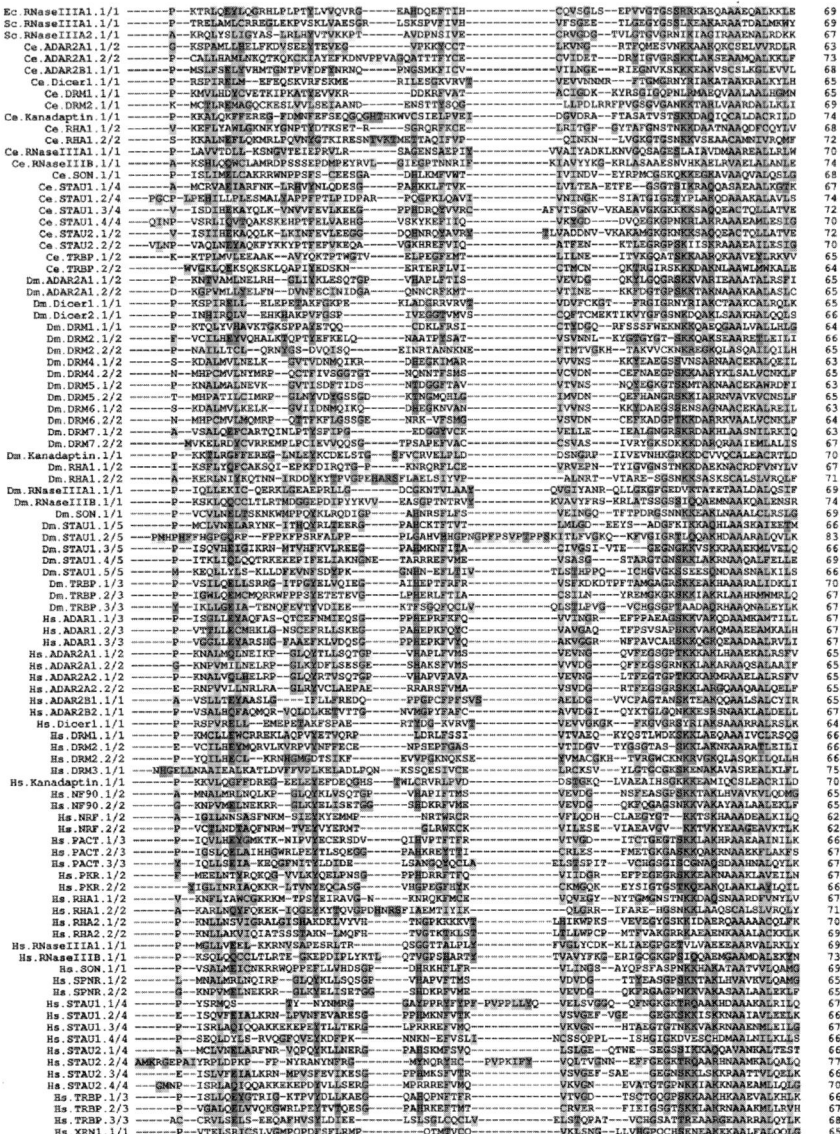

(Table IV). The spacing between the dsRBMs ranges from 22 aa (between the two dsRBMs in the nematode adenosine deaminase, Ce.ADAR2A1), with median of 48.5 aa and mean of 65.7 aa. No correlation was apparent between the spacing and any known feature or function, such as substrate specificity, of either the dsRBMs or proteins.

An additional seven domains are tethered to dsRBMs in plant and viral proteins. Thus, of the 23 recognized domains found in association with the dsRBM, four are uniquely present in plants and three in viruses (Table III). Of the remainder, four are well-characterized enzymatic domains, namely the RNase III domain (IPR000999), helicase domain (IPR001410 and IPR001650), ADAR domain (IPR002466), and protein kinase domain (IPR000719). These domains and the proteins that contain them have been investigated in considerable depth. A substantial amount of information is available on the functions of several other proteins in the dsRBM family, including a few (like Staufen) that lack additional protein domains. On the other hand, many dsRBM-containing proteins have no additional domains uncovered at the present time, and some possess domains that have yet to be thoroughly investigated: the functions of these proteins are generally less well understood, if they are appreciated at all. We outline the current state of knowledge about the functions of the dsRBM-containing proteins, and of the additional tethered domains that they contain, in the next section.

V. dsRBM Protein Families in Five Type Species

We used the human data as our main resource to group the dsRBM proteins into families. Based on domain structure and overall sequence criteria, 21 protein families were denominated (Table I and Fig. 2). These are RNase IIIA, RNase IIIB, Dicer, RHA, Staufen, ADAR1, ADAR2, NF90, TRBP, PKR,

---

FIG. 1. Alignment of double-stranded RNA-binding motif. The dsRBMs from five species, namely *Escherichia coli*, *Saccharomyces cerevisiae*, *Caenorhabditis elegans*, *Drosophila melanogaster*, and *Homo sapiens*, were aligned by ClustalX using parameters specified in Section II. dsRBMs from two putative pseudogenes are not included. Domain names are listed on the left of the body of the figure as abbreviations for the species (Ec, *Escherichia coli*; Sc, *Saccharomyces cerevisiae*; Ce, *Caenorhabditis elegans*; Dm, *Drosophila melanogaster*; Hs, *Homo sapiens*), followed by gene names and a number of the type n/N where N denotes the total number of dsRBMs in the protein and n denotes the particular dsRBM under consideration, counting from the protein N-terminus. A ruler is provided under the alignment, indicating the length of the domain. At the bottom of the figure is a homology plot generated by ClustalX, in which the height of the bar corresponds to the degree of conservation of the amino acid at each position in the motif.

TABLE III
Protein Domains Associated with dsRBMs[a]

InterPro accession	Name
IPR000051[b]	SAM (and some other nucleotide)-binding motif
253	Forkhead-associated (FHA) domain
467	D111/G-patch domain
605[c]	RNA helicase
607	Double-stranded RNA-specific adenosine deaminase (DRADA) repeat
694	Proline-rich region
719	Eukaryotic protein kinase
999	Ribonuclease III family
1159	Double-stranded RNA binding (DsRBD) domain
1202	WW/rsp5/WWP domain
1205[c]	RNA-dependent RNA polymerase (P3D)
1269	Uncharacterized protein family UPF0034
1374	Single-stranded nucleic acid binding (R3H)
1410	DEAD/DEAH box helicase
1650	Helicase C-terminal domain
2057[b]	Isopenicillin N synthetase
2466	Adenosine-deaminase (editase) domain
2873[c]	Rotavirus nonstructural protein NSP3
3009	Proteins binding FMN and related compounds core region
3100	PAZ domain
4844[b]	Serine/threonine-specific protein phosphatase
4843[b]	Metallophosphoesterase
5034	Protein of unknown function DUF283
DZF[d]	Domain in DSRM or ZnF_C2H2 domain-containing proteins

[a]Listed are the InterPro accession numbers of domains in dsRBM-containing proteins of the five type species (human, fly, worm, yeast, and bacterium), A. thaliana, and viruses. InterPro accession numbers and names were obtained from the InterPro website (http://www.ebi.ac.uk/interpro/). Domain numbers are abbreviated following the convention of Table I.
[b]Domains associated with dsRBM in A. thaliana only.
[c]Domains associated with dsRBM in viruses only.
[d]The DZF motif in NF90 has not been assigned an InterPro number.

SON, Kanadaptin, NRF, and XRN1, together with seven unnamed families, DRM1–7. Five of the families (RHA, ADAR2, Staufen, TRBP, and NF90) are represented by two or more genes in the human genome. Table V lists the genes with a brief description, their chromosomal loci, and some additional information. Locuslink identification numbers are presented for further

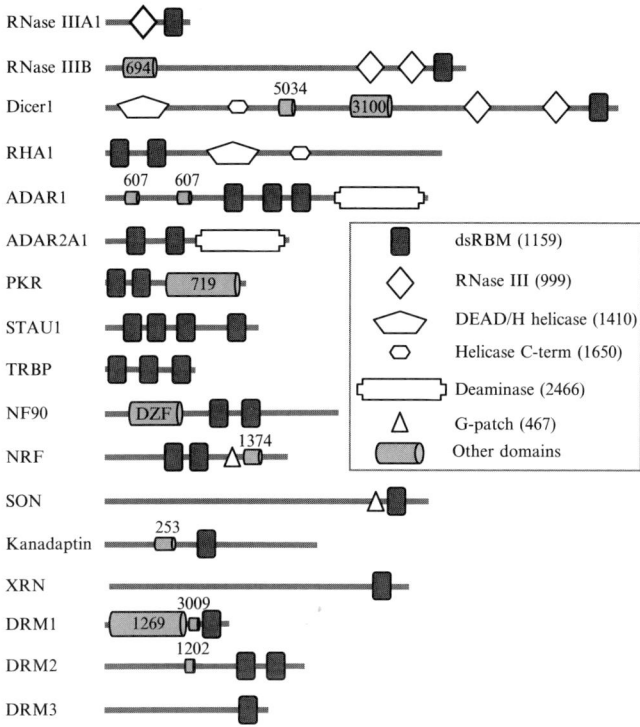

FIG. 2. Structures of human dsRBM-containing proteins. Depicted are the domain structures of representative human proteins, one from each family (Table I). Filled boxes represent dsRBMs. Distinctive open shapes are used for domains that occur in more than one protein. Other domains are represented as shaded cylinders marked with accession numbers as in Table I.

reference as they provide useful links to several data sources on gene names, maps, sequences, and functional annotation (22). For each gene, one Swiss-Prot, TrEMBL, or GenBank ID number was selected as representative; although the selected example may not be the longest one for the gene, it contains all the domains known for that gene.

The similarities among the dsRBM sequences (*not* the entire proteins) can be evaluated from the rectangular cladogram shown in Fig. 3. One notable feature is that, in most cases, the similarity between corresponding dsRBMs within a protein family is greater than that between families, reflecting sequence conservation among orthologues through evolution. RNase IIIA is exceptional in this regard, the *E. coli* dsRBM being an outlier for reasons that are not immediately apparent. On the other hand, multiple dsRBMs within a single protein often display considerable divergence, as clearly seen in the

TABLE IV
PLANT dsRBM-CONTAINING PROTEINS[a]

Domain structure	Representative accession
4844(4843)–1410(1650)–3100–999–999–1159	AAF26098
1410(1650)–5034–3100–999–999–1159–1159	AAF26461
1410(1650)–1159	AAG60124
1410(1650)–1159	BAB11511
1410(1650)–1159	AAD14515
1410(1650)–1159	AAM15307
999–1159–999–1159–1159	BAB11388
999–1159–1159	BAB02825
1159–2057–51	CAB45887
1159–1159	AAD20688
1159–1159	AAB60726
1159–1159	BAB09709
1159–1159	BAB01188
1159–1159	AAL67059
1159	CAB36811
1159	CAB69855
1159	AAF27126

[a]Domain layouts are shown for the dsRBM-containing proteins of *A. thaliana* using the convention of Table I. Domain numbers and names are as in Table III.

Staufen proteins. This presumably reflects either functional differences among the individual dsRBMs or their distinct evolutionary origins.

A. RNase III and Dicer: RNA Processing Enzymes

The RNase III domain (IPR000999) is found in three protein families: RNase IIIA, RNase IIIB, and Dicer (also known as class I, II, and III RNase III). The domain possesses endonucleolytic activity, cleaving each strand of dsRNA substrates and releasing fragments that carry 5'-phosphates and 3'-OH groups with 3' overhangs of two nucleotides (nt) (23).

The RNase IIIA proteins are dsRNA-dependent endoribonucleases implicated in the processing of preribosomal, messenger, and small nuclear RNAs (23, 24). These functions are essential for cell survival and, not surprisingly, the proteins are found in all five type species. The dsRBM sequences of the eukaryotic RNase IIIA proteins are very similar to each other and form a cluster in the cladogram (Fig. 3D). The dsRBM of bacterial RNase IIIA, however, differs from the rest of the group, possibly suggestive of functional as well as

TABLE V
HUMAN DSRBM-CONTAINING GENES[a]

Gene name	Accession	LocusID	Description	Map
RNase IIIA1	BAB14234	65080	MRPL44: mitochondrial ribosomal protein L44	2q36.3
RNase IIIB	AAF80558	29102	RNASE3L: putative ribonuclease III	5p13.3
Dicer1	BAA78691	23405	DICER1: Dicer1, Dcr-1 homolog (*Drosophila*)	14q32.2
RHA1	CAA71668	1660	DDX9: DEAD/H (Asp-Glu-Ala-Asp/His) box polypeptide 9 (RNA helicase A, nuclear DNA helicase II; leukophysin)	1q25
RHA1B	XP_062934	122056	Similar to ATP-dependent RNA helicase A (nuclear DNA helicase II) (NDH II) (DEAD-box protein 9)	13q22.1
RHA2	BAA74913	22907	DDX30: DEAD/H (Asp-Glu-Ala-Asp/His) box polypeptide 30	3p21.31
ADAR1	CAA55968	103	ADAR: adenosine deaminase, RNA specific	1q21.1–q21.2
ADAR2A1	CAB09392	104	ADARB1: adenosine deaminase, RNA specific, B1 (RED1 homolog rat)	21q22.3
ADAR2A2	AAF78094	55523	ADAR3: double-stranded RNA-specific adenosine deaminase	10p15
ADAR2B1	XP_091235	161931	Testis nuclear RNA-binding protein-like	16q24.1
ADAR2B2	BAB71416	132612	Tenr: testis nuclear RNA-binding protein	4q26
PKR	AAA36409	5610	PRKR: protein kinase, interferon-inducible double-stranded RNA dependent	2p22–p21
STAU1	AAD17531	6780	STAU: Staufen, RNA-binding protein (*Drosophila*)	20q13.1
STAU2	BAA92111	27067	STAU2: Staufen, RNA-binding protein, homolog 2 (*Drosophila*)	8q13–q21.1
TRBP	AAB50581	6895	TARBP2: TAR (HIV) RNA-binding protein 2	12q12–q13
TRBP2	XP_070394	137396	Similar to TAR RNA-binding protein 2, isoform a; TAR RNA-binding protein 2; transactivation responsive RNA-binding protein	8q22.3
PACT	AAC25672	8575	PRKRA: protein kinase, interferon-inducible double-stranded RNA-dependent activator	2q31.2
NF90	AAD51099	3609	ILF3: interleukin enhancer-binding factor 3, 90 kDa	19p13
SPNR	BAA92120	55342	STRBP: spermatid perinuclear RNA-binding protein	9q34.11

(*Continues*)

TABLE V (*Continued*)

Gene name	Accession	LocusID	Description	Map
NRF	XP_010360	55922	NRF: transcription factor NRF	Xq24
SON	CAA45282	6651	SON: SON DNA-binding protein	21q22.11
Kanadaptin	BAA91718	22950	SLC4A1AP: solute carrier family 4 (anion exchanger), member 1, adaptor protein	2p23.3
XRN1	CAB63749	54464	DKFZP434P0721: similar to mouse Xm1/Dhm2 protein	3q24
DRM1	BAA91143	54920	FLJ20399: hypothetical protein FLJ20399	16q22.2
DRM2	BAB83032	64790	FLJ22127: hypothetical protein FLJ22127	22q11.2
DRM3	AAF68967	55602	FLJ20036: hypothetical protein FLJ20036	4q35.1

[a] All human dsRBH-containing genes are listed, including two putative pseudogenes (RHA1B and TRBP2). Gene names and accession numbers are as in Table I. For each gene, the Locus Link ID and description and chromosomal location (Map) are given for reference.

evolutionary differences between prokaryotic and eukaryotic RNase IIIA proteins. Uniquely, yeast has two RNase IIIA proteins, Sc.RNase IIIA1 and Sc.RNase IIIA2, and the homology between them is not high, either in their dsRBMs or their overall sequence (data not shown). Because Sc.RNase IIIA1 is more homologous to other RNase IIIA proteins, Sc.RNase IIIA2 may be a paralogue of Sc.RNase IIIA1 with specialized functions.

Proteins of the RNase IIIB family are found only in the higher eukaryotes (metazoa). They differ from proteins of the RNase IIIA family in having two RNase III domains and a longer N-terminus. It is likely that this extended N-terminus carries uncharacterized domains that confer additional functions upon RNase IIIB proteins. So far, only a proline-rich region (IPR000694) has been detected in this region, and the significance of its presence is not clear.

Finally, the Dicer proteins (also known as MOI or drosha) have one helicase domain (IPR001410 and IPR001650; Section V.B), one PAZ domain (IPR003100), and one domain with unknown function (IPR005034), in addition to two RNase III domains and one dsRBM domain. These proteins play an important role in RNA interference (RNAi), which has been implicated in development and antiviral defense in several organisms (25, 26). Dicer enzymes digest long dsRNAs down to oligomers of 21–23 base pairs that are used by the cell to degrade other mRNAs in a sequence-specific manner (27, 28). The high homology among Dicer proteins suggests a function that is conserved across species. The plant homologue of Dicer, known as CAF (AAF26461 in Table IV), has similar biochemical functions and also plays a role in the RNAi pathway (Section VI.B). Dicer is also responsible for generating small temporal RNAs (stRNAs), 21–23 nt RNAs that play roles in the control of developmental timing acting through translational repression (29, 30). It is

dsRBM PHYLOGENETICS AND FUNCTIONS

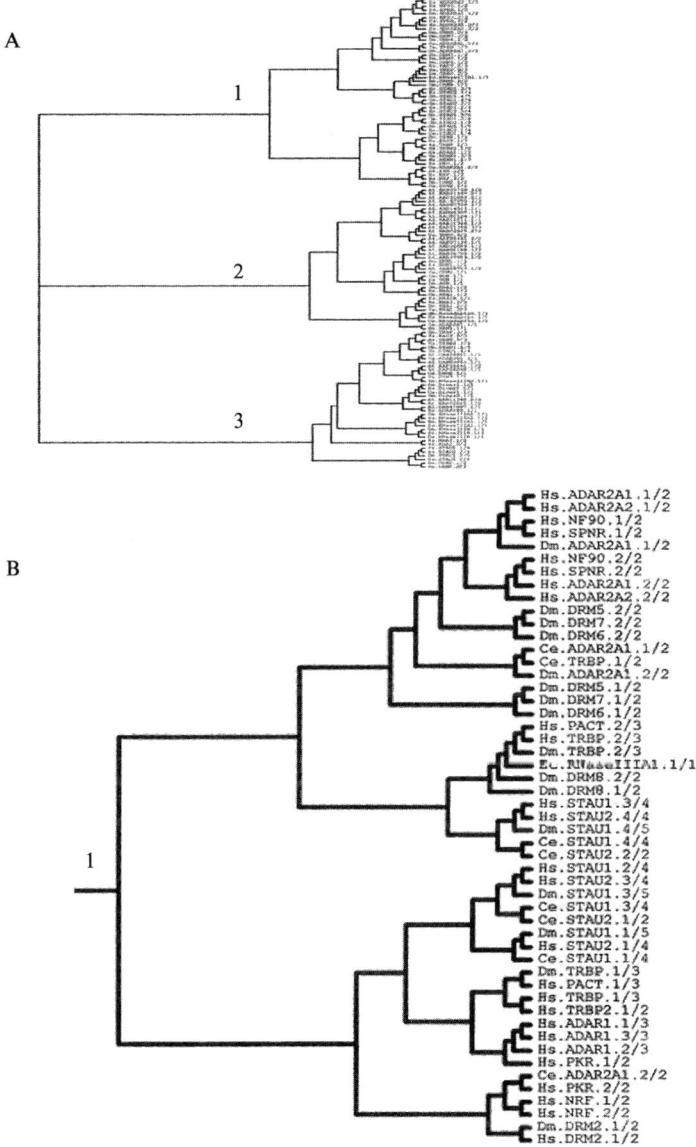

FIG. 3. (Continued)

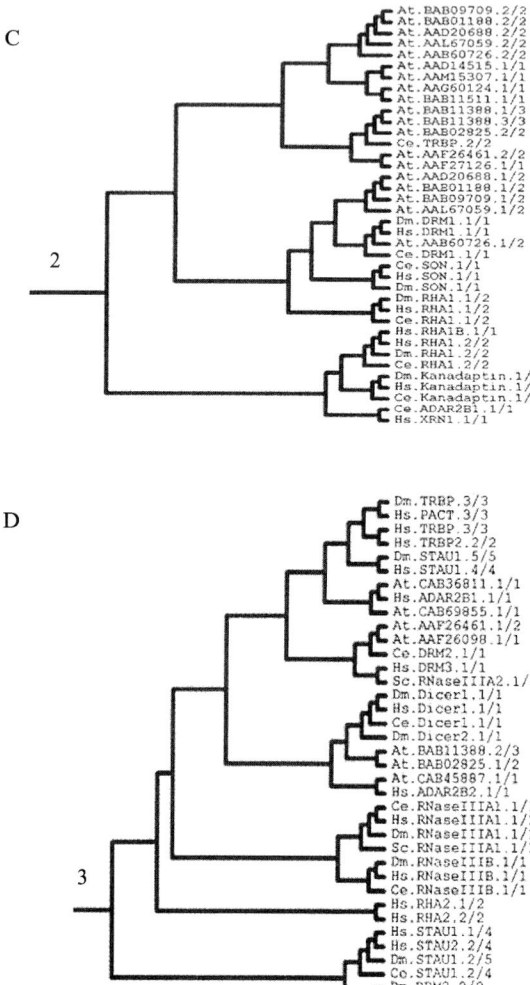

FIG. 3. Cladogram of dsRMBs. (A) dsRBMs from *Arabidopsis thaliana* and the five type species of Fig. 1 were used to generate the cladogram tree with TreeView software and an alignment made with ClustalX. Detailed parameters for the alignment are given in Section II. Domain names are abbreviated following the convention explained in Fig. 1 with the addition of At for *Arabidopsis thaliana*. Protein names, instead of gene names, are used for domains from *Arabidopsis thaliana*. (B–D) expansions of branches 1–3 of the cladogram. Note that the separation of dsRBMs into branches 1–3 was not consistent when different parameters were used, so we have not attempted to infer its meaning. The branches of the cladogram tree at the right side of the tree are much more stable, however, and are exploited in the text.

noteworthy that the PAZ domain of Dicer is found in another group of proteins, which also has a PIWI domain. Proteins of the PIWI family have been shown to work with Dicer, and are thought to play roles in controlling entry into the RNAi and stRNA pathways (25, 31).

The availability of the structure of RNase III domain from the bacterium *Aquifex aeolicus* has shed light on the mechanism by which this domain cleaves its dsRNA substrates (32, 33). However, it remains to be elucidated how the dsRBM and RNase III domains function in concert. A truncated form of *E. coli* RNase III lacking the dsRBM can specifically recognize a dsRNA substrate and accurately cleave it *in vitro*, leading to speculation that the dsRBM's role is to enhance the binding affinity of the enzyme for its substrates in the cell (34).

B. RHA: A DNA/RNA Helicase

Like Dicer, the RNA helicase A (RHA) family of proteins contains both dsRBM and DEAD/DEAH helicase domains. The canonical member of the family, RNA helicase A (RHA1), is present in human, fly, and worm. Its name notwithstanding, RHA unwinds both RNA and DNA in the 3' to 5' direction (35, 36). The fly orthologue Maleless (MLE_DROME) is required for dosage compensation, i.e., to up-regulate X chromosome transcription in the early stages of male fly development (37). It has been proposed that this process involves a ribonucleoprotein complex consisting of two noncoding RNAs (roX1 and roX2) and several proteins, and that Maleless is required at an early stage of complex assembly (38). Several lines of evidence suggest that mammalian RHA1 bridges between RNA polymerase II and transcriptional activators such as CBP/p300 and BRCA1 (39, 40). RHA1 can directly bind and activate the p16/INK4a promoter (41). Furthermore, RHA1 has been implicated in the transcriptional and posttranscriptional regulation of several retroviruses, and it is inhibited by binding adenovirus VA RNA_{II} (42–45).

A second gene, RHA1B, has very high homology with RHA1 both in its dsRBM and overall sequences. Although the predicted protein product (XP_062934) contains only one dsRBM, a BLAST search using the RHA1 mRNA sequence revealed a full-length RHA1B gene (data not shown). This gene aligns with the mRNA sequence of RHA1 without any insertion, indicating that RHA1B is probably a pseudogene derived from RHA1 mRNA.

The human genome also contains a third RHA helicase, RHA2, absent from the four other type species compared here. The overall domain layout is similar in RHA1 and RHA2, suggesting that the two genes may be paralogues that originated by duplication of a parental gene and then diverged. Alternatively, RHA2 may have evolved independently of RHA1, as indicated by the fact that their sequence homology is weak, even in the dsRBMs, and the two dsRBMs of RHA2 are more homologous to one another than to the corresponding dsRBMs of RHA1 (Fig. 3). Because the RHA1 knockout is lethal in mouse (46) despite the presence of RHA2 in the murine genome, RHA2 must behave differently than RHA1 either in its biochemical functions or in the timing, location, or level of its expression. The fact that RHA2 was first identified in brain hints that the two helicases may be expressed in different tissues (47).

C. ADARs: RNA Editases

The adenosine deaminase (editase) domain (IPR002466) confers the enzymatic ability to convert adenosine to inosine in RNA. Two families of proteins contain this domain, ADAR1 and ADAR2. Among the five type species surveyed, ADAR1 is restricted to humans whereas ADAR2 genes are present in all metazoa (human, fly, and worm). The human genome encodes four ADAR2 genes: ADAR2A1 (also known as ADAR2), ADAR2A2 (also known as ADAR3), ADAR2B1, and ADAR2B2.

Several lines of evidence suggest a role for ADAR1 in antiviral defense: it can be induced by interferon, an antiviral cytokine (48); viral RNAs from hepatitis δ are substrates (49); adenovirus VA RNA_1 can antagonize its activity (50); and the dsRNA-specific adenosine deaminase (DRADA) repeat or Z-DNA-binding domain (IPR000607) in the N-terminus of ADAR1, as well as a dsRBM, is found in the vaccinia virus protein E3L (Section VI. C). ADAR1 is also implicated in many other physiological pathways as discussed below.

ADAR1 and ADAR2A1 are ubiquitously expressed throughout the body, whereas ADAR2A2 has been detected only in brain (51). Despite some overlap in their expression patterns and biochemical activities, gene knockout studies indicate that ADAR1 and ADAR2A1 carry out different functions in the body (52–54). $ADAR1^{+/-}$ mice died before embryonic Day 14 with defects in the hematopoietic system, whereas $ADAR2A1^{-/-}$ mice were afflicted with a different set of defects as a result of impaired editing of glutamate receptor GluR-B premessenger RNA (53, 54). Although known endogenous ADAR substrates are limited to several signaling components of the nervous system, such as glutamate and serotonin receptors, it is expected that additional substrates await discovery (55).

Although ADARs use their dsRBMs to bind dsRNA, substrate specificity resides in the deaminase domain (56). The combination of these two features may explain why ADARs edit certain mRNAs only. ADAR1, ADAR2A1/2, and ADAR2B1/2 genes have three, two, and one dsRBMs, respectively; it remains to be seen whether the number of dsRBMs has an impact on substrate recognition or enzymatic function.

D. The Protein Kinase PKR

The dsRNA-activated protein kinase PKR (protein kinase, RNA dependent) has been studied particularly intensively. It is a serine/threonine kinase that is induced by interferon. It plays an important part in antiviral defense mechanisms as well as apoptosis, cell differentiation, and growth control (57, 58). Present in a latent form until activation, PKR appears to undergo a conformational change after binding to its activator, and autophosphorylation transforms the protein to its active state. The RNAs that activate PKR during viral infection are believed to be the viral genomes themselves or intermediates in viral replication or transcription, but several cellular RNAs with duplex structure have been found to activate or inhibit PKR (6, 59–62). The highly structured adenovirus VA RNA$_I$ is a well-studied inhibitor of PKR activation that serves to neutralize the antiviral action of PKR (63). In addition, PKR activity can be modulated by two dsRBM-containing proteins, TRBP and PACT (Section V. F).

The best known substrate for PKR is eukaryotic translation initiation factor 2 (eIF2). Phosphorylation of the α-subunit of eIF2 characteristically results in inhibition of translation. Other substrates include the inhibitor of NF-κB (I$_\kappa$B), the HIV-1 transcriptional activator Tat, and the tumor suppressor p53, suggesting that PKR is involved in signal transduction pathways as well as in translational control (64–66). Several dsRBM-containing proteins, such as NF90 and PACT, are phosphorylated by PKR (67, 68), although it is not clear whether this leads to any change in their activities, particularly dsRNA or PKR binding. The autophosphorylation of PKR, however, has been implicated in modulating its dsRNA-binding activity (69).

E. Staufen: mRNA Transport and Translation

Staufen proteins play important roles in mRNA transport and the regulation of protein synthesis. Staufen is involved in the localization of maternal mRNAs in the fly oocyte (70): it forms a complex with oskar mRNA, whose localization is important in establishing the anterior–posterior axis, and with kinesin I, a microtubule motor (71). Mammalian Staufen is expressed during oogenesis and spermatogenesis. It is also found in hippocampal neurons, associated with polysomes and the rough endoplasmic reticulum (72).

Staufen proteins have several dsRBMs (five in fly, four in human, and four or fewer in worm; Fig. 4) that have different functions in mRNA localization and translation. The fly Staufen dsRBMs 1, 3, and 4 bind dsRNA *in vitro*, whereas dsRBMs 2 and 5 do not; dsRBM 2 is required for the microtubule-dependent localization of oskar mRNA, and dsRBM 5 for the derepression of oskar mRNA translation once localized (73). The second motif is highly homologous among Staufen proteins from different species (Fig. 4), indicating that it may be responsible for transporting mRNAs in a microtubule-dependent manner in other species. The fifth domain, however, is found only in fly and human Staufen. Staufen protein has also been found to interact with telomerase RNA and HIV-1 genomic RNA (74, 75), but the significance of these interactions is unknown.

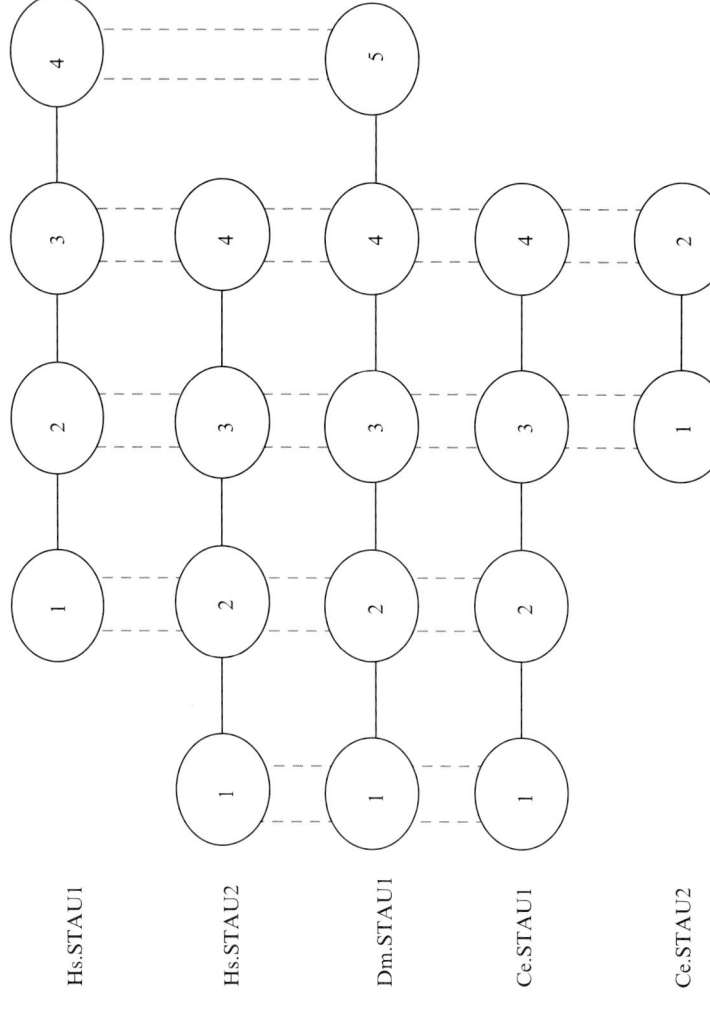

FIG. 4. Alignment of dsRBMs in Staufen proteins. The alignment was based on pair-wise BLAST searches as well as the cladogram generated from multiple alignments (Fig. 3). Corresponding domains are connected by pairs of dashed lines.

F. TRBP and PACT: Regulators of PKR

Like Staufen, the TAR RNA-binding protein (TRBP) has been implicated in spermatogenesis and the translational activation of repressed mRNAs (76, 77). First identified by its ability to bind HIV-1 TAR RNA, TRBP activates transcription from the viral LTR (78). Its *X. laevis* homologue, Xlrbpa, associates with ribosomes and heterogeneous nuclear RNA (10). TRBP interacts with PKR and inhibits its kinase activity (79). In contrast, a close family member, the cellular protein activator of PKR (PACT), functions as an activator of PKR in response to a number of stress stimuli (67, 80, 81). Although both proteins have been shown to bind PKR, the mechanisms by which they regulate its activity are not clear.

Another gene, TRBP2, has been predicted in the human genome (Table V). Sequence alignment indicates that it has high homology with TRBP, and it appears to be a pseudogene evolved from TRBP mRNA (data not shown).

G. NF90: A Multifunctional Regulator

The nuclear factor 90 (NF90) family has two members, NF90 itself (forms of which are known as ILF3, MPP4, DRBP76, and NFAR1 and 2) and spermatid perinuclear protein (SPNR). Among the species surveyed, these proteins are found only in human, but they are present in other vertebrates including mouse and *Xenopus*. Several lines of evidence indicate that NF90 and SPNR are paralogues and that the gene duplication happened quite recently: first, the proteins share high homology in overall sequence; second, the dsRBMs of these two proteins form clusters in the cladogram (Fig. 3); finally, the two dsRBMs are equally spaced in both proteins.

NF90 is implicated in a number of different biological processes (82). It has both DNA- and RNA-binding activities; it can regulate the translation of some transcripts; it interacts with the protein–arginine methyltransferase PRMT1, and DNA-dependent protein kinase (DNA-PK) complex; it is differentially expressed in nasopharyngeal carcinoma cells; and it can be phosphorylated by PKR and possibly regulates the activity of PKR (45, 68, 83–88). The *Xenopus* homologues are involved in developmental processes (89). NF90 gene transcripts are found in various tissues including germ cells, and they are alternatively spliced, generating proteins that differ at their C-termini (90). The functional significance of this heterogeneity remains to be determined.

Mouse SPNR mRNA, also present in multiple forms, is expressed at high levels in the testis, ovary, and brain, and the protein localizes in cytoplasmic microtubules (91). Mice deficient for SPNR show neurological, spermatogenic, and sperm morphological abnormalities, indicating a role in neural function and spermatogenesis (92).

At their N-termini, both NF90 and SPNR have a DZF (domain in DSRM or ZnF_C2H2 domain-containing proteins) domain. Although this domain is not yet categorized in the InterPro database, it is also present in nuclear factor 45 (NF45), a protein that forms heterodimers with NF90, and several ZnF_C2H2 or Zinc finger-containing proteins, which have also been found to form homo- or heterodimers.

H. NRF: A Transcriptional Repressor

The NF-κB repressing factor (NRF) has two dsRBMs, neither of which was previously recognized. Their degree of homology to other dsRBMs is relatively low and their dsRNA-binding properties remain to be investigated. NRF is a constitutive and position-independent silencer of NF-κB-binding sites (93). It inhibits the activation of interferon-β (IFN-β) genes prior to viral activation. The presence of dsRBMs in this protein suggests that it may be involved in a coordinated pathway by which cells respond to viral infection: dsRNAs produced during viral infection could bind NRF and relieve the inhibition of NF-κB, thereby activating the IFN-β genes. It may be significant that the genes encoding the antiviral proteins PKR and ADAR1 are both transcriptionally induced by IFN-β.

NRF also has two other known domains, the G-patch (IPR000467) and R3H (IPR001374). The G-patch is a short region of about 40 aa that is found in a number of putative RNA-binding proteins (94). R3H has characteristic spacing of conserved arginine (R) and histidine (H) residues capable of binding single-stranded nucleic acids (95). The importance of the presence of these domains is not yet established.

I. SON: Another Transcriptional Regulator

The SON protein (also known as NREBP) was shown to be a nuclear protein, and to inhibit the transcription of human hepatitis B virus (HBV) by binding to a negative regulatory element (NRE) in the viral genome (96). Through this binding, it represses HBV core promoter activity, transcription of HBV genes, and production of HBV virions. In addition to the dsRBM, a G-patch (IPR000467) domain is also found in SON. Interestingly, the dsRBM-containing protein NRF (Section V.H), also has a G-patch and acts as a transcriptional repressor.

J. Kanadaptin

Little is known about kanadaptin (kidney anion exchanger adaptor protein). Given that it exists in human, fly, and worm, and that all orthologues have conserved residues throughout their sequences, it is reasonable to speculate that it is involved in some basic cellular function(s) conserved in metazoa. The mouse homologue of kanadaptin was identified as a protein that interacts with the cytoplasmic domain of kAE1, an anion exchanger, and is possibly involved in the localization of kAE1 (97). Kanadaptin is present in the nuclei of various epithelial and nonepithelial cells, raising the possibility that it has other functions as well (98).

The forkhead-associated (FHA) domain (IPR000467) present in the protein is a phosphopeptide-binding domain found in several prokaryotic and eukaryotic proteins (99). Many FHA-containing proteins in yeast and human nuclei are involved in DNA repair, the cell cycle, and mRNA processing.

K. XRN1: RNA Degradation

Although XRN1 proteins are present in several species, their homology is confined to certain regions. The C-terminus of human and murine XRN1, where the dsRBM resides, is missing in lower organisms (data not shown). The *S. cerevisiae* homologue, Xrn1p, is a 5'-3' exoribonuclease localized in the cytoplasm and is involved in multiple mRNA decay pathways. However, the 5'-3' exonuclease activity, which has been mapped to the N-terminus of the protein, is capable of degrading a variety of substrates including single-stranded RNA (ssRNA), single-stranded DNA (ssDNA), and double-stranded DNA (dsDNA) (100). Besides the defects in RNA metabolism, a yeast mutant lacking Xrn1p exhibited meiotic arrest and defects in microtubule-related processes, suggesting many roles for the protein (100). Mouse XRN1 (mXRN1p) displays homology with human XRN1 throughout the entire protein, as expected, indicating that they have a greater degree of functional similarity than the XRN1 proteins of lower species, mXRN1p, like Xrn1p in yeast, localizes to cytoplasm, but the purified protein exhibits a substrate preference for G4 RNA tetraplexes (101). Whether this novel ability is due to the presence of a dsRBM has yet to be determined.

L. DRMs: Unassigned dsRBM-Containing Proteins

The functions of some dsRBM family proteins are still entirely obscure even though their primary sequences are available. Because no official or meaningful names have been given to these proteins, we provisionally refer to them as dsRBM-containing protein families DRMs 1–7. DRMs 1–3 are found in the human genome, and DRM1 and DRM2 are also present in the fly and worm genomes. DRM1 contains two additional domains, IPR001269 and IPR003009. DRM2 contains the forkhead-associated (FHA) domain (IPR000253), a putative nuclear signaling domain found in a variety of otherwise unrelated proteins. DRM3 contains a serine-rich region and has some homology with NRF.

Four other genes, DRMs 4–7, are found exclusively in the *Drosophila* genome. Among them, DRMs 4–6 all have two dsRBMs and their dsRBM sequences are highly homologous to one another as indicated in the cladogram

(Fig. 3), raising the possibility that they belong to the same gene family and have similar functions. Circumstantial evidence indicates that DRM5 (also called DIP1 or DISCO interacting protein 1) has chromatin-binding activity. It remains to be seen whether DRMs 4–7 are truly unique to the fly or have homologues in other species that are yet to be discovered.

VI. dsRBMs in Other Taxa

Although a detailed analysis of dsRBMs in other organisms lies beyond the scope of this review, we have expanded our survey to take stock of the dsRBMs in three additional, highly diverse, taxa—the archebacteria, a plant, and viruses.

A. Archaea

No dsRBM has been discovered in archaeal sequences. HMM and BLAST searches failed to disclose any dsRBMs in several archaea whose genome sequences are complete, namely *Aeropyrum pernix*, *Archaeoglobus fulgidus*, *Halobacterium* sp., *Methanobacterium thermoautotrophicum*, *Methanococcus jannaschii*, *Pyrococcus horikoshii*, and *Thermoplasma acidophilum*. Although the possibility cannot be excluded that dsRBMs will be found in other archaeal genomes, the occurrence of dsRBMs presently appears to be limited to eubacteria and eukaryota. It is especially notable that the RNase III domain, present in all other species surveyed, has also not been found in archaeal genomes.

B. Plant

In the mouse-ear cress *Arabidopsis thaliana*, for which the complete genome sequence is available, there are at least 17 dsRBM-containing proteins (Table IV). The *Arabidopsis* dsRBMs tend to form a cluster in the multiple alignment, as shown in the cladogram (Fig. 3), suggesting that plant dsRBMs may have evolved from dsRBM ancestor(s) that diverged from those of the other species surveyed at a very early time.

Helicase (IPR001410 and IPR001650) and RNase III (IPR000999) domains are very well represented among the set of plant dsRBM-containing proteins. Two proteins contain dsRBM and RNase III domains and are presumably homologues of RNase III; four contain dsRBM and helicase domains, and may serve functions similar to that of RHA; and two contain dsRBM, helicase, and RNase III domains, including one (AAF26461) that resembles Dicer. A notable feature of the plant proteins is that some of them appear to be reorganized versions of their animal counterparts, differing in the number and order of the domains in the protein. Compared to other metazoa, *Arabidopsis* also exhibits a distinct set of enzymatic domains associated with the dsRBM, namely the phosphatase domain (IPR004844 and IPR004843), SAM-binding motif (IPR000051), and isopenicillin N synthetase signature (IPR002057).

The existence of a large number of plant dsRBM-containing proteins is not surprising, given the fact that dsRNAs have been implicated in many vital cellular pathways in plants. These include development, the containment of mobile genetic elements, and antiviral defense (26). In plants, dsRNA can function by sequence-specific RNA degradation (RNAi), or transcriptional gene silencing (TGS), which involves RNA-directed DNA methylation (102). Consistent with the deployment of different mechanisms of antiviral defense in plants and animals, kinase and adenosine deaminase domains, two enzymatic domains implicated in fending off viral infection in vertebrates, are absent from the *Arabidopsis* set of dsRBM-containing proteins.

C. Viruses

Although dsRBMs are widespread in the virus world, in no case has more than one dsRBM been found in any viral protein (Table VI) implying that a single dsRBM is sufficient to carry out viral activities. This presumably reflects

TABLE VI
VIRAL DsRBM-CONTAINING PROTEINS[a]

Domain structure	Representative accession	Virus	Taxonomy
607–1159	AAA48040	Vaccinia virus	dsDNA viruses, no RNA stage; Poxviridae; Chordopoxvirinae Orthopoxvirus.
1159	AAF14917	Myxoma virus	dsDNA viruses, no RNA stage; Poxviridae Chordopoxvirinae Leporipoxvirus
2873–1159	CAB52751	Human rotavirus (group C/strain Bristol)	dsRNA viruses; Reoviridae Rotavirus
1159	AAC72057	Banna virus	dsRNA viruses; Reoviridae; Coltivirus
999–1159	AAB94459	Chilo iridescent virus	dsDNA viruses, no RNA stage; Iridoviridae Iridovirus
1159–605–1205	AAC58807	Drosophila C virus	ssRNA positive–strand viruses, no DNA stage; picornaviridae Cricket paralysis-like viruses
605–1205–1159	AAc58718	Acyrthosiphon pisum virus	ssRNA positive-strand viruses, no DNA stage

[a]Domain layouts are shown for viral dsRBM-containing proteins following the convention of Table I. Viral names and taxonomy are also listed. Domain numbers and names are as in Table III.

the compactness and efficiency of most viral genomes, and it may also signify that these proteins serve a limited range of functions among viruses. These functions may be crucial for viral survival, however. In mammalian cells, endowed with potent dsRNA-dependent antiviral mechanisms, proteins that bind dsRNA and prevent its exposure to cellular components of the defense pathway are important for successful viral infection. The vaccinia virus E3L protein (AAA48040) is the best studied example. This protein can inhibit PKR, either by sequestering dsRNA or by forming heterodimers with PKR. The presence of the adenosine deaminase domain (IPR000607) suggests that E3L has some connection with ADAR1, another gene thought to play an antiviral role (Section V. C). Other poxvirus proteins, such as that of myxoma virus (AAF14917), lack the ADAR1-like domain and presumably function simply by dsRNA sequestration (*103*). Not surprisingly, dsRBMs are also found in dsRNA viruses, such as the reoviruses (CAB52751 and AAC72057).

Less predictably, dsRBMs are found in insect viruses, whose hosts have not been shown to possess anti-dsRNA machinery although they do have RNAi mechanisms. In these viruses, the presence of additional protein domains (RNase III domain in AAB94459 and RNA-dependent RNA polymerase and RNA helicase domains in AAC58807 and AAC58718) may point to roles in RNA processing and replication. Strikingly, proteins from two insect viruses (AAC58807 and AAC58718) contain the same domains arranged in a different order, with the dsRBM at the N-terminus or C-terminus of the protein, respectively. This supports the view of the dsRBM as a modular element of protein structure.

VII. Concluding Remarks

We have exploited computational approaches to catalogue and compare the dsRBM-containing proteins in a number of model systems. Because of the recent availability of many entire genome sequences, we were able to assess the full set of proteins that contains the dsRBM in representatives of a variety of taxonomic groups. Therefore, the present survey is the most comprehensive one assembled to date, both in terms of the range of organisms examined and the completeness of the genomes scrutinized. In addition to disclosing the existence of previously unidentified dsRBM-containing proteins, this study sheds light on the evolution and function of the motif, as well as on the cellular roles of dsRNA.

A. Origin and Evolution of the dsRBM

dsRNA is likely to be an extremely ancient form of nucleic acid. If an "RNA World" presaged the current era, RNA's capacity for self-annealing and the relatively high stability of RNA duplexes would argue that dsRNA is younger than ssRNA but probably older than DNA. It is, therefore, not surprising that the dsRBM, a protein motif able to interact with dsRNA, is widespread in nature and seems to have a long lineage. The existence of intriguing clusters in the cladogram of Fig. 3, taken together with the wide distribution of the dsRBM, invites speculation that the motif might have had a single evolutionary origin. This distribution of dsRBMs among the major branches of the tree may not be reliable, however, as it does not remain constant when the alignment parameters are varied. Consequently, on the basis of present data and methods of analysis we cannot make a compelling case to distinguish between the single-origin hypothesis and the alternative that the dsRBM arose more than once in the evolution of protein families.

Apart from the archaea, the dsRBM is represented in every group of living organisms studied, encompassing the eubacterial, yeast, animal, and plant taxa, as well as viruses. Remarkably, no dsRBM was detected in the seven genomes belonging to the archebacterial kingdom that were surveyed. How can this be explained? Because these seven species are drawn from different genera, it does not seem likely that they are simply aberrant microorganisms and that other representatives of the kingdom will turn out to possess the motif. More plausibly, in view of the flexibility of the dsRBM consensus sequence, the motif might exist in archebacterial genomes but in a form sufficiently distinct to escape detection by our search tools. Alternatively, the archaea may have evolved an entirely different motif for interacting with dsRNA. Finally, the possibility that no such motif exists in the archaea cannot be dismissed out of hand, even though this conclusion would lead to a reevaluation of some entrenched ideas.

One approach toward resolving the issue is to attempt the biochemical purification and characterization of dsRNA-binding proteins from archebacterial extracts. The sequence of any proteins that emerged from such an analysis would be immediately illuminating. In this connection, it should be borne in mind that a small number of cellular and viral proteins are known to bind dsRNA without benefit of a dsRBM; examples include the cellular $2',5'$-oligoadenylate synthetases, the reovirus $\sigma 3$ protein, and the influenza virus NS1 protein (104–106). Comparison of the RNA-binding motifs of these proteins with those of archaeal dsRNA-binding proteins, if any, might provide clues as to the origin of the dsRBM.

It is striking that the RNase III domain, present in all other species surveyed, is also missing in archaea (data not shown). This observation may be significant in two respects. First, it implies that dsRNA is probably handled differently in these organisms, underlining the need for thorough biochemical investigation. Second, because the RNase III domain has both dsRNA-binding and cleavage activities (Section V.A), the coordinate absence of the RNase III and dsRBM domains prompts us to speculate that the dsRBM came into existence because of its role in enhancing substrate recognition and binding by RNase III.

In contrast to the three major branches of the cladogram, the relationships among the dsRBMs that populate the minor branches are stably conserved (Fig. 3 and data not shown). This finding is consistent with the view that protein genes evolve as a unit once their motifs are assembled and protein families are established. Moreover, it is evident from our analysis that evolutionarily newer proteins, such as NRF, RHA2, and ADAR1, which are absent from lower species, tend to display a greater degree of homology between multiple dsRBM domains within a single protein than evolutionarily older proteins such as Staufen, RHA1, and TRBP. We infer that the multiple dsRBMs within each protein stem from identical sequences, originating either by motif duplication or possibly as the result of repeated insertion of the same dsRBM sequence. Subsequently, the individual dsRBMs diverged over time, presumably as a result of differing evolutionary pressures. The fact that corresponding dsRBMs in the same protein family are generally similar, as evident from their proximity in the cladogram, may indicate that the several dsRBMs in a protein subserve distinct functions.

B. The dsRBM and Protein Function

Although there are several proteins in which it is the only motif recognized, the dsRBM is more typically tethered to one or more other types of motif in a protein. These associations increase the repertoire of cellular functions and responses to dsRNA. In human cells, dsRNA is a potent signal that can be either beneficial or deleterious depending upon its length and the context. Judging by their exquisite sensitivity to low concentrations of dsRNA, cells and enzymes seem poised to respond to long dsRNA (>25–30 bp) as a challenge or alarm signal, presumably because such molecules are commonly produced during virus infection. Because recognition of dsRNA by the dsRBM is predominantly structure based, the motif is well adapted for processes that involve interactions with generic dsRNAs free of any specific sequence restrictions. Recent studies have shown that short dsRNAs (<25–30 bp) do not trigger the antiviral alarm machinery in human cells but, as in several organisms, can regulate gene expression manner via the RNAi response. This phenomenon involves the dsRBM-containing enzyme Dicer at an early stage. Although RNA interference is sequence specific, Dicer's role in the process appears to be sequence nonspecific.

Our analysis reveals an evolutionary trend toward increased numbers of dsRBMs per protein, and toward a greater number and broader range of motifs of different sorts in dsRBM-containing proteins. The presence of varied functional elements equips the proteins to fulfill additional and more sophisticated roles, and to receive regulatory signals from more directions. Through activities of the dsRBM itself (e.g., RNA annealing and RNA binding) and activities brought in by other domains that are associated with it (catalyzing dsRNA unwinding, base modification, and cleavage), dsRBM-containing proteins enable cells to react to dsRNA in a variety of ways. The inventory of responses ranges from cell death, which can be elicited via the PKR pathway, to measures that neutralize dsRNA as a signal in consequence of helicase, adenosine deaminase, and RNase III activities.

The presence of multiple dsRBMs in a protein can enhance the motif's affinity for dsRNA through cooperative interactions, as seen in PKR (7). Moreover, dsRBMs are capable of carrying out activities other than binding to dsRNA, which expand the function of the protein that harbors them. For example, they can participate in protein:protein interactions, as in the heterodimerization of PKR with TRBP or PACT (14, 81); they can play a part in protein subcellular localization, as suggested by the finding that dsRBMs from several dsRBM-containing proteins can bind the nuclear export receptor exportin-5 (107), they can anneal RNA helices (8), and they can bind ribosomes (108). Some dsRBMs that have lost most of their dsRNA-binding activity, including the second motif of PKR and the third motif of PACT (7, 81), have been clearly implicated in functions other than RNA binding, such as protein dimerization and kinase activation. On the other hand, the exact determinants of the sequence specificity in RNA binding displayed by Staufen remains to be established.

The spectrum of biochemical activities displayed by dsRBMs undoubtedly reflects the diverse evolutionary pressures imposed on them in different proteins. From another perspective, dsRBMs confer a variety of properties upon the domains with which they are associated. Despite the motif's considerably relaxed sequence conservation, nearly identical three-dimensional folding is adopted by all the dsRBMs whose structures have been solved to date. It seems likely, therefore, that the broad spectrum of activities is made possible by virtue of the high tolerance for sequence changes within the framework of a relatively fixed dsRBM structure.

C. Scope of the dsRBM Protein Family

Although our compilation of sequences that contain dsRBMs is comprehensive and probably close to complete for the five species that form the primary focus of this study, it is far from exhaustive. In terms of phylogenetic space, we have merely scratched the surface. What is more, with a few notable exceptions the functional characterization of the dsRBM family proteins is in an early phase, and in many cases has not even begun. This is especially true in the plant kingdom, where—to judge from the solitary species in our survey—dsRBMs occur in a range of proteins substantially distinct from that seen in animals. The identification and understanding of proteins and their domains represent daunting tasks in the postgenomic era when function is at center stage. With the development of new computational tools and the accumulation of further experimental data, including the elucidation of higher order structures, we anticipate that domains will continue to be defined and that their interplay during protein function will be increasingly amenable to study.

Acknowledgments

This work is supported by Grant AI34552 from the National Institutes of Health to M.B.M.

References

1. Nanduri, S., Carpick, B. W., Yang, Y., Williams, B. R., and Qin, J. (1998). Structure of the double-stranded RNA-binding domain of the protein kinase PKR reveals the molecular basis of its dsRNA-mediated activation. *EMBO J.* **17,** 5458–5465.
2. Ryter, J. M., and Schultz, S. C. (1998). Molecular basis of double-stranded RNA-protein interactions: Structure of a dsRNA-binding domain complexed with dsRNA. *EMBO J.* **17,** 7505–7513.
3. Kharrat, A., Macias, M. J., Gibson, T. J., Nilges, M., and Pastore, A. (1995). Structure of the dsRNA binding domain of E. coli RNase III. *EMBO J.* **14,** 3572–3584.
4. Bycroft, M., Grunert, S., Murzin, A. G., Proctor, M., and St Johnston, D. (1995). NMR solution structure of a dsRNA binding domain from Drosophila staufen protein reveals homology to the N-terminal domain of ribosomal protein S5. *EMBO J.* **14,** 3563–3571.
5. Bevilacqua, P. C., George, C. X., Samuel, C. E., and Cech, T. R. (1998). Binding of the protein kinase PKR to RNAs with secondary structure defects: Role of the tandem A-G mismatch and noncontiguous helixes. *Biochemistry* **37,** 6303–6316.
6. Tian, B., White, R. J., Xia, T., Welle, S., Turner, D. H., Mathews, M. B., and Thornton, C. A. (2000). Expanded CUG repeat RNAs form hairpins that activate the double-stranded RNAdependent protein kinase PKR. *RNA* **6,** 79–87.
7. Green, S. R., Manche, L., and Mathews, M. B. (1995). Two functionally distinct RNA-binding motifs in the regulatory domain of the protein kinase DAI. *Mol. Cell. Biol.* **15,** 358–364.
8. Hitti, E., Neunteufl, A., and Jantsch, M. F. (1998). The double-stranded RNA-binding protein Xlrbpa promotes RNA strand annealing. *Nucleic Acids Res.* **26,** 4382–4388.
9. Tian, B., and Mathews, M. B. (2001). Functional characterization of and cooperation between the double-stranded RNA-binding motifs of the protein kinase PKR. *J. Biol. Chem.* **276,** 9936–9944.
10. Eckmann, C. R., and Jantsch, M. F. (1997). Xlrbpa, a double-stranded RNA-binding protein associated with ribosomes and heterogeneous nuclear RNPs. *J. Cell Biol.* **138,** 239–253.
11. Kiebler, M. A., Hemraj, I., Verkade, P., Kohrmann, M., Fortes, P., Marion, R. M., Ortin, J., and Dotti, C. G. (1999). The mammalian staufen protein localizes to the somatodendritic domain of cultured hippocampal neurons: Implications for its involvement in mRNA transport. *J. Neurosci.* **19,** 288–297.
12. Marion, R. M., Fortes, P., Beloso, A., Dotti, C., and Ortin, J. (1999). A human sequence homologue of Staufen is an RNA-binding protein that is associated with polysomes and localizes to the rough endoplasmic reticulum. *Mol. Cell. Biol.* **19,** 2212–2219.

13. Wu, S., Kumar, K. U., and Kaufman, R. J. (1998). Identification and requirement of three ribosome binding domains in dsRNA-dependent protein kinase (PKR). *Biochemistry* **37**, 13816–13826.
14. Daher, A., Longuet, M., Dorin, D., Bois, F., Segeral, E., Bannwarth, S., Battisti, P. L., Purcell, D. F., Benarous, R., Vaquero, C., Meurs, E. F., and Gatignol, A. (2001). Two dimerization domains in the TAR RNA binding protein, TRBP, individually reverse the protein kinase R inhibition of HIV-1 long terminal repeat expression. *J. Biol. Chem.* **276**, 33899–33905.
15. Ung, T. L., Cao, C., Lu, J., Ozato, K., and Dever, T. E. (2001). Heterologous dimerization domains functionally substitute for the double-stranded RNA binding domains of the kinase PKR. *EMBO J.* **20**, 3728–3737.
16. Apweiler, R., Attwood, T. K., Bairoch, A., Bateman, A., Birney, E., Biswas, M., Bucher, P., Cerutti, L., Corpet, F., Croning, M. D., Durbin, R., Falquet, L., Fleischmann, W., Gouzy, J., Hermjakob, H., Hulo, N., Jonassen, I., Kahn, D., Kanapin, A., Karavidopoulou, Y., Lopez, R., Marx, B., Mulder, N. J., Oinn, T. M., Pagni, M., and Servant, F. (2001). The InterPro database, an integrated documentation resource for protein families, domains and functional sites. *Nucleic Acids Res.* **29**, 37–40.
17. Jeanmougin, F., Thompson, J. D., Gouy, M., Higgins, D. G., and Gibson, T. J. (1998). Multiple sequence alignment with Clustal X. *Trends Biochem. Sci.* **23**, 403–405.
18. Karplus, K., Barrett, C., and Hughey, R. (1998). Hidden Markov models for detecting remote protein homologies. *Bioinformatics* **14**, 846–856.
19. Page, R. D. (1996). TreeView: An application to display phylogenetic trees on personal computers. *Comput. Appl. Biosci.* **12**, 357–358.
20. Gonnet, G. H., Cohen, M. A., and Benner, S. A. (1992). Exhaustive matching of the entire protein sequence database. *Science* **256**, 1443–1445.
21. Fierro-Monti, I., and Mathews, M. B. (2000). Proteins binding to duplexed RNA: One motif, multiple functions. *Trends Biochem. Sci.* **25**, 241–246.
22. Pruitt, K. D., and Maglott, D. R. (2001). RefSeq and LocusLink: NCBI gene-centered resources. *Nucleic Acids Res.* **29**, 137–140.
23. Nicholson, A. W. (1999). Function, mechanism and regulation of bacterial ribonucleases. *FEMS Microbiol. Rev.* **23**, 371–390.
24. Conrad, C., and Rauhut, R. (2002). Ribonuclease III: New sense from nuisance. *Int. J. Biochem. Cell Biol.* **34**, 116–129.
25. Hutvagner, G., and Zamore, P. D. (2002). RNAi: Nature abhors a double-strand. *Curr. Opin. Genet. Dev.* **12**, 225–232.
26. Vance, V., and Vaucheret, H. (2001). RNA silencing in plants—defense and counterdefense. *Science* **292**, 2277–2280.
27. Elbashir, S. M., Lendeckel, W., and Tuschl, T. (2001). RNA interference is mediated by 21- and 22-nucleotide RNAs. *Genes Dev.* **15**, 188–200.
28. Bernstein, E., Caudy, A. A., Hammond, S. M., and Hannon, G. J. (2001). Role for a bidentate ribonuclease in the initiation step of RNA interference. *Nature* **409**, 363–366.
29. Grishok, A., Pasquinelli, A. E., Conte, D., Li, N., Parrish, S., Ha, I., Baillie, D. L., Fire, A., Ruvkun, G., and Mello, C. C. (2001). Genes and mechanisms related to RNA interference regulate expression of the small temporal RNAs that control C. elegans developmental timing. *Cell* **106**, 23–34.
30. Banerjee, D., and Slack, F. (2002). Control of developmental timing by small temporal RNAs: A paradigm for RNA-mediated regulation of gene expression. *BioEssays* **24**, 119–129.
31. Cerutti, L., Mian, N., and Bateman, A. (2000). Domains in gene silencing and cell differentiation proteins: The novel PAZ domain and redefinition of the Piwi domain. *Trends Biochem. Sci.* **25**, 481–482.

32. Blaszczyk, J., Tropea, J. E., Bubunenko, M., Routzahn, K. M., Waugh, D. S., Court, D. L., and Ji, X. (2001). Crystallographic and modeling studies of RNase III suggest a mechanism for double-stranded RNA cleavage. *Structure (Camb.)* **9**, 1225–1236.
33. Zamore, P. D. (2001). Thirty-three years later, a glimpse at the ribonuclease III active site. *Mol. Cell* **8**, 1158–1160.
34. Sun, W., Jun, E., and Nicholson, A. W. (2001). Intrinsic double-stranded-RNA processing activity of Escherichia coli ribonuclease III lacking the dsRNA-binding domain. *Biochemistry* **40**, 14976–14984.
35. Lee, C. G., and Hurwitz, J. (1992). A new RNA helicase isolated from HeLa cells that catalytically translocates in the 3′ to 5′ direction. *J. Biol. Chem.* **267**, 4398–4407.
36. Zhang, S., and Grosse, F. (1994). Nuclear DNA helicase II unwinds both DNA and RNA. *Biochemistry* **33**, 3906–3912.
37. Lee, C. G., Chang, K. A., Kuroda, M. I., and Hurwitz, J. (1997). The NTPase/helicase activities of Drosophila maleless, an essential factor in dosage compensation. *EMBO J.* **16**, 2671–2681.
38. Meller, V. H., Gordadze, P. R., Park, Y., Chu, X., Stuckenholz, C., Kelley, R. L., and Kuroda, M. I. (2000). Ordered assembly of roX RNAs into MSL complexes on the dosage-compensated X chromosome in Drosophila. *Curr. Biol.* **10**, 136–143.
39. Anderson, S. F., Schlegel, B. P., Nakajima, T., Wolpin, E. S., and Parvin, J. D. (1998). BRCA1 protein is linked to the RNA polymerase II holoenzyme complex via RNA helicase A. *Nat. Genet.* **19**, 254–256.
40. Nakajima, T., Uchida, C., Anderson, S. F., Lee, C. G., Hurwitz, J., Parvin, J. D., and Montminy, M. (1997). RNA helicase A mediates association of CBP with RNA polymerase II. *Cell* **90**, 1107–1112.
41. Myohanen, S., and Baylin, S. B. (2001). Sequence-specific DNA binding activity of RNA helicase A to the p16INK4a promoter. *J. Biol. Chem.* **276**, 1634–1642.
42. Fujii, R., Okamoto, M., Aratani, S., Oishi, T., Ohshima, T., Taira, K., Baba, M., Fukamizu, A., and Nakajima, T. (2001). A role of RNA helicase A in cis-acting transactivation response element-mediated transcriptional regulation of human immunodeficiency virus type 1. *J. Biol. Chem.* **276**, 5445–5451.
43. Li, J., Tang, H., Mullen, T. M., Westberg, C., Reddy, T. R., Rose, D. W., and Wong-Staal, F. (1999). A role for RNA helicase A in post-transcriptional regulation of HIV type 1. *Proc. Natl. Acad. Sci. USA* **96**, 709–714.
44. Tang, H., Gaietta, G. M., Fischer, W. H., Ellisman, M. H., and Wong-Staal, F. (1997). A cellular cofactor for the constitutive transport element of type D retrovirus. *Science* **276**, 1412–1415.
45. Liao, H. J., Kobayashi, R., and Mathews, M. B. (1998). Activities of adenovirus virus-associated RNAs: Purification and characterization of RNA binding proteins. *Proc. Natl. Acad. Sci. USA* **95**, 8514–8519.
46. Lee, C. G., da Costa Soares, V., Newberger, C., Manova, K., Lacy, E., and Hurwitz, J. (1998). RNA helicase A is essential for normal gastrulation. *Proc. Natl. Acad. Sci. USA* **95**, 13709–13713.
47. Nagase, T., Ishikawa, K., Suyama, M., Kikuno, R., Hirosawa, M., Miyajima, N., Tanaka, A., Kotani, H., Nomura, N., and Ohara, O. (1998). Prediction of the coding sequences of unidentified human genes. XII. The complete sequences of 100 new cDNA clones from brain which code for large proteins in vitro. *DNA Res.* **5**, 355–364.
48. Samuel, C. E. (2001). Antiviral actions of interferons. *Clin. Microbiol. Rev.* **14**, 778–809.
49. Polson, A. G., Bass, B. L., and Casey, J. L. (1996). RNA editing of hepatitis delta virus antigenome by dsRNA-adenosine deaminase. *Nature* **380**, 454–456.
50. Lei, M., Liu, Y., and Samuel, C. E. (1998). Adenovirus VAI RNA antagonizes the RNA-editing activity of the ADAR adenosine deaminase. *Virology* **245**, 188–196.

51. Chen, C. X., Cho, D. S., Wang, Q., Lai, F., Carter, K. C., and Nishikura, K. (2000). A third member of the RNA-specific adenosine deaminase gene family, ADAR3, contains both single- and double-stranded RNA binding domains. *RNA* **6,** 755–767.
52. Lehmann, K. A., and Bass, B. L. (2000). Double-stranded RNA adenosine deaminases ADAR1 and ADAR2 have overlapping specificities. *Biochemistry* **39,** 12875–12884.
53. Wang, Q., Khillan, J., Gadue, P., and Nishikura, K. (2000). Requirement of the RNA editing deaminase ADAR1 gene for embryonic erythropoiesis. *Science* **290,** 1765–1768.
54. Higuchi, M., Maas, S., Single, F. N., Hartner, J., Rozov, A., Burnashev, N., Feldmeyer, D., Sprengel, R., and Seeburg, P. H. (2000). Point mutation in an AMPA receptor gene rescues lethality in mice deficient in the RNA-editing enzyme ADAR2. *Nature* **406,** 78–81.
55. Reenan, R. A. (2001). The RNA world meets behavior: A→I pre-mRNA editing in animals. *Trends Genet.* **17,** 53–56.
56. Wong, S. K., Sato, S., and Lazinski, D. W. (2001). Substrate recognition by ADAR1 and ADAR2. *RNA* **7,** 846–858.
57. Williams, B. R. (1997). Role of the double-stranded RNA-activated protein kinase (PKR) in cell regulation. *Biochem. Soc. Trans.* **25,** 509–513.
58. Kaufman, R. J. (2000). Double-stranded RNA-activated protein kinase (PKR). In "Translational Control of Gene Expression" (N. Sonenberg, J. W. B. Hersey, and M. B. Mathews eds.), pp. 503–527. Cold Spring Harbor Laboratory Press, Cold Spring Harbor, NY.
59. Chu, W. M., Ballard, R., Carpick, B. W., Williams, B. R., and Schmid, C. W. (1998). Potential Alu function: Regulation of the activity of double-stranded RNA-activated kinase PKR. *Mol. Cell Biol.* **18,** 58–68.
60. Bommer, U. A., Borovjagin, A. V., Greagg, M. A., Jeffrey, I. W., Russell, P., Laing, K. G., Lee, M., and Clemens, M. J. (2002). The mRNA of the translationally controlled tumor protein P23/TCTP is a highly structured RNA, which activates the dsRNA-dependent protein kinase PKR. *RNA* **8,** 478–496.
61. Nussbaum, J. M., Gunnery, S., and Mathews, M. B. (2002). The 3′-untranslated regions of cytoskeletal muscle mRNAs inhibit translation by activating the double-stranded RNA-dependent protein kinase PKR. *Nucleic Acids Res.* **30,** 1205–1212.
62. Ben-Asouli, Y., Banai, Y., Pel-Or, Y., Shir, A., and Kaempfer, R. (2002). Human interferon-gamma mRNA autoregulates its translation through a pseudoknot that activates the interferon-inducible protein kinase PKR. *Cell* **108,** 221–232.
63. Mathews, M. B., and Shenk, T. (1991). Adenovirus virus-associated RNA and translation control. *J. Virol.* **65,** 5657–5662.
64. Cuddihy, A. R., Wong, A. H., Tam, N. W., Li, S., and Koromilas, A. E. (1999). The double-stranded RNA activated protein kinase PKR physically associates with the tumor suppressor p53 protein and phosphorylates human p53 on serine 392 in vitro. *Oncogene* **18,** 2690–2702.
65. Kumar, A., Haque, J., Lacoste, J., Hiscott, J., and Williams, B. R. (1994). Double-stranded RNA-dependent protein kinase activates transcription factor NF-kappa B by phosphorylating I kappa B. *Proc. Natl. Acad. Sci. USA* **91,** 6288–6292.
66. Brand, S. R., Kobayashi, R., and Mathews, M. B. (1997). The Tat protein of human immunodeficiency virus type 1 is a substrate and inhibitor of the interferon-induced, virally activated protein kinase, PKR. *J. Biol. Chem.* **272,** 8388–8395.
67. Patel, C. V., Handy, I., Goldsmith, T., and Patel, R. C. (2000). PACT, a stress-modulated cellular activator of interferon-induced double-stranded RNA-activated protein kinase, PKR. *J. Biol. Chem.* **275,** 37993–37998.
68. Parker, L. M., Fierro-Monti, I., and Mathews, M. B. (2001). Nuclear factor 90 is a substrate and regulator of the eukaryotic initiation factor 2 kinase double-stranded RNA-activated protein kinase. *J. Biol. Chem.* **276,** 32522–32530.

69. Jammi, N. V., and Beal, P. A. (2001). Phosphorylation of the RNA-dependent protein kinase regulates its RNA-binding activity. *Nucleic Acids Res.* **29,** 3020–3029.
70. St. Johnston, D., Beuchle, D., and Nusslein-Volhard, C. (1991). Staufen, a gene required to localize maternal RNAs in the Drosophila egg. *Cell* **66,** 51–63.
71. Brendza, R. P., Serbus, L. R., Duffy, J. B., and Saxton, W. M. (2000). A function for kinesin I in the posterior transport of oskar mRNA and Staufen protein. *Science* **289,** 2120–2122.
72. Roegiers, F., and Jan, Y. N. (2000). Staufen: A common component of mRNA transport in oocytes and neurons? *Trends Cell Biol.* **10,** 220–224.
73. Micklem, D. R., Adams, J., Grunert, S., and St Johnston, D. (2000). Distinct roles of two conserved Staufen domains in oskar mRNA localization and translation. *EMBO J.* **19,** 1366–1377.
74. Le, S., Sternglanz, R., and Greider, C. W. (2000). Identification of two RNA-binding proteins associated with human telomerase RNA. *Mol. Biol. Cell* **11,** 999–1010.
75. Mouland, A. J., Mercier, J., Luo, M., Bernier, L., DesGroseillers, L., and Cohen, E. A. (2000). The double-stranded RNA-binding protein Staufen is incorporated in human immunodeficiency virus type 1: Evidence for a role in genomic RNA encapsidation. *J. Virol.* **74,** 5441–5451.
76. Siffroi, J. P., Pawlak, A., Alfonsi, M. F., Troalen, F., Guellaen, G., and Dadoune, J. P. (2001). Expression of the TAR RNA binding protein in human testis. *Mol. Hum. Reprod.* **7,** 219–225.
77. Zhong, J., Peters, A. H., Lee, K., and Braun, R. E. (1999). A double-stranded RNA binding protein required for activation of repressed messages in mammalian germ cells. *Nat. Genet.* **22,** 171–174.
78. Gatignol, A., Buckler-White, A., Berkhout, B., and Jeang, K. T. (1991). Characterization of a human TAR RNA-binding protein that activates the HIV-1 LTR. *Science* **251,** 1597–1600.
79. Cosentino, G. P., Venkatesan, S., Serluca, F. C., Green, S. R., Mathews, M. B., and Sonenberg, N. (1995). Double-stranded-RNA-dependent protein kinase and TAR RNA-binding protein form homo- and heterodimers in vivo. *Proc. Natl. Acad. Sci. USA* **92,** 9445–9449.
80. Patel, R. C., and Sen, G. C. (1998). PACT, a protein activator of the interferon-induced protein kinase, PKR. *EMBO J.* **17,** 4379–4390.
81. Peters, G. A., Hartmann, R., Qin, J., and Sen, G. C. (2001). Modular structure of PACT: Distinct domains for binding and activating PKR. *Mol. Cell. Biol.* **21,** 1908–1920.
82. Reichman, T. W., and Mathews, M. B. NF90 family of double-stranded RNA-binding proteins: Regulators of viral and cellular function. *In* "Handbook of Cellular Signaling" (R. A. Bradshaw and E. Dennis eds.). 3, Ch. 321 Academic Press, San Diego, In press.
83. Reichman, T. W., Muniz, L. C., and Mathews, M. B. (2002). The RNA binding protein nuclear factor 90 functions as both a positive and negative regulator of gene expression in mammalian cells. *Mol. Cell. Biol.* **22,** 343–356.
84. Fung, L. F., Lo, A. K., Yuen, P. W., Liu, Y., Wang, X. H., and Tsao, S. W. (2000). Differential gene expression in nasopharyngeal carcinoma cells. *Life Sci.* **67,** 923–936.
85. Sakamoto, S., Morisawa, K., Ota, K., Nie, J., and Taniguchi, T. (1999). A binding protein to the DNase I hypersensitive site II in HLA-DR alpha gene was identified as NF90. *Biochemistry* **38,** 3355–3361.
86. Tang, J., Kao, P. N., and Herschman, H. R. (2000). Protein-arginine methyltransferase I, the predominant protein-arginine methyltransferase in cells, interacts with and is regulated by interleukin enhancer-binding factor 3. *J. Biol. Chem.* **275,** 19866–19876.
87. Ting, N. S., Kao, P. N., Chan, D. W., Lintott, L. G., and Lees-Miller, S. P. (1998). DNA-dependent protein kinase interacts with antigen receptor response element binding proteins NF90 and NF45. *J. Biol. Chem.* **273,** 2136–2145.

88. Xu, Y. H., and Grabowski, G. A. (1999). Molecular cloning and characterization of a translational inhibitory protein that binds to coding sequences of human acid beta-glucosidase and other mRNAs. *Mol. Genet. Metab.* **68,** 441–454.
89. Brzostowski, J., Robinson, C., Orford, R., Elgar, S., Scarlett, G., Peterkin, T., Malartre, M., Kneale, G., Wormington, M., and Guille, M. (2000). RNA-dependent cytoplasmic anchoring of a transcription factor subunit during Xenopus development. *EMBO J.* **19,** 3683–3693.
90. Duchange, N., Pidoux, J., Camus, E., and Sauvaget, D. (2000). Alternative splicing in the human interleukin enhancer binding factor 3 (ILF3) gene. *Gene* **261,** 345–353.
91. Schumacher, J. M., Lee, K., Edelhoff, S., and Braun, R. E. (1995). Spnr, a murine RNA-binding protein that is localized to cytoplasmic microtubules. *J. Cell Biol.* **129,** 1023–1032.
92. Pires-daSilva, A., Nayernia, K., Engel, W., Torres, M., Stoykova, A., Chowdhury, K., and Gruss, P. (2001). Mice deficient for spermatid perinuclear RNA-binding protein show neurologic, spermatogenic, and sperm morphological abnormalities. *Dev. Biol.* **233,** 319–328.
93. Nourbakhsh, M., and Hauser, H. (1999). Constitutive silencing of IFN-beta promoter is mediated by NRF (NF-kappaB-repressing factor), a nuclear inhibitor of NF-kappaB. *EMBO J.* **18,** 6415–6425.
94. Aravind, L., and Koonin, E. V. (1999). G-patch: A new conserved domain in eukaryotic RNA-processing proteins and type D retroviral polyproteins. *Trends Biochem. Sci.* **24,** 342–344.
95. Grishin, N. V. (1998). The R3H motif: A domain that binds single-stranded nucleic acids. *Trends Biochem. Sci.* **23,** 329–330.
96. Sun, C. T., Lo, W. Y., Wang, I. H., Lo, Y. H., Shiou, S. R., Lai, C. K., and Ting, L. P. (2001). Transcription repression of human hepatitis B virus genes by negative regulatory element-binding protein/SON. *J. Biol. Chem.* **276,** 24059–24067.
97. Chen, J., Vijayakumar, S., Li, X., and Al-Awqati, Q. (1998). Kanadaptin is a protein that interacts with the kidney but not the erythroid form of band 3. *J. Biol. Chem.* **273,** 1038–1043.
98. Hubner, S., Jans, D. A., Xiao, C. Y., John, A. P., and Drenckhahn, D. (2002). Signal- and importin-dependent nuclear targeting of the kidney anion exchanger 1-binding protein kanadaptin. *Biochem. J.* **361,** 287–296.
99. Li, J., Lee, G. I., Van Doren, S. R., and Walker, J. C. (2000). The FHA domain mediates phosphoprotein interactions. *J. Cell Sci.* **113,** 4143–4149.
100. Solinger, J. A., Pascolini, D., and Heyer, W. D. (1999). Active-site mutations in the Xrn 1p exoribonuclease of Saccharomyces cerevisiae reveal a specific role in meiosis. *Mol. Cell. Biol.* **19,** 5930–5942.
101. Bashkirov, V. I., Scherthan, H., Solinger, J. A., Buerstedde, J. M., and Heyer, W. D. (1997). A mouse cytoplasmic exoribonuclease (mXRN1p) with preference for G4 tetraplex substrates. *J. Cell Biol.* **136,** 761–773.
102. Aufsatz, W., Mette, M. F., Van Der Winden, J., Matzke, A. J., and Matzke, M. (2002). RNA-directed DNA methylation in Arabidopsis. *Proc. Natl. Acad. Sci. USA* **99**(Suppl. 4), 16499–16506.
103. Jacobs, B. L. (2000). Translational control in poxvirus-infected cells. *In* "Translational Control of Gene Expression" (N. Sonenberg, J. W. B. Hersey, and M. B. Mathews eds.), pp. 951–971. Cold Spring Harbor Laboratory Press, Cold Spring Harbor, NY.
104. Rebouillat, D., and Hovanessian, A. G. (1999). The human 2′, 5′-oligoadenylate synthetase family: Interferon-induced proteins with unique enzymatic properties. *J. Interferon Cytokine Res.* **19,** 295–308.

105. Yue, Z., and Shatkin, A. J. (1996). Regulated, stable expression and nuclear presence of retrovirus double-stranded RNA-binding protein sigma3 in HeLa cells. *J. Virol.* **70**, 3497–3501.
106. Hatada, E., and Fukuda, R. (1992). Binding of influenza A virus NS1 protein to dsRNA in vitro. *J. Gen. Virol.* **73**, 3325–3329.
107. Brownawell, A. M., and Macara, I. G. (2002). Exportin-5, a novel karyopherin, mediates nuclear export of double-stranded RNA binding proteins. *J. Cell Biol.* **156**, 53–64.
108. Kumar, K. U., Srivastava, S. P., and Kaufman, R. J. (1999). Double-stranded RNA-activated protein kinase (PKR) is negatively regulated by 60S ribosomal subunit protein L18. *Mol. Cell. Biol.* **19**, 1116–1125.

Mending the Break: Two DNA Double-Strand Break Repair Machines in Eukaryotes

Lumir Krejci, Ling Chen,
Stephen Van Komen, Patrick
Sung, and Alan Tomkinson

*Department of Molecular Medicine and
Institute of Biotechnology, University of
Texas Health Science Center at San
Antonio, San Antonio, Texas 78245*

I. Introduction	160
II. Biological Relevance of the DSB	161
A. Immunoglobulin Gene Switching	161
B. Meiotic Breaks	161
C. Mating Type Switching	162
III. Genetic Pathways for Homologous Recombination	162
A. Initiation of Homologous Recombination	162
B. Processing of the D-Loop: Recombination Models	164
IV. Recombination Genes: The *RAD52* Epistasis Group	166
A. DNA End Processing	167
B. Heteroduplex DNA Formation	169
C. Break-Induced DNA Replication	177
D. Single-Strand Annealing	178
E. Recombination Machinery in Higher Organisms	178
V. General Introduction to NHEJ	178
VI. Mechanisms and Function of NHEJ in Eukaryotes	179
A. *S. cerevisiae* NHEJ Genes	181
B. Molecular Mechanisms of NHEJ in *S. cerevisiae*	182
C. Mammalian NHEJ Genes	185
D. Molecular Mechanisms of NHEJ in Mammals	187
References	191

DNA double-strand breaks (DSBs) pose a special challenge for cells in the maintenance of genome stability. In eukaryotes, the removal of DSBs is mediated by two major pathways: homologous recombination (HR) and nonhomologous endjoining (NHEJ). Capitalizing on existing genetic frameworks, biochemical reconstitution studies have begun to yield insights into the mechanistic underpinnings of these DNA repair reactions.

I. Introduction

Our genome is vulnerable to injury inflicted by high-energy radiations, chemical agents, and also reactive intermediates that arise during cellular metabolism. In addition, stalling of DNA replication forks at certain DNA structures (e.g., DNA palindrome) or at bulky DNA lesions (e.g., ultraviolet-light-induced photoproducts) and slippage of DNA polymerases will give rise to discontinuities in the DNA template that pose a potential danger to genome integrity (1, 2). Furthermore, a preexisting nick in DNA, which arises when cells attempt to remove oxidative or other types of DNA base damage, can lead to the formation of a double-strand break upon encountering a DNA replication fork (3). Under these unsavory circumstances, cells promptly activate DNA damage checkpoints to momentarily halt cell cycle progression and summon DNA repair or replication restart machinery to fix the damaged DNA or rescue the stalled DNA replication forks (4–7). As seen in genetic studies in the yeast *Saccharomyces cerevisiae*, the failure to activate the checkpoint or repair mechanisms in these times of crisis results in high mutation rates and gross chromosome rearrangements characteristic of cancer cells (8). In fact, several well-documented cancer-prone diseases, including xeroderma pigmentosum, ataxia telangiectasia, Nijmegen breakage syndrome, and certain forms of hereditary breast and colon cancers, have been linked to defects in DNA damage checkpoint or repair mechanisms (8–10). Taken together, the available evidence makes a compelling case that the coordinated efforts of DNA damage checkpoint responses and DNA repair/replication fork restart pathways are critical for the suppression of chromosomal rearrangements and the prevention of cell death, cell transformation, and cancer formation.

Nucleotide base damages and intrastrand DNA cross-links can simply be excised from the DNA strand. The resulting gap is filled in by a DNA polymerase without much danger of alteration in the genetic information. However, the DSB presents a particular challenge for cells in terms of preservation of genomic integrity, because in this particular lesion, both of the DNA strands are damaged simultaneously. Indeed, mishandling of DSBs can lead to chromosome fragmentation, deletion, and translocation, causing irreversible alterations to a cell's genomic configuration and possibly cell death (8).

A series of reviews on the checkpoint mechanisms and the role of DNA checkpoints and DNA repair pathways in the maintenance of genomic stability have recently appeared (1, 8, 11, 12). This article focuses on the two major DSB repair pathways employed in eukaryotic cells—homologous recombination (HR) and nonhomologous DNA end joining (NHEJ). We shall first briefly review the genetic and conceptual frameworks for HR and NHEJ and then discuss the mechanistic aspects of these two repair pathways.

II. Biological Relevance of the DSB

As mentioned above, DSBs are induced by ionizing radiation, arise as intermediates of repair reactions, and occur during the course of DNA replication. The DSBs in these circumstances are perceived as lesions and dealt with promptly using one of the two repair mechanisms available. Interestingly, as briefly outlined below, DSBs also appear in a programmed fashion as an obligatory intermediate during certain biological processes.

A. Immunoglobulin Gene Switching

The vertebrate immune system uses V(D)J recombination to generate the diversity of immunoglobulins and T cell receptors. During this process, DSBs are formed at specific recombination signal sequences (RSSs) by the Rag1/Rag2 protein complex (13). The cleavage results in blunt signal ends and hairpin coding ends. Whereas the signal ends can be joined directly, the coding ends need to be opened first. A pair of NHEJ factors, Artemis/DNA-PKcs, likely carry out this hairpin opening process, which we will discuss in more detail later. The subsequent processing at the opened coding ends further increases the variety of joined products. The completion of V(D)J recombination is achieved through functions of NHEJ factors (14). Upon the interaction with antigen, the affinity and specificity of the different immunoglobulins generated by V(D)J recombination are increased through somatic hypermutation, a process termed "affinity maturation" (15). Furthermore, another event, class switching, alters the constant region of the immunoglobulin (16). Recent studies have shown the creation of DSBs in somatic hypermutation (17, 18); whether these DSBs are definite intermediates in this process remains to be determined. In contrast, class switching recombination clearly involves the concerted generation of two DSBs at the switch regions (16). The recently identified activation-induced cytidine deaminase (AID) has been suggested to function in initiating DSB formation in class switching recombination (19, 20).

B. Meiotic Breaks

In *S. cerevisiae*, DSBs appear during meiosis after DNA replication. These meiotic DSBs are found at hotspots that are present in each of the 16 chromosomes (21). Formation of these breaks is mediated by a protein complex containing the topoisomerase II-like protein, Spo11, which is thought to introduce the DSBs at meiotic hotspots via a transesterification mechanism. Formation of the meiotic DSBs leads to recombination between chromosomal homologues to provide stable linkage between them until time for their segregation in meiosis I. Accordingly, inactivation of the machinery that makes the meiotic DSBs not only abolishes recombination, but also results in chromosomal nondisjunction during meiosis I. The mechanism for the formation of meiotic DSBs and the function of meiotic recombination are both highly conserved through evolution (22–26). The role of DSBs in meiotic chromosome metabolism is the subject of a number of recent reviews (21, 23, 27–29).

C. Mating Type Switching

The switching of mating type in the yeast *S. cerevisiae* is triggered by the introduction of a site-specific DSB at the *MAT* locus on chromosome 3. Once formed, the DSB break is utilized by the recombination machinery for conducting a gene conversion event that results in the switching of one mating type to the other (a or α) via one of the gene cassettes at *HML* and *HMR*, which store the a and α mating type information, respectively. The high specificity of the HO endonuclease that makes the site-specific DSB at *MAT* has been used with great success in model systems for delineating the temporal order of events during DSB repair (30).

III. Genetic Pathways for Homologous Recombination

A. Initiation of Homologous Recombination

Studies in *S. cerevisiae* have revealed that in recombination events induced by a DSB, the ends of the DNA break are processed by a nuclease activity to expose single-stranded (ss) tails of a considerable length, typically a few hundred bases (Step 1 in Fig. 1). The ssDNA tails serve as the substrate for the recruitment of the recombination machinery, which assembles on these tails and then conducts a search for a chromosomal homologue. Subsequently, a DNA joint that links the recombining DNA molecules is formed and genetic information is transferred from the donor (the intact DNA molecule) to the recipient (the DNA molecule that has sustained the DSB). During mitotic growth, the chromosomal donor is most frequently the sister chromatid, and hence as a DNA repair tool, homologous recombination is most useful during the S and G_2 phases of the cell cycle. In fact, the cell cycle stage appears to be a major determinant as to whether a DSB is repaired by a homology-directed or an end-joining mechanism (31). In contrast, meiotic cells almost exclusively use the homologous chromosome as the donor to repair the Spo11-associated DSBs. Genetic loci that help determine meiotic recombination partner choice (i.e., exclusive usage of the homologous chromosome as recombination partner) have been described, and mutations in these genes compromise meiotic chromosome transmission, resulting in aneuploidy and death (29).

Regardless of whether the sister chromatid or homologous chromosome is used as the recombination partner, the first DNA joint molecule that forms

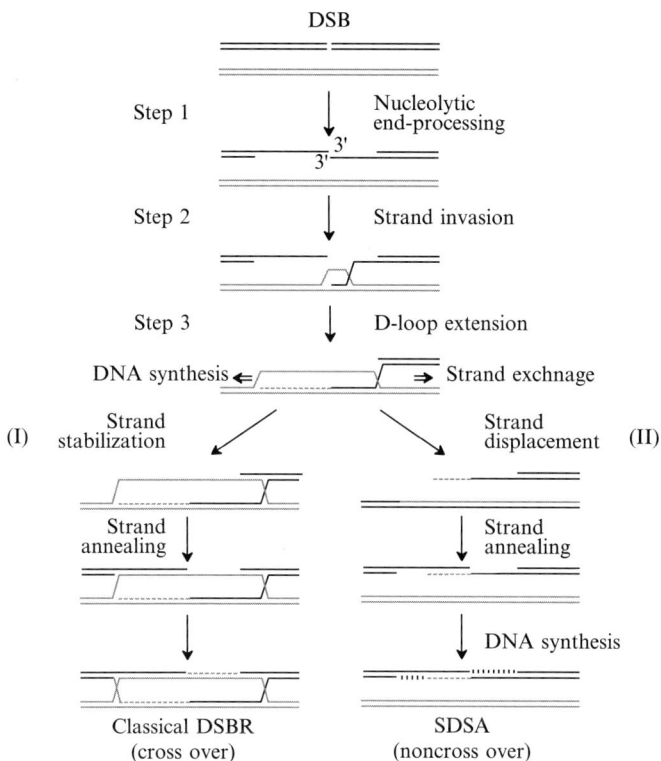

FIG. 1. Recombination models. The DSB is resected to yield 3′ single-stranded (ss) tails (Step 1). One of the ssDNA tails invades the DNA homologue to form a D-loop (Step 2). The D-loop is extended by continual strand exchange and by *de novo* DNA synthesis. The D-loop can be processed through two separate pathways. In one case (classic DSB repair mechanism in Panel I), the second ssDNA tail anneals with the DNA strand displaced by the expanding D-loop structure, leading to the formation of a second Holliday junction. Alternatively (SDSA mechanism in Panel II), the invading DNA strand dissociates from the donor DNA molecule and then anneals with the second ssDNA tail. Gap filling DNA synthesis and ligation then complete the process.

between one of the initiating ssDNA tails (product of the nucleolytic end-processing reaction: Step 1 in Fig. 1) and the undamaged DNA homologue is a structure called D-loop (Step 2 in Fig. 1). Subsequent to its formation, the size of the D-loop is expanded by continual uptake of the initiating ssDNA tails into the DNA homologue, with concomitant displacement of the like strand in the DNA homologue (Step 3 in Fig. 1). In widely accepted enzymological terms, the process responsible for formation of the initial DNA joint (D-loop) is commonly referred to as "homologous DNA pairing," while the extension of the D-loop is called "DNA strand exchange" or "DNA branch migration." Accordingly, the overall biochemical reaction that leads to the formation and extension of DNA joints during recombination has been termed "homologous DNA pairing and strand exchange."

B. Processing of the D-Loop: Recombination Models

Concomitant with homologous DNA pairing and strand exchange, DNA synthesis initiates from the extremity of the invading single strand that is now paired with the homologous strand in the donor molecule. As a result, extension of the D-loop occurs not only by way of DNA strand exchange, but also as a consequence of *de novo* DNA synthesis (Fig. 1, Step 3). Most current recombination models picture the initial step of DNA–DNA interactions as being one-ended, i.e., only one of the ssDNA tails that arise through nucleolytic end processing is utilized for promoting DNA joint formation. As the D-loop expands in size because of DNA synthesis, the other single-stranded DNA tail may also become engaged in the recombination process, resulting in a DNA intermediate that contains two crossover structures, better known as Holliday junctions. This sequence of events is envisioned in the recombination model proposed by Szostak *et al.* in 1983 (32). The presence of the Holliday structure in the DNA joint molecule allows crossover recombination products (i.e., recombinants that harbor reciprocal exchange of genetic information), as opposed to simple gene conversion products wherein the information is nonreciprocally transferred. The available evidence is consistent with the premise that crossovers are critical for proper disjunction (i.e., segregation) of the homologous chromosome pairs during the first meiotic division (i.e., meiosis I). For additional information on the subject of meiotic recombination, the reader is referred to a number of highly informative articles (23, 27, 29, 33–35).

1. SDSA

Interestingly, investigations in yeast and other organisms have provided compelling evidence that the D-loop intermediate is not always processed via the reaction sequence envisioned in the 1983 Szostak *et al.* recombination model (32). For instance, it is now well established that crossover recombination products are relatively rare in mitotic cells. In fact, even during meiosis, only a fraction the DNA joint molecules mature into the double crossover structure depicted in panel I of Fig. 1. The available data point to a prevalent mitotic recombination pathway that involves the dissociation of the invading single strand from the homologue and its hybridization to the other single strand derived from the end-processing reaction. Gap filling DNA synthesis and ligation then complete the recombination process. This type of recombination has been termed DNA synthesis-dependent single-strand annealing, or SDSA (Fig. 1, panel II). Variants of the SDSA model have been discussed (36).

2. BREAK-INDUCED REPLICATION (BIR)

Here, a D-loop is again made but is extended primarily by DNA synthesis without much heteroduplex being formed. The length of the newly synthesized DNA can cover the entire length of the donor chromosome, resulting in really long gene conversion tracts (Fig. 2). It is important to emphasize that the BIR mode of recombination is not merely an extension or variation of the SDSA pathway, as it can occur in the absence of some recombination proteins, most notably Rad51, that are indispensable for the latter pathway. In fact, BIR seems to rely on a trio of proteins, Rad50, Mre11, and Xrs2, which do not appear to have any significant involvement in the Rad51-dependent pathway of D-loop formation. Even though the majority of cellular recombination events are carried out by Rad51-mediated pathways, the Rad50/Mre11/Xrs2-dependent BIR reaction appears to make important contributions to certain biological processes. For instance, Kolodner has proposed that BIR, but not Rad51-mediated recombination, is chiefly responsible for the suppression of gross chromosomal rearrangements that could arise due to delinquent DNA replication forks (8, 37). In addition, the available evidence has implicated BIR as an important mechanism (as is the Rad51-dependent

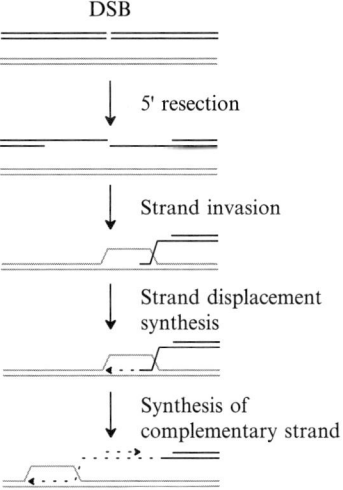

FIG. 2. Recombination by BIR. A 3′ ssDNA tail invades the donor chromosome to form a short DNA joint, which is then stabilized by DNA synthesis. The invading DNA strand continually dissociates from the donor chromosome as DNA synthesis continues. Second strand DNA synthesis then initiates on the nondissociated invading strand. This recombination process can result in the copying of a substantial portion of the donor chromosome, resulting in gene conversion tracts that are many kilobases in length. Alternative models of BIR have been discussed (107).

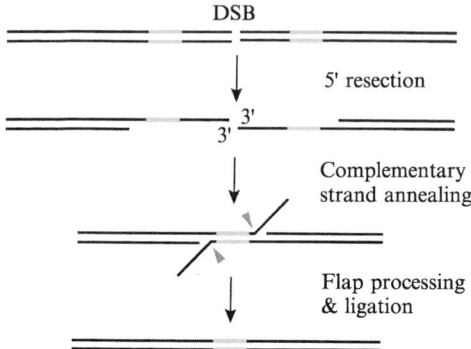

FIG. 3. Recombination between direct repeats by single-strand annealing. Resection of the ends of a DSB located between two direct DNA repeats (light boxes) results in 3′ ssDNA tails that can hybridize within the homologous region. Following trimming of the flap structure, gap-filling DNA synthesis and ligation lead to a recombinant that has one of the DNA repeats and the intervening DNA sequence deleted.

recombination pathway) for the elongation of shortened telomeres in cells that are defective in telomerase function (30, 38–40). The biology of BIR and variants of the BIR model depicted in Fig. 2 are discussed in a recent review by Paques and Haber (30). The players in BIR, some of which also function in the Rad51-mediated recombination pathways, will be mentioned below.

3. RECOMBINATION BY SINGLE-STRAND ANNEALING

Recombination by single-strand annealing (SSA) is an efficient process that occurs between directly repeated DNA sequences. In this type of recombination, regions of homology in the 3′ single-stranded tails originating from the end-processing reaction hybridize to form a DNA joint, followed by the trimming of the nonhomologous overhangs, fill-in DNA synthesis, and ligation (Fig. 3). The end result is a recombinant that has one of the DNA repeats and the intervening nonhomologous DNA sequence deleted. Genetic analyses have revealed that only a subset of the factors that function in general homologous recombination events is needed for single-strand annealing. We shall return to this subject when discussing the biochemical requirements of the SSA reaction below.

IV. Recombination Genes: The *RAD52* Epistasis Group

Defects in homologous recombination ablate a major pathway of DSB repair and hence render cells sensitive to break-inducing agents, e.g., ionizing radiation. This phenotypic manifestation has been exploited with great success in the isolation of the majority of genes that are essential for homologous recombination and homology-directed DNA repair. Complementation assays using specific recombination substrates, yeast two-hybrid analyses, and protein homology-based computer searches have helped identify additional recombination genes. The recombination/DSB repair genes isolated using the aforementioned approaches—*RAD50, RAD51, RAD52, RAD54, RAD55, RAD57, RAD59, RDH54/TID1, MRE11*, and *XRS2*—are collectively known as the *RAD52* epistasis group.

Based on the conservation of these factors during evolution, we anticipate that the mechanistic studies with yeast recombination factors will provide a conceptual framework for the same pathways in higher eukaryotes. Below is a summary of our current knowledge on the biochemical properties of the *RAD52* group proteins and the hierarchy of physical and functional interactions among them. The biochemical attributes of functionally homologous recombination factors from *E. coli*, yeast, and mammals are summarized in Table I.

A. DNA End Processing

The trio of genes—*RAD50, MRE11*, and *XRS2*—are unique among the *RAD52* group because of their multifunctional nature. Rad50 protein is a member of the SMC protein family, possessing telltale coiled coils, a weak ATPase activity, and ATP-stimulated DNA-binding activity (*12, 41, 42*). Mre11 protein has 3′ to 5′ exonuclease and DNA structure-specific endonuclease activities (*43–45*). The least characterized component of this trio is Xrs2. Through interactions of Mre11 with Rad50 and Xrs2, these three proteins form a stable complex with a stoichiometry of Rad50:Mre11:Xrs2 of 2:2:1 (*46, 47*). Optimal nuclease activity of Mre11 is contingent upon complex formation with Rad50 and Xrs2/Nbsl (*44, 45*).

Null mutants of the *RAD50, MRE11*, and *XRS2* genes are not only defective in the nucleolytic processing of meiotic DSBs made by Spo11, but also in the formation of these breaks. Thus the nuclease activity of Mre11 is critical for meiotic DSB formation and processing. In mitotic cells, other nucleases appear to substitute for this activity, as nucleolytic processing of DNA ends still occurs, albeit at a reduced rate in the *mre11* null mutant. Because the 3′ to 5′ exonuclease activity of Mre11 cannot be solely responsible for the creation of 3′ single-stranded tails from DNA ends, it has been postulated that its DNA structure-specific endonuclease activity is germane for this processing reaction. In this regard, the Rad50/Mre11/Xrs2 complex may target a DNA helicase to the DNA ends to initiate DNA strand separation, followed by clippage of the 5′ overhanging single strand by the DNA structure-specific endonuclease activity of Mre11 (*45*). Another, perhaps more popular, premise is that the Rad50/Mre11/Xrs2 complex recruits a nuclease (or nucleases) to help

TABLE I
BIOCHEMICAL PROPERTIES OF RECOMBINATION FACTORS

Protein	Biochemical function	E. coli homologue	Human homologue	Features
Rad50	DNA-binding and ATPase	SbcC	hRad50	Member of SMC family; interacts with Mre11 and Xrs2
Mre11	Endonuclease and 3' to 5' exonuclease	SbcD	hMre11	Homology to phosphoesterases; interacts with Rad50 and Xrs2
Xrs2	Not known	None	NBS1	Interacts with Mre11, Rad50, and Lif1
Rad51	ATP-dependent homologous DNA pairing and strand exchange	RecA	hRad51	Forms nucleoprotein filaments; interacts with Rad52, Rad54, and Rad55/Rad57
Rad52	ssDNA binding and annealing	None	hRad52	Mediator of strand exchange; single-strand annealing activity
Rad55 and Rad57	ssDNA binding	None	XRCC2 XRCC3 Rad51B Rad51C Rad51D	Form heterodimer; mediator of strand exchange
Rad54	DNA-dependent ATPase	None	hRad54 hRad54B	Member of Swi2/Snf2 family; promotes homologous DNA pairing by Rad51; DNA tracking/supercoiling function
Rdh54/Tid1	DNA-dependent ATPase	None	hRad54 hRad54B	Member of Swi2/Snf2 family; promotes homologous DNA pairing by Rad51; interacts with Dmc1 and Rad51
Rad59	ssDNA binding and annealing	None	Not known	Homology to Rad52; required for single-strand annealing and BIR
RPA	ssDNA binding	SSB	hRPA	Removes secondary structure in ssDNA; sequesters ssDNA during homologous pairing and strand exchange

process the ends of DSBs. This latter notion stems from the observations that in cells harboring certain nuclease null alleles of *MRE11*, nucleolytic end processing remains unaffected. However, it will be quite important to examine the products of the various presumed nuclease null alleles of *MRE11* for nucleolytic activities in conjunction with Rad50 and Xrs2, to be sure that they are indeed completely defective in nuclease function. Until more data become available, the *in vivo* role of the Mre11 nuclease function in DNA end processing in mitotic cells will undoubtedly remain controversial.

Remarkably, in addition to its role in DNA end processing in mitotic and meiotic cells, this trio of genes has been shown to participate in the tolerance of specific DNA structures, DNA damage checkpoint signaling, telomere maintenance, BIR, and NHEJ (*12, 48, 49*). Mutations in two components of the equivalent complex in humans, namely, Mre11 and Nbs1 (Xrs2 equivalent), give rise to the ataxia telangiectasia-like disease and Njimegen breakage syndrome, respectively. These ailments are marked by abnormal DNA damage checkpoint responses, radiation sensitivity, chromosomal fragility, and an elevated cancer incidence in afflicted individuals (*12*). Considerable structural insights into Rad50, Mre11, and the Rad50/Mre11 complex have been garnered through crystallographic studies and analyses involving electron microscopy and scanning force microscopy (*50–54*). The role of the yeast Rad50/Mre11/Xrs2 complex in the DNA end-joining reaction mediated by the Dnl4/Lif1 and Hdf1/Hdf2 complexes will be discussed below. A number of recent articles have exhaustively reviewed the general biology and advances in understanding the structure and function of this trio of factors and their complexes (*12, 42, 49*).

B. Heteroduplex DNA Formation

1. Rad51, the RecA Homologue

Mutants of *RAD51* display the classic phenotype of a gene critical for recombination, including sensitivity to ionizing radiation and alkylating agents, and also defects in mitotic and meiotic recombination. Three groups reported the cloning and characterization of *RAD51* in 1992 (*55–57*). Importantly, sequence alignment revealed that the *RAD51* gene product is structurally related to the *E. coli* RecA protein, which plays a central role in recombination via its ability to promote the homologous DNA pairing and strand-exchange reaction. Rad51 was later shown to form helical filaments on dsDNA (*58*) and ssDNA (*59, 60*). The demonstration of Rad51 nucleoprotein filaments reinforced the presumption that Rad51 is a true functional homologue of RecA, as the latter carries out its biochemical functions in the context of a nucleoprotein filament (*61*).

That Rad51 possesses a homologous DNA pairing and strand-exchange activity was first reported in 1994 (*62*). This study utilized Rad51 purified from yeast cells, the heterotrimeric single-strand DNA-binding protein replication protein A (RPA, whose role will be discussed below), and an assay system first devised for use in RecA studies (62). Subsequently, recombinase activity was found in the human Rad51 protein as well (63–65). Together, the biochemical studies with yeast and human Rad51 have firmly established that this protein is a bona fide homologue of RecA.

2. Assay Systems for Homologous DNA Pairing and Strand Exchange

During homologous recombination, a single-stranded DNA tail is used by the recombination machinery to invade a homologous duplex to form a DNA joint (Step 2 in Fig. 1). This principle has guided the development of *in vitro* recombination assays, which all entail the use of a single-stranded DNA molecule as initiating substrate and a homologous duplex molecule as the pairing partner. The three most commonly used *in vitro* systems germane for our discussions are described in Fig. 4. Extensive biochemical studies with *E. coli* RecA have been instrumental for defining three kinetically distinct phases in

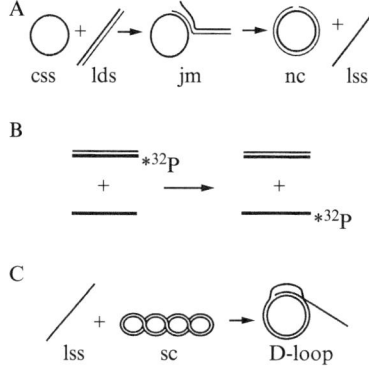

FIG. 4. *In vitro* systems for examining homologous DNA pairing and strand exchange. (A) In this system, circular single-stranded DNA (css) from a bacteriophage, typically ϕX174, is paired with the homologous linear duplex (lds) to yield a joint molecule (jm). DNA strand exchange, if successful over the entire length of the DNA molecules (5.4 kilobase pairs for ϕX DNA), generates a nicked circular duplex (nc) and a linear ssDNA (lss) as products. (B) The second system for characterizing the recombinase activity of Rad51 utilizes an oligonucleotide as the initiating substrate. The joint molecule that results from pairing between the oligonucleotide and the homologous duplex is rapidly resolved by DNA strand exchange, leading to the release of the ^{32}P-labeled strand from the duplex. Because the substrates are short, this system has its utility limited to studying homologous pairing. (C) This system has been specifically designed to study D-loop formation. Herein, a linear single strand (lss), either an oligonucleotide or a plasmid-length DNA molecule, invades a covalently closed, typically negatively supercoiled, duplex molecule (sc) to form a D-loop. Biochemical studies have revealed that Rad51 needs the cooperation of Rad54 or Rdh54 to make D-loop efficiently.

the homologous DNA pairing and strand-exchange reaction: presynapsis, synapsis, and DNA branch migration (2, 66). The homologous DNA pairing and strand-exchange characteristics of Rad51 will be discussed in the context of these three reaction phases.

a. The Presynaptic Phase. Rad51 nucleates onto ssDNA to form a helical protein filament, often referred to as the Rad51–ssDNA nucleoprotein filament or presynaptic filament. The presynaptic filament holds the DNA in an extended conformation. Each helical repeat in the presynaptic filament contains ~18 bases of ssDNA and ~6 Rad51 molecules (58, 59). Because Rad51 has an ATPase activity that is greatly stimulated by ssDNA (62), the filament assembly process can be conveniently followed by measuring ATP hydrolysis (67). Alternatively, presynaptic filament assembly has been studied by direct visualization with the electron microscope and by DNA mobility shift in agarose gels (59, 60, 68, 69). There are two distinct DNA-binding sites in the presynaptic filament: the ssDNA is situated within the primary site and the incoming duplex molecule is bound within the secondary site. The ability to hold three DNA strands in proximity underlies the ability of the presynaptic filament to mediate homologous DNA pairing and strand exchange.

Formation of the presynaptic filament needs adenosine triphosphate (ATP) (58–60). However, analogues of ATP—ATP-γ-S and AMP-PNP—that are non-hydrolyzable (or only slowly hydrolyzable) can support a substantial level of homologous DNA pairing and strand exchange (70), indicating that assembly of a functional presynaptic filament can occur with little or no ATP hydrolysis. The role of ATP binding and hydrolysis in presynaptic filament assembly has been further investigated by making mutants of yeast Rad51 that either binds but does not hydrolyze ATP (rad51 K191R) or fails to bind ATP (rad51 K191A). The rad51 K191R mutant protein can perform homologous DNA pairing and strand exchange *in vitro* and complements the DNA repair defects of a *rad51* null mutant (70, 71). However, an increased quantity of the rad51 K191R mutant protein is needed (57, 70, 72). Similarly, when an ATP analogue (ATP-γ-S or AMP-PNP) is used, more wild-type Rad51 is required to achieve a significant level of homologous pairing and strand exchange (70). It thus remains possible that ATP hydrolysis influences the efficiency of assembly of the presynaptic filament or its maintenance without being an absolute requirement (70). As expected, the rad51 K191A mutant protein is incapable of DNA binding and is therefore defective in DNA pairing and strand exchange. Genetically, the *rad51 K191A* allele behaves like the null mutant, thus firmly establishing the requirement for nucleotide binding in Rad51 functions. It should be noted that studies on variants of human Rad51 (hrad51 K133A and hrad51

K133R) harboring the equivalent mutations reached the same conclusion that ATP binding alone is sufficient for the recombinase activity of this factor (73).

b. Synapsis. This reaction phase entails the search for DNA homology in the donor duplex and the formation of nascent DNA joints between the initiating ssDNA and the duplex. In the case of RecA, and likely with Rad51 also, the search for DNA homology in the duplex molecule appears to proceed by way of reiterative collisions between the presynaptic filament and the duplex (66). In other words, the duplex molecule is incorporated into the presynaptic filament through multiple contact points and lingers within the secondary DNA-binding site of the presynaptic filament for a finite amount of time. The duplex molecule is released if homology is not found and another segment of the duplex is sampled. This kiss-then-release process continues until homology is located in the duplex to initiate the homologous pairing process. Conceptually, for the DNA homology search process to proceed efficiently, the duplex–presynaptic filament interactions should be transient. This is clearly illustrated in the case of the presynaptic filament formed with human Rad51, where weakening of its interactions with the duplex DNA by the inclusion of relatively high levels of salts greatly stimulates the efficiency of homologous DNA pairing and strand exchange (65).

Once homology is located within the duplex molecule, DNA joint formation with the ssDNA becomes possible. The presynaptic filament is capable of making DNA joints that are either paranemic or plectonemic in nature. The paranemic joints occur within the internal regions of the recombining DNA molecules, whereas the plectonemic joints are formed at a free end located either in the ssDNA or the duplex substrate (see later). There is considerable evidence to indicate that the paranemic joints are held together by canonical Watson–Crick hydrogen bonds, but, because the two strands involved in joint formation are not topologically wound around each other, these joints are transient and will rapidly dissociate upon deproteinization. Nonetheless, the paranemic joints are thought to play an important role in bringing the recombining ssDNA and duplex substrate in homologous registry to enhance the likelihood for the formation of a plectonemic joint. Studies in the Radding group have revealed that recognition of homology, helix destabilization, and initiation of DNA joint formation are integral parts of a concerted mechanism in which A:T base pairs play a critical role (74).

With RecA, there is no evidence that the duplex slides along the presynaptic filament during the homology search process, and it seems reasonable to assume the same for Rad51. However, as discussed below, Rad54, which tracks on DNA and physically interacts with Rad51, has been suggested to actively pump the duplex DNA through the fold of the Rad51 presynaptic filament to facilitate the search for DNA homology (see later).

c. DNA Branch Migration (Strand Exchange). After its formation, the nascent plectonemic joint is lengthened in a unidirectional fashion. The extent of this is determined by the length of the Rad51–ssDNA nucleoprotein filament.

3. FACTORS THAT FACILITATE THE RAD51-MEDIATED HOMOLOGOUS DNA PAIRING AND STRAND-EXCHANGE REACTION

a. RPA. RPA, through its ability to bind and facilitate the removal of secondary structure in ssDNA, plays a central role in just about all the DNA metabolic pathways. RPA is indispensable for homologous DNA pairing and strand-exchange efficiency when plasmid-length DNA substrates are used (62, 67). It was originally thought that the only significant role of RPA in homologous pairing and strand exchange was in the promotion of Rad51 presynaptic filament assembly by helping eliminate DNA secondary structure (67, 75). However, recent studies have implicated RPA in at least two other capacities in the formation and stabilization of DNA joints in recombination reactions.

Eggler *et al.* (76) have recently reported that when the Rad51–ssDNA nucleoprotein filament is assembled under low magnesium buffer conditions to minimize secondary structure in ssDNA, RPA is still needed for maximal DNA strand-exchange efficiency with plasmid-length DNA substrates (System A in Fig. 4). Additional studies involving the use of exonuclease digestion of the DNA joint molecules and electron microscopy have yielded compelling evidence that RPA ensures DNA strand-exchange efficiency by sequestering the noncomplementary DNA strand displaced from the duplex substrate as the result of DNA strand exchange. Eggler *et al.* (76) suggested that the RPA-mediated sequestration of the displaced ssDNA ensures that the DNA strand-exchange process is not reversed upon deproteinization of the reaction mixture for gel analysis.

A third role for RPA was recently revealed in the studies of Van Komen *et al.* (77). It has been known for quite some time that in the presence of Rad54, robust homologous DNA pairing can occur with amounts of Rad51 well below what is needed to saturate the ssDNA (i.e., three nucleotides of ssDNA/Rad51 monomer) (69, 78). Interestingly, under these conditions, RPA is still indispensable for pairing efficiency (77). It was demonstrated in this work that protein-free ssDNA inhibits the ability of Rad51 to promote homologous DNA pairing, likely by occupying the secondary DNA binding site in the presynaptic filament, and it competes with duplex DNA for binding to Rad54. By sequestering free ssDNA, RPA alleviates its inhibitory effects on Rad51 and Rad54.

Thus, RPA promotes Rad51-mediated homologous DNA pairing and strand exchange by facilitating the assembly of the presynaptic filament on long ssDNA molecules (59, 67), sequestering protein-free ssDNA (77), and preventing the reversal of DNA strand exchange (76). Paradoxically, because RPA has high avidity for ssDNA, an excess of RPA added with (or before)

Rad51 to the ssDNA substrate results in exclusion of Rad51 and marked suppression of presynaptic filament assembly (65, 67, 79). Specific recombination mediator proteins that promote the nucleation of Rad51 onto ssDNA and even onto an RPA-coated ssDNA template have been identified. The presence of these mediator proteins in the homologous DNA pairing and strand-exchange reaction can effectively overcome the competitive effect imposed by RPA (Fig. 5; see below).

b. Rad52. Rad52 is a multimeric ring-shaped molecule (80, 81) that binds ssDNA avidly (82) and physically interacts with Rad51 (57, 83). In reactions wherein homologous pairing and strand exchange would be otherwise compromised by coincubation of the ssDNA substrate with RPA and Rad51, an amount of Rad52 about one-tenth that of Rad51 fully restores the reaction efficiency (84, 85). Importantly, Rad52 allows Rad51 to utilize an RPA-coated ssDNA template for recombination reactions (75, 85, 86). This effect likely involves a specific interaction between Rad52 and RPA (87). In addition to overcoming the suppressive effect of RPA, Rad52 exerts modest stimulation on the Rad51 recombinase activity in the absence of RPA (75, 86, 88). However, even with Rad52, RPA is still needed for maximal homologous pairing and strand exchange (75, 86). The physical interaction between Rad51 and Rad52 is indispensable for the recombination mediator function of the latter (89).

It should be emphasized that deletion of *RAD52* engenders defects in recombination more severe than those observed in a *rad51* null mutant.

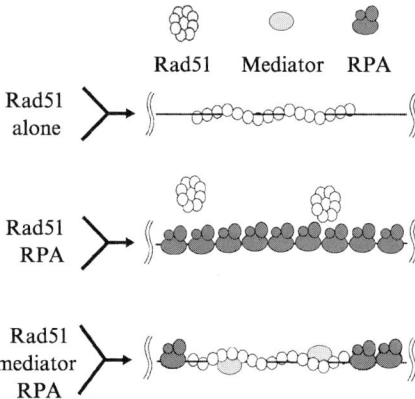

FIG. 5. Function of recombination mediators. RPA has high affinity for ssDNA and can therefore effectively compete with Rad51 for binding to the DNA, resulting in the suppression of presynaptic filament assembly. The recombination mediators help overcome the suppressive effect of RPA.

In addition, mutants of Rad52 lacking the Rad51 interaction domain (89–91) are less impaired for DSB repair and meiosis than the rad52 null mutant (92). These observations provide evidence that Rad52 possesses Rad51-independent functions in recombination reactions. Consistent with this premise, detailed genetic analyses have revealed a key role for Rad52 in SSA and BIR (see below).

c. *Rad55/Rad57.* *RAD55* and *RAD57* genes show a tight epistatic relationship and their encoded products interact in the yeast two hybrid system (93–95). Consistent with these observations, Rad55 and Rad57 have been shown to form a stable heterodimer by coimmunoprecipitation and copurification through a number of chromatographic column steps. Like Rad52, Rad55–Rad57 also has a recombination mediator function, capable of overcoming the competition posed by RPA for binding sites on the initiating ssDNA substrate (Fig. 6). Through a Rad51–Rad55 association (93–95), the Rad55–Rad57 heterodimer interacts with Rad51 in the absence of ssDNA (our unpublished results). Rad55–Rad57 has an ssDNA binding function and a weak ATPase activity (our unpublished results). It seems likely that Rad55–Rad57 delivers Rad51 to the ssDNA template to facilitate the assembly of the presynaptic filament. Whether the Rad55–Rad57 heterodimer can also interact with RPA and specifically recognize an RPA-coated ssDNA template in performing its mediator function has not yet been tested.

d. *Rad54.* Despite its structural and functional similarities to RecA, Rad51 is poorly adept at making a D-loop by itself (78, 96, 97). The addition of Rad54 protein significantly stimulates D-loop formation by Rad51 (78). Rad54 binds Rad51 in both the yeast two-hybrid system and *in vitro* (78, 98, 99). This interaction is likely important for DNA joint formation, as Rad54 has no effect on RecA-mediated recombination reactions *in vitro*.

Rad54 belongs to the Swi2/Snf2 protein family and, like other members of this protein family, has a DNA-dependent ATPase function. This ATPase activity prefers dsDNA as cofactor. Further insights into the role of ATP has been derived from studying Walker mutants that are either defective in nucleotide binding (rad54 K341A) or bind ATP but are inactivated for ATP hydrolysis (rad54 K341R). These *rad54* mutant alleles are defective in haploid-specific mitotic intrachromosomal gene conversion and DNA repair in both haploid and diploid cells (78, 98, 99). Interestingly, the two *rad54* Walker mutants are still capable of carrying out interchromosomal gene conversion in diploid cells (*100*), suggesting that Rad54 has another recombination function that is independent of its ATPase activity.

Rad54 protein has been shown to utilize the free energy from ATP hydrolysis to track along duplex DNA. This tracking motion generates a positively

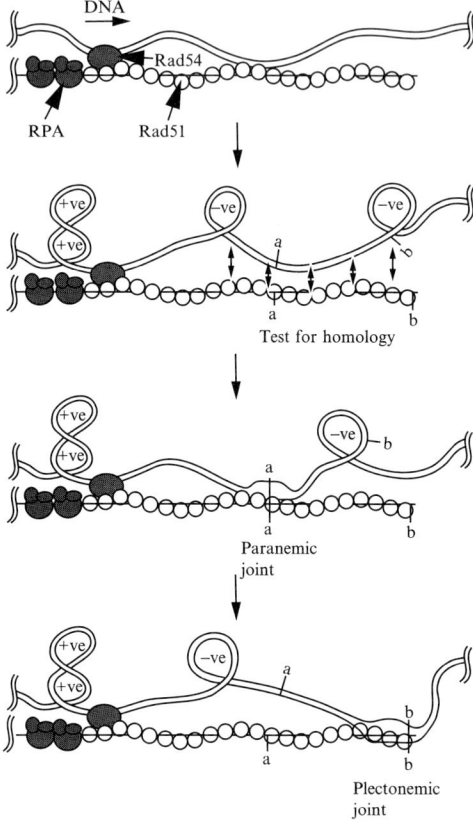

FIG. 6. Cooperation between Rad51 and Rad54 in DNA joint formation. In this model, the incoming duplex is actively pumped through the presynaptic complex (consisting of Rad51, Rad54, RPA, and ssDNA) by Rad54, creating positively and negatively supercoiled domains as depicted. The Rad51–ssDNA nucleoprotein filament samples the negatively supercoiled domain for DNA homology. Once homology is located, formation of a DNA joint molecule ensues. The nascent joint molecule formed will be paranemic in nature if it occurs away from the end of the initiating ssDNA molecule. However, if DNA joint formation occurs near the end of the ssDNA molecule, it has the potential of being converted into a plectonemic linkage.

supercoiled domain ahead of protein movement and a compensatory, negatively supercoiled domain behind (69, 101). The negative supercoils that accumulate lead to transient DNA strand opening. Interestingly, the ATPase, DNA supercoiling, and DNA strand opening activities of Rad54 are greatly stimulated via an interaction with Rad51 (69, 96, 97).

We envision that Rad54 exerts two major effects on the efficiency of the homologous DNA pairing reaction. First, an ability of Rad54 associated with the Rad51–ssDNA nucleoprotein complex to pull the duplex through its fold (i.e., tracking) is expected to enhance the rate at which the duplex molecule is sampled for homology. Second, the transient DNA strand opening within the negatively supercoiled domain generated by Rad54 very likely facilitates the formation of a nascent DNA joint upon DNA homology location in the duplex. This latter suggestion is supported by the demonstration that Rad54 enables Rad51 to utilize even a topologically relaxed DNA substrate for efficient DNA joint formation (69). These ideas are summed up in Fig. 5.

Notably, the DNA tracking/supercoiling function first demonstrated in Rad54 appears to be a conserved property of the Swi2/Snf2 family of proteins (102, 103). It has been suggested that the ability to track on and supercoil DNA underlies the chromatin remodeling function of the various Swi2/Snf2-Snf2-like proteins and complexes that contain them (102, 103).

e. Rdh54/Tid1. A *RAD54*-related gene, *RDH54* (*RAD* Homologue 54), was identified through computer searches of yeast databases (104, 105). The *RDH54* gene was also independently isolated via a yeast two-hybrid screen for genes whose products interact with the meiotic specific recombinase Dmc1 (106), and was named *TID1* (*T*wo-hybrid *I*nteraction with *D*mc1 I). In the same study (106), the *RDH54/TID1*-encoded protein was also found to bind Rad51 in the two-hybrid assay. Consistent with this observation, purified Rdh54/Tid1 physically interacts with Rad51 (69). Like Rad54, Rdh54 has a robust dsDNA-activated ATPase function and an ability to track on and supercoil DNA (69). Importantly, Rdh54/Tid1 greatly stimulates Rad51-mediated D-loop formation (69).

C. Break-Induced DNA Replication

Genetic analyses have revealed that BIR is dependent on *RAD52, RAD59, RDH54,* and the trio of *RAD50, MRE11,* and *XRS2.* Kolodner *et al.* (8) have suggested that BIR is chiefly responsible for the suppression of gross chromosomal rearrangements originating from delinquent DNA replication forks. Furthermore, the same set of BIR factors appears to participate in telomere elongation to yield type II survivors in cells that lack telomerase (107).

Rad59 is related in sequence to the amino-terminal portion of Rad52 and physically interacts with the latter (108, 109). It is therefore likely that Rad52 and Rad59 work in the BIR reaction as a complex. Exactly how Rad52–Rad59 and the products of the other aforementioned genes establish a D-loop structure to initiate the DNA synthesis reaction in BIR is unknown at the moment. However, it seems reasonable to consider the possibility that the combination of Rad52–Rad59, Rad50–Mre11–Xrs2, and Rdh54 can utilize a ssDNA template to invade a duplex template, much like what has been shown for classic recombination events mediated by Rad51, Rad52, Rad55–Rad57, and Rad54.

D. Single-Strand Annealing

SSA requires Rad52 and Rad59 and its efficiency is modulated by RPA in both *in vitro* and *in vivo* settings (*82, 86, 110, 111*). The involvement of Rad52 and Rad59 in SSA is easily explained by the ability of these factors to promote the annealing of complementary DNA strands (*82, 110*). Rad52 is capable of annealing DNA strands coated with RPA (*86, 87*), and it has been suggested that a specific interaction between Rad52 and RPA is necessary for the annealing reaction (*87*). By contrast, the single-strand annealing activity of Rad59 is inhibited by RPA (*110*). Rad52 and Rad59 appear to functionally cooperate in the strand annealing reaction (*109*). Smith and Rothstein (*111*) have described an RPA mutant, *rfa1-D228Y*, that allows SSA to occur in a *rad52* mutant. SSA is seen more frequently in the *rfa1-D228Y* mutant strain, suggesting that RPA normally suppresses SSA in wild-type cells.

E. Recombination Machinery in Higher Organisms

In single-cell eukaryotes, deleting genes of the *RAD52* epistasis group does not in general affect mitotic viability unless cells are challenged with a DNA-damaging agent. In contrast, null mutations in the *RAD52* group genes often engender a defect in cell proliferation in vertebrate cells and embryonic lethality in mice. The cell proliferation defect and embryonic inviability likely reflect the fact that recombination is indispensable for the repair of spontaneous lesions that arise during DNA replication. Biochemical studies with human recombination proteins have revealed a hierarchy of physical and functional interactions among these proteins that largely follow the paradigms established with the equivalent yeast factors (*65, 97, 112–114*). Importantly, the breast tumor suppressors BRCA1 and BRCA2 both affect the efficiency of recombination and recombinational DNA repair, and BRCA2 binds Rad51 (*10, 115–118*). A recent study reports the finding that BRCA2 cooperates with Rad51/RPA in DNA joint formation (*119*). These observations provide compelling evidence that recombination plays a major role in the maintenance of genome stability and cancer avoidance in mammals; several recent reviews on this topic are available (*10, 114, 120, 121*).

V. General Introduction to NHEJ

In contrast to homologous recombination, the repair of DSBs by nonhomologous end joining is conceptually much simpler. The ends of broken DNA molecules are brought together and joined in the absence of significant DNA sequence homology. An unusual feature of this repair pathway is that many of the early groundbreaking studies were carried out in the mammalian

TABLE II
IR-SENSITIVE RODENT MUTANTS DEFECTIVE IN NHEJ

IR group	Gene defective	Protein	Identified mutants
4	XRCC4	XRCC4	XR-1
5	XRCC5	Ku80	xrs1–7
6	XRCC6	Ku70	
7	XRCC7	DNA-PKcs	Mouse scid cell line, V-3

system and it is only recently that it has been examined in lower eukaryotes such as S. cerevisiae. One of the major reasons for this was the isolation and characterization of X-ray-sensitive mutant mammalian cell lines that are defective in NHEJ (Table II). This led to the cloning of the genes comprising NHEJ factors such as the DNA-dependent protein kinase (DNA-PK), which is composed of the Ku70/Ku80 DNA-binding subunit and the DNA-PKcs catalytic subunit, and XRCC4 (122).

The availability of the complete sequence of the S. cerevisiae genome (123) has facilitated the search for homologues of mammalian NHEJ genes and stimulated the study of NHEJ mechanisms in this model eukaryote. In contrast to mammals, inactivation of S. cerevisiae NHEJ genes does not usually confer a significant increase to killing by agents that cause DSBs because, in this organism, these lesions are predominantly repaired by homologous recombination. Nonetheless, these studies have shown that the fundamental mechanisms of this repair pathway are conserved among eukaryotes. The biochemical features of proteins involved in NHEJ are summarized in Table III. In the next section, we will describe the generally accepted model for NHEJ and the role of this repair pathway in maintaining genome integrity and stability.

VI. Mechanisms and Function of NHEJ in Eukaryotes

As mentioned previously, NHEJ can simply be described as the bringing together and joining of broken DNA molecules ends. Because these events are not mediated by long tracts of DNA sequence homology, it has been proposed that the ends are held together primarily by protein–protein interactions. These so-called end-bridging factors have been central to most models of NHEJ and the focus of many investigations. However, the end-bridging model raises the following question: If a cell suffers more than one DNA double strand break, can it distinguish between ends that were once linked and ends from different DNA molecules? Studies in mammalian cells have shown that although the joining of ends from different DNA molecules

TABLE III
BIOCHEMICAL PROPERTIES OF NHEJ FACTORS

Protein	Biochemical function	S. cerevisiae homologue	Features
Ku70/Ku80	DNA end binding	Hdf1/Hdf2	Interacts with DNA-PKcs
DNA-PKcs	DNA activated serine-threonine protein kinase	None	Interacts with Ku70/Ku80 and Artemis
Artemis	5′ to 3′ exonuclease; acquires structure-specific endonuclease activity when complexed with DNA-PKcs	Not known	Interacts with DNA-PKcs
DNA ligase IV	ATP-dependent DNA ligase	Dnl4	Interacts with XRCC4; Dnl4 also interacts with Pol4
XRCC4	DNA binding	Lif1	Interacts with DNA ligase IV; Lif1 also interacts with Xrs2
Pol μ	DNA polymerase	Pol4	
Fen-1	Flap endonuclease	Rad27	
hRad50/hMre11/NBS1[a]			

[a]See Table I.

to generate chromosomal translocations does occur, these events are rare and the cell usually rejoins the previously linked ends. One plausible explanation for these observations is that the arrangement of chromatin in loops attached to the nuclear matrix restricts the mobility of broken DNA ends and favors their rejoining to reconstitute the loop. A prediction of this idea is that chromosomal translocations will begin to occur when the cell suffers more than one DSB per chromatin loop.

The majority of DSBs caused by ionizing radiation cannot be joined directly by a DNA ligase because they have inappropriate termini. Thus, after end-bridging, the ends need to be processed to generate a ligatable structure. Analysis of *in vivo* DNA joining has revealed that many of the events occur at sites of short DNA sequence homology, ranging from 2 to 4 nucleotides in length (*124–126*). This suggests that DNA ends are brought together in a sequence-independent manner and then the DNA sequences adjacent to the ends are sampled in some manner to identify short complementary sequences, so-called microhomologies. Presumably these sequences are used to align the DNA molecules and then ligatable structures are generated from the aligned DNA molecules by further processing. In the following sections of this review we will summarize recent advances in our understanding of the mechanisms of NHEJ in *S. cerevisiae* and mammalian cells.

A. S. cerevisiae NHEJ Genes

DSBs in yeast are predominantly repaired by homologous recombination. Consequently, mutation of any one of the yeast NHEJ genes generally has little effect on the sensitivity to DNA damage of proliferating cells unless the homologous recombination pathway is also inactivated. To directly examine the defect in NHEJ, several in vivo end-joining assays have been developed, including the most frequently used method of measuring recircularization of plasmid DNA molecules previously linearized with a restriction endonuclease.

As mentioned above, key players in NHEJ identified by biochemical and genetic studies with mammalian cells include DNA-PK, XRCC4, and, by extension, DNA ligase IV because of its stable interaction with XRCC4 (127). Although Hdf1 and Hdf2 have been identified as the yeast homologs of Ku70 and Ku80 (128, 129), respectively, yeast appears to lack a functional homologue of DNA-PKcs. Besides NHEJ, Hdf1 and Hdf2 likely participate in other cellular functions as hdf1 and hdf2 strains exhibit phenotypes that appear to be independent of the defect in NHEJ, such as temperature-sensitive growth and pronounced telomere shortening (128–132). The counterpart of the mammalian LIG4 gene, DNL4, was identified in the yeast genome by sequence homology search (133–136). Like DNA ligase IV, Dnl4 also forms a stable complex with a partner protein Lif1, which is likely to be the functional homologue of XRCC4 (137). Consistent with their role in NHEJ, inactivation of HDF1, HDF2, DNL4, or LIF1 results in a similar reduction of plasmid rejoining (130, 134–139).

As mentioned before, the products of the RAD50, MRE11, and XRS2 genes are indispensable for NHEJ. In this regard, rad50, mre11, and xrs2 strains exhibit the same reduction in transformation efficiency of linearized plasmid DNA as the other NHEJ mutants (125, 131, 139–141). These genetic studies indicate that the Hdf1/Hdf2, Rad50/Mre11/Xrs2, and Dnl4/Lif1 complexes are key components of the major NHEJ pathway. Although inactivation of any one of these NHEJ factors causes a similar reduction in transformation efficiency, there are significant differences in the type of residual end joining. Specifically, the residual end-joining activity in dnl4, rad50, mre11, or xrs2 cells generates a significant number of accurately repaired products, whereas the residual repair products in hdf cells are mostly imprecise and suffer large deletions (125, 130, 131, 135, 137–139). Epistasis studies have shown that the end-joining phenotype conferred by inactivation of the Hdf1/Hdf2 complex predominates over the mutant phenotypes caused by mutating other genes, suggesting that Hdf1/Hdf2 acts earlier than the other NHEJ factors (131, 135). In support of this hypothesis, the recruitment of the Dnl4/Lif1 complex to in vivo DSBs is dependent upon Hdf1/Hdf2 (142).

Although the Hdf1/Hdf2, Rad50/Mre11/Xrs2, and Dnl4/Lif1 complexes are critical for the efficient recircularization of linear plasmid DNA molecules with cohesive ends *in vivo*, the majority of DSBs induced by DNA-damaging agents require additional factors to process the DNA ends. Genetic studies have implicated the nuclease encoded by the *FEN1* (*RAD27*) gene, the DNA helicase encoded by the *SRS2* gene, and the DNA polymerase encoded by the *POL4* gene in NHEJ events (*143–145*).

Four laboratories have recently identified a novel NHEJ gene, *NEJ1* (*LIF2*), whose inactivation reduces transformation efficiency of linearized plasmid DNA to the same extent as the other key NHEJ genes (*146–149*). The residual end-joining products in *nej1* cells suffer large deletions similar to those seen in *hdf* cells (*148*). The *NEJ1* gene is expressed in haploids but not diploids, providing a possible mechanism for the previously suggested regulation of NHEJ by mating type (*150, 151*). In support of this notion, the deletion of *NEJ1* inhibited NHEJ in haploid cells, whereas its constitutive expression alleviated the repression of NHEJ in diploid cells (*147, 148*). The identification of *NEJ1* also provides a molecular explanation for the indirect role of the *SIR* genes in NHEJ. Mutation in the *SIR* genes results in a/α mating-type heterozygosity that inhibits *NEJ1* expression and causes defective NHEJ (*147, 148, 150, 151*). Because all naturally a/α expressing cells are diploid, it appears that NHEJ is inhibited when a homologous chromosome is present, presumably reflecting the preference for the accurate repair pathway, homologous recombination. However, haploid cells in the G_1 phase of the cell cycle cannot repair DSBs by homologous recombination, providing an explanation for the activation of NHEJ in these cells.

B. Molecular Mechanisms of NHEJ in *S. cerevisiae*

Genetic studies have provided a conceptual framework for NHEJ, but the molecular mechanisms of this pathway have not been investigated to the same extent. As mentioned previously, the Mre11 subunit of the Rad50/Mre11/Xrs2 complex is a nuclease. However, mutations that inactivate nuclease activity have no effect on the efficiency of the recircularization of linear plasmid DNA molecules with cohesive ends, suggesting that Rad50/Mre11/Xrs2 has another role in NHEJ (*152*). In a recent study, it was shown that the Rad50/Mre11/Xrs2 complex not only stimulates the catalytic activity of Dn14/Lif1 but also alters the mechanism of ligation from intra- to intermolecular (*47*). Consistent with this latter observation, atomic force microscopy studies have revealed that Rad50/Mre11/Xrs2 has robust end-bridging activity, forming oligomers from linear DNA molecules (*47*). In many of the oligomers, Rad50/Mre11/Xrs2 complexes were bound at internal sites corresponding to the junctions between DNA molecules. Because these complexes were the same size as end-bound and free protein complexes, it appears that a single Rad50/Mre11/Xrs2 complex, which is composed of two Rad50 molecules, two Mre11 molecules, and one Xrs2 molecule, can bind two DNA ends simultaneously. In addition to end-bridging, the Rad50/Mre11/Xrs2 complex also specifically recruits the Dn14/Lif1 complex via an interaction between the Xrs2 and Lif1 subunits (47).

Similar to mammalian Ku70/Ku80, the Hdf1/Hdf2 complex binds avidly to DNA ends and can translocate inward along the DNA duplex from the end (47, 128, 129). It has been shown that Ku70/Ku80 can also bridge DNA ends (153), but this activity appears to be much weaker than that of the Rad50/Mre11/Xrs2 complex. In accord with genetic studies, Hdf1/Hdf2 is required for efficient intermolecular ligation by Rad50/Mre11/Xrs2 and Dnl4/Lif1 at physiological salt concentrations (47). The inability of mammalian Ku70/Ku80 to substitute for the yeast complex suggests that there are specific functional interactions between the Hdf1/Hdf2, Rad50/Mre11/Xrs2, and Dnl4/Lif1 complexes. In the model shown in Fig. 7, we propose that after DSB formation, Hdf1/Hdf2 binds to the ends, protecting them from degradation. The presence of Hdf1/Hdf2 at or near the DNA end enhances the recruitment of Rad50/Mre11/Xrs2, possibly in a manner similar to the recruitment of DNA-PKcs by Ku70/Ku80, and promotes end-bridging by this complex. Dnl4/Lif1 is recruited to the nucleoprotein complex containing the two DNA ends via the Xrs2 subunit of the Rad50/Mre11/Xrs2 complex. On occasions when the ends can be aligned and ligated, the reaction will be completed by the coordinated actions of these three complexes without additional factors. The newly identified *NEJ1* gene product is also required for the efficient recirculization of DNA molecules with cohesive ends (146–149). Although Nej1 has been shown to interact with Lif1 and is necessary for the nuclear localization of Dnl4/Lif1 (146–149), it is not known whether Nej1 participates directly in the end joining reaction.

In most cases, the ends of DSBs will be either resected or unwound to expose microhomologies, with the Hdf1/Hdf2 complex functioning to limit the extent of resection. The nuclease(s) and helicase(s) involved in this step have not been definitively identified. Based on biochemical studies with hMre11, the Mre11 subunit of the Rad50/Mre11/Xrs2 complex is an attractive candidate for the nuclease whereas genetic studies have suggested a possible role for the Srs2 DNA helicase (145). Once the ends of the DNA molecules have been aligned via short tracts of complementary DNA sequences, nuclease and DNA polymerase activities remove single-strand flaps and fill in the resulting gaps, respectively, to generate a ligatable structure. Genetic studies have implicated the Fen-1 (Rad27) nuclease and the Pol4 DNA polymerase in these final end-processing reactions (143, 144). In support of the suggested role of Pol4 in NHEJ, a recent biochemical study has shown that Pol4 efficiently catalyzes DNA synthesis on small gaps generated by the alignment of

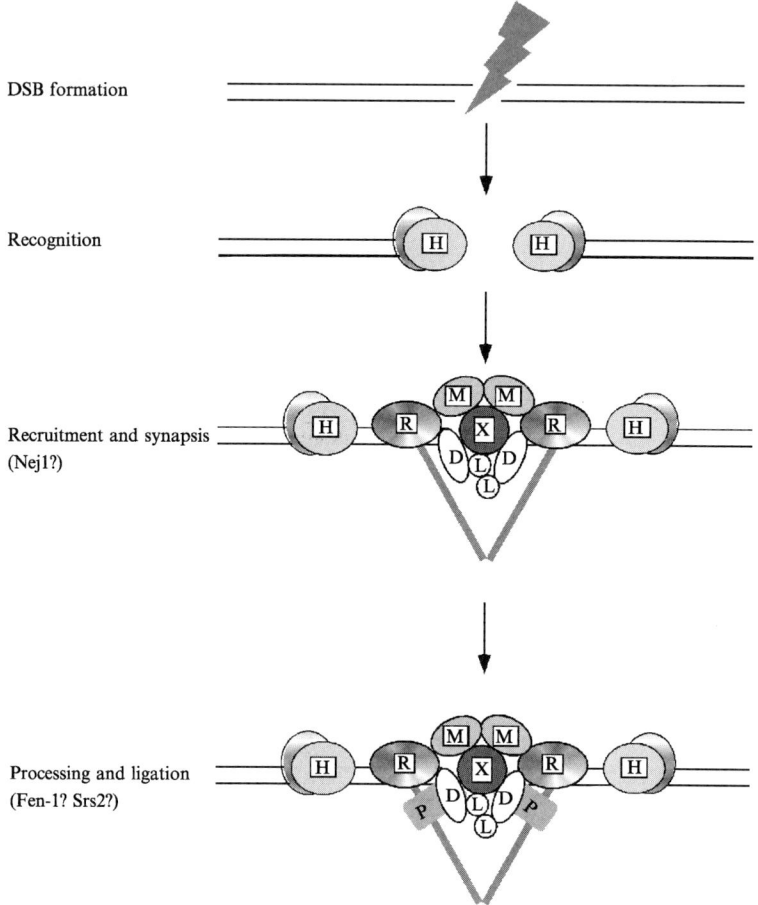

FIG. 7. A model for NHEJ in *S. cerevisiae*. After DSB formation, Hdf1/Hdf2 binds to the DNA ends. The presence of Hdf1/Hdf2 at or near DNA ends enhances the recruitment of Rad50/Mre11/Xrs2 and promotes its end-bridging activity. Dnl4/Lif1 is in turn recruited to the nucleoprotein complex via the Xrs2 subunit. If necessary, end processing by factors such as Pol4 may take place within the nucleoprotein complex. H, Hdf1/Hdf2; R, Rad50; M, Mre11; X, Xrs2; D, Dnl4; L, Lif1; and P, Pol4.

linear duplex DNA molecules with complementary ends. Furthermore, Pol4 specifically interacts with Dnl4/Lif1, resulting in the stimulation of DNA synthesis by Pol4 as well as DNA joining by Dnl4/Lif1 (154). Because Hdf1/Hdf2, Hdf2, Rad50/Mre11/Xrs2, and Dnl4/Lif1 are all required for the efficient recircularization of DNA molecules with cohesive ends *in vivo*, we propose that the resection, the alignment of DNA molecules via microhomologies, and the end-processing reactions that generate a ligatable structure occur within the context of the nucleoprotein complex formed by these three core NHEJ factors.

C. Mammalian NHEJ Genes

NHEJ makes a much larger contribution to cell survival in response to DSBs in mammalian cells compared with yeast cells. Although mammalian somatic cells are diploid, the size and complexity of the genome, in particular the large amounts of repetitive DNA, make the identification of homologous sequences for recombinational repair a much more difficult proposition than in yeast. Interestingly, it was observed that the X-ray sensitivity of the Chinese hamster ovary cell line XR-1 varied as a function of the cell-cycle stage (155). The later identification of XRCC4 as the missing component in XR-1 (156) as well as a crucial NHEJ factor revealed cell-cycle regulation of NHEJ. Subsequent studies by many investigators have confirmed and extended this study and have led to the generally accepted notion that NHEJ is the major DSB repair pathway in the G_1 phase of the cell cycle and in quiescent cells, whereas recombinational repair is more effective in late S and G_2 phases of the cell cycle when sister chromatids are present (157).

The X-ray-sensitive rodent cells have been a valuable resource for the identification of human genes (X-Ray Cross Complementing, XRCC) that participate in DSB repair. Unlike yeast cells, defects in NHEJ result in X-ray sensitivity, so human genes involved in each of these pathway have been cloned by functional complementation. The mutant rodent cell lines have been used for the cloning of the *XRCC4* gene, which encodes the partner protein of DNA ligase IV, and the functional dissection of the Ku70/Ku80 and DNA-PKcs subunits of DNA-PK (122) ARTEMIS is a very recent addition to the human NHEJ gene group (158). This gene, which is mutated in an inherited form of immunodeficiency, was cloned by linkage analysis. Although the predicted amino acid sequence of this gene exhibited homology with β-lactamases, the X-ray sensitivity of Artemis cell lines and the identification of additional mammalian and yeast DNA repair genes encoding β-lactamase-like proteins (159, 160) indicate that *ARTEMIS* is an NHEJ gene.

Although genetic and biochemical studies in yeast have firmly established the role of the Rad50/Mre11/Xrs2 complex in NHEJ, the participation of its human equivalent, hRad50/hMre11/NBS1, in NHEJ is less clear. As mentioned above, mutations in the *hMRE11* and *NBS1* genes have been identified as the causative factors in the cancer-prone human syndromes ataxia telangiectasia-like disorder (ATLD) and Nijmegen breakage syndrome (NBS), respectively (161–163). However, ATLD and NBS1 cell lines do not exhibit an obvious defect in either NHEJ or V(D)J recombination (161, 164). Unfortunately, the genes encoding subunits of the hRad50/hMre11/NBS1 complex are required for cell viability, hindering more detailed analysis of the involvement of this complex in NHEJ (165–167). Intriguingly, individuals with *LIG4* mutations have been identified and exhibit the developmental abnormalities and mental retardation characteristic of NBS (168). One possible explanation of these observations is that the hRad50/hMre11/NBS1 complex functions in an NHEJ subpathway that plays a critical role during embryogenesis and/or development.

Other than DSB repair, NHEJ factors also appear to participate in other cellular functions such as V(D)J recombination and retroviral integration (14, 169, 170). Cell lines from the immunodeficient mouse strain scid (severe combined immunodeficiency) are X-ray sensitive (171–173) and defective in DNA-PK activity (174–176), revealing the link between NHEJ and immunoglobulin gene rearrangements. Specifically, NHEJ factors are required to complete V(D)J recombination, which is initiated by the site-specific Rag1/Rag2 endonuclease (14). Thus it was surprising that the first human individual identified with *LIG4* mutations was not immunodeficient (177, 178). This is presumably because the residual levels of DNA ligase IV activity in this patient are sufficient for V(D)J recombination but not for NHEJ. Indeed, a recent examination of patients with uncharacterized immunodeficiency led to the identification of *LIG4* mutations that do associate with this syndrome (168).

The mouse DNA-PKcs, Ku70, Ku80, *LIG4*, and *XRCC4* genes have all been inactivated by conventional gene targeting to generate mouse models of NHEJ deficiency. As expected, all of the mutant cell lines are X-ray sensitive and defective in V(D)J recombination. However, there are significant differences in the severity of the phenotype both at the cellular and organismal level. Apart from immunodeficiency, the DNA-PKcs null animal has no overt defects (179–181). Surprisingly, there are differences in the phenotype of *Ku70* and *Ku80* null animals that include cancer predisposition and premature aging, consistent with the premise that these proteins have independent functions (182–184). In contrast, the *XRCC4* and *LIG4* null animals have essentially identical phenotypes, exhibiting embryonic lethality at around Day 16.5 (185–187). The mutant embryos appear to die because of abnormally high levels of apoptosis in the developing central nervous system (CNS) (185, 187). Embryonic lethality can be rescued by a second mutation inactivating either Atm function or p53 function (188–191). This suggests that unrepaired DSBs resulting from the NHEJ defect trigger apoptosis. Neuronal cells in the developing CNS appear to be particularly sensitive to apoptotic triggers, an effect that is suppressed by mutations inactivating signal transduction pathways linking DSBs to apoptosis.

The milder phenotype of cells deficient in DNA-PK compared with cells deficient in DNA ligase IV activity possibly reflects the sequestration of DNA ends into the DNA ligase IV-dependent pathway by DNA-PK assembly, whereas in the absence of functional DNA-PK, the ends can be repaired by other DNA repair pathways. This model is supported by studies in DT40 chicken cells showing that inactivation of the Ku70/Ku80 heterodimer in *lig4* mutant cells reduces the severity of the phenotype to the same level as *ku* mutant cells (*192*). This observation is reminiscent of the dominant nature of the end-joining defect caused by *hdf* mutations compared with inactivation of other NHEJ factors in yeast.

D. Molecular Mechanisms of NHEJ in Mammals

Insights into the mechanisms of several different DNA transactions have been based on the development of assays with cell-free extracts. For many years analysis of NHEJ by this approach was complicated by the presence of robust end-joining activity in mutant cell extracts that appeared to be independent of NHEJ factors. This was a particular problem in extracts from rodent cells because they contain significantly lower levels of DNA-PK than human cells (*193*). Baumann and West (*194*) have developed an assay to detect end joining in an extract from human lymphoblastoid cells that was, in accord with genetic studies, dependent upon Ku, DNA-PKcs, and DNA ligase IV/XRCC4. In subsequent fractionation studies, inositol 6-phosphate was unexpectedly identified as an important cofactor for efficient end joining (*195–197*). Inositol 6-phosphate binds to Ku and could regulate DNA-PK activity (*198, 199*).

The recent determination of the structure of the Ku70/Ku80 heterodimer complexed with DNA by X-ray crystallography has provided a molecular explanation for the DNA-binding properties of this complex (*200*). Specifically, the Ku70/Ku80 complex forms an asymmetric ring around the DNA helix, suggesting a mechanism for both its ability to translocate along the DNA molecule from an end and to align and bridge DNA ends. However, the topological linking of Ku70/Ku80 to DNA creates the problem of how to remove it after repair is completed. In contrast to the contacts between the subunits of the PCNA homotrimer, the extensive and intertwined nature of the interactions between the Ku70 and Ku80 subunits argue against the dissociation of the Ku70/Ku80 heterodimer from DNA by separation of the subunits. In fact, the thin section of the asymmetric ring encircling the DNA suggests that the Ku70/Ku80 heterodimer may be released by proteolysis.

Two laboratories have reported interactions between Ku70/Ku80 and DNA ligase IV/XRCC4, but the effects of Ku on DNA joining were different (*201, 202*). In one study, Ku stimulated DNA joining whereas in the other study Ku inhibited joining by DNA ligase IV/XRCC4. The reason for the discrepancy in these studies is not clear. It is possible that the stimulatory effect is a consequence of DNA end-bridging because Ku not only stimulated DNA ligase IV/XRCC4 but also DNA ligases I and III (*202*). However, Hdf1/Hdf2, the yeast homologue of Ku70/Ku80, inhibited DNA joining by Dnl4/Lif1 (*47*).

The effect of Ku on the DNA joining activity of DNA ligase IV/XRCC4 may not be biologically relevant because, in the yeast system, it appears that end-bridging is mediated by the Rad50/Mre11/Xrs2 complex whereas the requirement for Hdf1/Hdf1 is only apparent at ionic conditions close to physiological levels (47). Similarly, the efficient interaction of DNA-PKcs with DNA ends only requires Ku70/Ku80 at ionic conditions close to physiological levels (203). Interestingly, when DNA-PKcs loads onto DNA ends bound by Ku70/Ku80, the heterodimer is moved inward along the DNA helix, leaving DNA-PKcs at the end (204).

Recently, it was observed that DNA-PKcs was autophosphorylated in response to IR and that this autophosphorylation was necessary for DSB repair (205). In vitro studies have shown that autophosphorylation of DNA-PKcs led to its dissociation from Ku and the loss of kinase activity (206), leading to the proposal that DNA-PK plays a regulatory role in NHEJ. However, there is accumulating evidence that DNA-PKcs plays an important structural role in NHEJ. The structure of DNA-PKcs determined by electron crystallography revealed the presence of channels, suggesting that the DNA molecule is threaded through DNA-PKcs with the single-strand ends protruding from the molecule (207). Such an arrangement is intriguing as it suggests plausible mechanisms for DNA end-bridging, alignment, and processing but, as with Ku70/Ku80, it raises questions about the removal of DNA-PKcs after repair is completed. DNA-PKcs associates with DNA ligase IV/XRCC4 at DNA ends and promotes intermolecular joining by DNA ligase IV/XRCC4 (201). Indeed, under conditions compatible with DNA-PK assembly, Ku70/Ku80 still inhibited intramolecular joining by DNA ligase IV/XRCC4, so the majority of DNA joining events were intermolecular (201). This change in the type of ligation resembles the effect of Rad50/Mre11/Xrs2 on Dnl4/Lif1 activity and suggests that DNA-PKcs has end-bridging activity. Interestingly, the efficient activation of kinase activity appears to require both the occupation of the open channel by double-strand DNA and the simultaneous interaction with two single-strand DNA ends (208, 209). This provides a mechanism for DNA-PKcs to distinguish DSBs from SSBs as well as to enable the synapsis of two DNA ends. A recent study using electron microscopy has provided direct evidence for end-bridging by DNA-PKcs (210). In contrast to the end-bridging by a single yeast Rad50/Mre11/Xrs2 complex, end-bridging by DNA-PKcs occurs between DNA ends each bound by a DNA-PKcs molecule and involves interactions between these DNA-PKcs molecules (210).

Recently, physical and functional interactions between DNA-PKcs and Artemis have been described (211). The assembly of a DNA-PKcs/Artemis complex on a DNA end results in the phosphorylation of Artemis and the activation of the cryptic endonuclease activity of Artemis. The DNA-PKcs/Artemis complex is able to open DNA hairpins, the unique reaction intermediate generated by the Rag1/Rag2 endonuclease during V(D)J recombination, providing a molecular explanation for the severe combined immunodeficiency of Artemis patients. The sensitivity of Artemis-deficient cell lines to ionizing radiation suggests that the Artemis nuclease is also a key player in NHEJ, presumably either contributing to exposing microhomologies for alignment or removing flaps after the alignment of microhomologies. Thus, the DNA-PKcs/Artemis complex is similar to the yeast Rad50/Mre11/Xrs2 complex in that it posseses end-bridging and nuclease activities.

Although we have argued that Ku70/Ku80 is not directly involved in end-bridging and alignment, it likely remains associated with the nucleoprotein complex formed by NHEJ factors and may recruit additional NHEJ factors. In support of this idea, a functional interaction between Ku70/Ku80 and the product of the Werner's syndrome (WS) gene has been characterized (212–214). The helicase and nuclease activities of the WS gene product may contribute to end processing. In the model shown in Fig. 8, we propose that the DSBs are initially bound by Ku70/Ku80, which in turn recruits the DNA-PKcs/Artemis complex. DNA ends are then brought together by interactions between the DNA-PKcs molecules. DNA-PK phosphorylates Artemis to activate its nuclease activity. Although the identities and roles of the nucleases that expose microhomologies and remove single-strand flaps after alignment remain to be determined, a recent study has provided evidence for a functional link between the Pol X family members, Pol μ and terminal deoxynucleotidyltransferase (TdT), and NHEJ factors (215). Because Pol4, the only member of the Pol X family in S. cerevisiae, interacts with Dnl4/Lif1 (154), it appears that the link between Pol X DNA polymerases and NHEJ factors is conserved among eukaryotes. Presumably, TdT has evolved to play a specialized role in adding untemplated nucleotides during V(D)J recombination after the removal of Rag proteins and assembly of the NHEJ nucleoprotein complex containing Ku70/Ku80, DNA-PKcs, Artemis, and DNA ligase IV/XRCC4, whereas Pol μ presumably interacts with the same NHEJ complex in nonlymphoid cells and fills in small gaps prior to ligation.

Huang and Dynan (216) fractionated HeLa extracts and identified fractions that stimulated end joining by purified Ku70/Ku80 and DNA ligase IV/XRCC4. hRad50/hMre11/NBS1 but not DNA-PKcs copurified with the stimulatory activity. This result resembles the observation in the yeast system where Rad50/Mre11/Xrs2 stimulates the end-joining activity of Dnl4/Lif1 (47), suggesting that hRad50/hMre11/NBS1, like the yeast counterpart, has a role in NHEJ. This idea is supported by the reported interaction between Ku and hMre11 (217). Further evidence for the participation of the hRad50/hMre11/NBS1 complex in NHEJ comes from biochemical studies. The hMre11 nuclease is activated by nonhomologous DNA ends but is inhibited by complementary ends, suggesting that it has the ability to identify

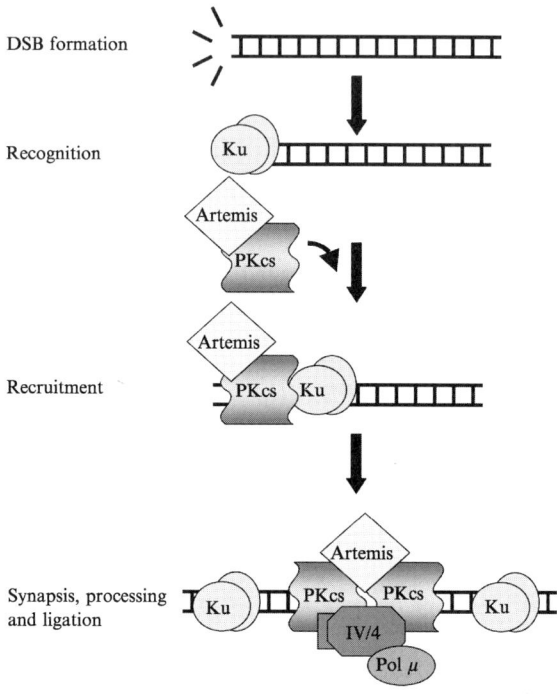

FIG. 8. A model for NHEJ in mammals. After DSB formation, Ku binds to DNA ends and recruits DNA-PKcs/Artemis. The binding of DNA-PKcs to DNA ends causes the inward translocation of Ku, leaving DNA-PKcs at the ends. Protein–protein interactions between end-bound DNA-PKcs molecules mediate the synapsis of the ends. Assembly of the DNA-PK complex on DNA ends activates the kinase activity that results in the phosphorylation of Artemis and XRCC4. DNA ends may be processed by Artemis and/or Pol μ. Finally, DNA ends are joined by DNA ligase IV/XRCC4. PKcs, DNA-PKcs; IV/4, DNA ligase IV/XRCC4.

and align microhomologies (218). Indeed, the processing of noncomplementary DNA ends by hMre11 generated ligatable structures that were aligned at microhomologies.

In the *in vitro* system, addition of purified DNA-PKcs inhibited the stimulatory activity by hRad50/hMre11/NBS1, suggesting that hRad50/hMhMre11/NBS1 and DNA-PKcs compete for Ku-bound DNA ends (216). This competition, together with the fact that both protein complexes possess end-bridging and nucleases activities, suggests that parallel NHEJ pathways may exist in the mammalian system to achieve efficient DSB repair. Interestingly, the repair of DSBs appears to occur by two kinetically distinct mechanisms in mammalian cells. Inactivation of DNA-PKcs eliminates the rapid mechanism whereas the rapid and slow mechanisms are both lost in the absence of Ku function (219). It is possible that the DNA-PKcs and Artemis have evolved in higher eukaryotes to increase the efficiency of NHEJ, giving rise to the rapid pathway. We suggest that the slow pathway may correspond to the yeast pathway involving the Rad50/Mre11/Xrs2 complex. The identification of *LIG4* mutations in patients with NBS-like symptoms also supports the existence of the hRad50/hMre11/NBS1-mediated NHEJ pathway and suggests that such a pathway is critical during early development (*168*).

In summary, it appears that DNA end joining in mammalian cells may occur by several distinct pathways. Future studies are needed to elucidate the biological roles of the NHEJ subpathways.

Acknowledgments

The studies in the laboratories of the authors have been supported by research grants from the U.S. National Institutes of Health. L.K. was supported in part by a NATO Science Fellowship and S.V.K. was supported in part by a U.S. Army fellowship. We are grateful to Michael Sehorn and Jana Villemain for reading the manuscript.

References

1. Norbury, C. J., and Hickson, I. D. (2001). Cellular responses to DNA damage. *Annu. Rev. Pharmacol. Toxicol.* **41**, 367–401.
2. Cox, M. M. (2001). Recombinational DNA repair of damaged replication forks in Escherichia coli: Questions. *Annu. Rev. Genet.* **35**, 53–82.
3. Michel, B. *et al.* (2001). Rescue of arrested replication forks by homologous recombination. *Proc. Natl. Acad. Sci. USA* **98**, 8181–8188.
4. Foiani, M. *et al.* (2000). DNA damage checkpoints and DNA replication controls in *Saccharomyces cerevisiae*. *Mutat. Res.* **451**, 187–196.
5. Greenwood, J., Costanzo, V., Robertson, K., Hensey, C., and Gautier, J. (2001). Responses to DNA damage in Xenopus: Cell death or cell cycle arrest. *Novartis Found. Symp.* **237**, 221–230; discussion 230–234.
6. Kaliraman, V., Mullen, J. R., Fricke, W. M., Bastin-Shanower, S. A., and Brill, S. J. (2001). Functional overlap between Sgs1-Top3 and the Mms4-Mus81 endonuclease. *Genes Dev.* **15**, 2730–2740.
7. Rothstein, R., Michel, B., and Gangloff, S. (2000). Replication fork pausing and recombination or "gimme a break." *Genes Dev.* **14**, 1–10.
8. Kolodner, R. D., Putnam, C. D., and Myung, K. (2002). Maintenance of genome stability in *Saccharomyces cerevisiae*. *Science* **297**, 552–557.
9. Lindahl, T., and Wood, R. D. (1999). Quality control by DNA repair. *Science* **286**, 1897–1905.
10. Pierce, A. J. *et al.* (2001). Double-strand breaks and tumorigenesis. *Trends Cell Biol.* **11**, S52–59.
11. Cromie, G. A., Connelly, J. C., and Leach, D. R. (2001). Recombination at double-strand breaks and DNA ends: Conserved mechanisms from phage to humans. *Mol. Cell* **8**, 1163–1174.

12. D'Amours, D., and Jackson, S. P. (2002). The Mre11 complex: At the crossroads of DNA repair and checkpoint signalling. *Nat. Rev. Mol. Cell Biol.* **3**, 317–327.
13. McBlane, J. F. *et al.* (1995). Cleavage at a V(D)J recombination signal requires only RAG1 and RAG2 proteins and occurs in two steps. *Cell* **83**, 387–395.
14. Grawunder, U., West, R. B., and Lieber, M. R. (1998). Antigen receptor gene rearrangement. *Curr. Opin. Immunol.* **10**, 172–180.
15. Siskind, G. W., and Benacerraf, B. (1969). Cell selection by antigen in the immune response. *Adv. Immunol.* **10**, 1–50.
16. Stavnezer, J. (1996). Immunoglobulin class switching. *Curr. Opin. Immunol.* **8**, 199–205.
17. Bross, L. *et al.* (2000). DNA double-strand breaks in immunoglobulin genes undergoing somatic hypermutation. *Immunity* **13**, 589–597.
18. Papavasiliou, F. N., and Schatz, D. G. (2000). Cell-cycle-regulated DNA double-stranded breaks in somatic hypermutation of immunoglobulin genes. *Nature* **408**, 216–221.
19. Muramatsu, M. *et al.* (1999). Specific expression of activation-induced cytidine deaminase (AID), a novel member of the RNA-editing deaminase family in germinal center B cells. *J. Biol. Chem.* **274**, 18470–18476.
20. Petersen, S. *et al.* (2001). AID is required to initiate Nbs1/gamma-H2AX focus formation and mutations at sites of class switching. *Nature* **414**, 660–665.
21. Martini, E., and Keeney, S. (2002). Sex and the single (double-strand) break. *Mol. Cell* **9**, 700–702.
22. Smith, K. N., Penkner, A., Ohta, K., Klein, F., and Nicolas, A. (2001). B-type cyclins CLB5 and CLB6 control the initiation of recombination and synaptonemal complex formation in yeast meiosis. *Curr. Biol.* **11**, 88–97.
23. Keeney, S. (2001). Mechanism and control of meiotic recombination initiation. *Curr. Top. Dev. Biol.* **52**, 1–53.
24. Romanienko, P. J., and Camerini-Otero, R. D. (1999). Cloning, characterization, and localization of mouse and human SPO11. *Genomics* **61**, 156–169.
25. Keeney, S. *et al.* (1999). A mouse homolog of the *Saccharomyces cerevisiae* meiotic recombination DNA transesterase Spo11p. *Genomics* **61**, 170–182.
26. Dernburg, A. F. *et al.* (1998). Meiotic recombination in C. elegans initiates by a conserved mechanism and is dispensable for homologous chromosome synapsis. *Cell* **94**, 387–398.
27. Villeneuve, A. M., and Hillers, K. J. (2001). Whence meiosis?. *Cell* **106**, 647–650.
28. Zickler, D., and Kleckner, N. (1999). Meiotic chromosomes: Integrating structure and function. *Annu. Rev. Genet.* **33**, 603–754.
29. Roeder, G. S. (1997). Meiotic chromosomes: It takes two to tango. *Genes Dev.* **11**, 2600–2621.
30. Paques, F., and Haber, J. E. (1999). Multiple pathways of recombination induced by double-strand breaks in *Saccharomyces cerevisiae*. *Microbiol. Mol. Biol. Rev.* **63**, 349–404.
31. Takata, M. *et al.* (1998). Homologous recombination and non-homologous end-joining pathways of DNA double-strand break repair have overlapping roles in the maintenance of chromosomal integrity in vertebrate cells. *EMBO J.* **17**, 5497–5508.
32. Szostak, J. W., Orr-Weaver, T. L., Rothstein, R. J., and Stahl, F. W. (1983). The double-strand-break repair model for recombination. *Cell* **33**, 25–35.
33. Allers, T., and Lichten, M. (2001). Intermediates of yeast meiotic recombination contain heteroduplex DNA. *Mol. Cell* **8**, 225–231.
34. Kleckner, N. (1996). Meiosis: How could it work? *Proc. Natl. Acad. Sci. USA* **93**, 8167–8174.
35. Roeder, G. S., and Bailis, J. M. (2000). The pachytene checkpoint. *Trends Genet.* **16**, 395–403.
36. Allers, T., and Lichten, M. (2001). Differential timing and control of noncrossover and crossover recombination during meiosis. *Cell* **106**, 47–57.

37. Myung, K., Chen, C., and Kolodner, R. D. (2001). Multiple pathways cooperate in the suppression of genome instability in *Saccharomyces cerevisiae*. *Nature* **411**, 1073–1076.
38. Lundblad, V. (2000). DNA ends: Maintenance of chromosome termini versus repair of double strand breaks. *Mutat. Res.* **451**, 227–240.
39. Lundblad, V. (2002). Telomere maintenance without telomerase. *Oncogene* **21**, 522–531.
40. Zakian, V. A. (1996). Structure, function, and replication of *Saccharomyces cerevisiae* telomeres. *Annu. Rev. Genet.* **30**, 141–172.
41. Sung, P., Trujillo, K. M., and Van Komen, S. (2000). Recombination factors of *Saccharomyces cerevisiae*. *Mutat. Res.* **451**, 257–275.
42. Connelly, J. C., and Leach, D. R. (2002). Tethering on the brink: The evolutionarily conserved Mre11-Rad50 complex. *Trends Biochem. Sci.* **27**, 410–418.
43. Paull, T. T., and Gellert, M. (1998). The 3′ to 5′ exonuclease activity of Mre 11 facilitates repair of DNA double-strand breaks. *Mol. Cell* **1**, 969–979.
44. Paull, T. T., and Gellert, M. (1999). Nbs1 potentiates ATP-driven DNA unwinding and endonuclease cleavage by the Mre11/Rad50 complex. *Genes Dev.* **13**, 1276–1288.
45. Trujillo, K. M., Yuan, S. S., Lee, E. Y., and Sung, P. (1998). Nuclease activities in a complex of human recombination and DNA repair factors Rad50, Mre11, and p95. *J. Biol. Chem.* **273**, 21447–21450.
46. Usui, T. *et al.* (1998). Complex formation and functional versatility of Mre11 of budding yeast in recombination. *Cell* **95**, 705–716.
47. Chen, L., Trujillo, K., Ramos, W., Sung, P., and Tomkinson, A. E. (2001). Promotion of Dn14-catalyzed DNA end-joining by the Rad50/Mre11/Xrs2 and Hdf1/Hdf2 complexes. *Mol. Cell* **8**, 1105–1115.
48. Lobachev, K. S., Gordenin, D. A., and Resnick, M. A. (2002). The Mre11 complex is required for repair of hairpin-capped double-strand breaks and prevention of chromosome rearrangements. *Cell* **108**, 183–193.
49. Hopfner, K. P., Putnam, C. D., and Tainer, J. A. (2002). DNA double-strand break repair from head to tail. *Curr. Opin Struct. Biol.* **12**, 115–122.
50. Hopfner, K. P. *et al.* (2002). The Rad50 zinc-hook is a structure joining Mre11 complexes in DNA recombination and repair. *Nature* **418**, 562–566.
51. Hopfner, K. P. *et al.* (2001). Structural biochemistry and interaction architecture of the DNA double-strand break repair Mre11 nuclease and Rad50-ATPase. *Cell* **105**, 473–485.
52. Hopfner, K. P. *et al.* (2000). Mre11 and Rad50 from Pyrococcus furiosus: Cloning and biochemical characterization reveal an evolutionarily conserved multiprotein machine. *J. Bacteriol.* **182**, 6036–6041.
53. Anderson, D. E., Trujillo, K. M., Sung, P., and Erickson, H. P. (2001). Structure of the Rad50 × Mre11 DNA repair complex from *Saccharomyces cerevisiae* by electron microscopy. *J. Biol. Chem.* **276**, 37027–37033.
54. de Jager, M. *et al.* (2001). Human Rad50/Mre11 is a flexible complex that can tether DNA ends. *Mol. Cell* **8**, 1129–1135.
55. Basile, G., Aker, M., and Mortimer, R. K. (1992). Nucleotide sequence and transcriptional regulation of the yeast recombinational repair gene RAD51. *Mol. Cell. Biol.* **12**, 3235–3246.
56. Aboussekhra, A., Chanet, R., Adjiri, A., and Fabre, F. (1992). Semidominant suppressors of Srs2 helicase mutations of *Saccharomyces cerevisiae* map in the *RAD51* gene, whose sequence predicts a protein with similarities to procaryotic RecA proteins. *Mol. Cell. Biol.* **12**, 3224–3234.
57. Shinohara, A., Ogawa, H., and Ogawa, T. (1992). Rad51 protein involved in repair and recombination in S. cerevisiae is a RecA-like protein. *Cell* **69**, 457–470.
58. Ogawa, T., Yu, X., Shinohara, A., and Egelman, E. H. (1993). Similarity of the yeast RAD51 filament to the bacterial RecA filament. *Science* **259**, 1896–1899.

59. Sung, P., and Robberson, D. L. (1995). DNA strand exchange mediated by a RAD51-ssDNA nucleoprotein filament with polarity opposite to that of RecA. *Cell* **82**, 453–461.
60. Benson, F. E., Stasiak, A., and West, S. C. (1994). Purification and characterization of the human Rad51 protein, an analogue of E. coli RecA. *EMBO J.* **13**, 5764–5771.
61. Yu, X., Jacobs, S. A., West, S. C., Ogawa, T., and Egelman, E. H. (2001). Domain structure and dynamics in the helical filaments formed by RecA and Rad51 on DNA. *Proc. Natl. Acad. Sci. USA* **98**, 8419–8424.
62. Sung, P. (1994). Catalysis of ATP-dependent homologous DNA pairing and strand exchange by yeast RAD51 protein. *Science* **265**, 1241–1243.
63. Gupta, R. C., Bazemore, L. R., Golub, E. I., and Radding, C. M. (1997). Activities of human recombination protein Rad51. *Proc. Natl. Acad. Sci. USA* **94**, 463–468.
64. Baumann, P., Benson, F. E., and West, S. C. (1996). Human Rad51 protein promotes ATP-dependent homologous pairing and strand transfer reactions in vitro. *Cell* **87**, 757–766.
65. Sigurdsson, S., Trujillo, K., Song, B., Stratton, S., and Sung, P. (2001). Basis for avid homologous DNA strand exchange by human Rad51 and RPA. *J. Biol. Chem.* **276**, 8798–8806.
66. Bianco, P. R., Tracy, R. B., and Kowalczykowski, S. C. (1998). DNA strand exchange proteins: A biochemical and physical comparison. *Front. Biosci.* **3**, D570–603.
67. Sugiyama, T., Zaitseva, E. M., and Kowalczykowski, S. C. (1997). A single-stranded DNA-binding protein is needed for efficient presynaptic complex formation by the *Saccharomyces cerevisiae* Rad51 protein. *J. Biol. Chem.* **272**, 7940–7945.
68. Zaitseva, E. M., Zaitsev, E. N., and Kowalczykowski, S. C. (1999). The DNA binding properties of *Saccharomyces cerevisiae* Rad51 protein. *J. Biol. Chem.* **274**, 2907–2915.
69. Van Komen, S., Petukhova, G., Sigurdsson, S., Stratton, S., and Sung, P. (2000). Superhelicity-driven homologous DNA pairing by yeast recombination factors Rad51 and Rad54. *Mol. Cell* **6**, 563–572.
70. Sung, P., and Stratton, S. A. (1996). Yeast Rad51 recombinase mediates polar DNA strand exchange in the absence of ATP hydrolysis. *J. Biol. Chem.* **271**, 27983–27986.
71. Fortin, G. S., and Symington, L. S. (2002). Mutations in yeast Rad51 that partially bypass the requirement for Rad55 and Rad57 in DNA repair by increasing the stability of Rad51-DNA complexes. *EMBO J.* **21**, 3160–3170.
72. Morgan, E. A., Shah, N., and Symington, L. S. (2002). The requirement for ATP hydrolysis by Saccharomyces cerevisiae Rad51 is bypassed by mating-type heterozygosity or *RAD54* in high copy. *Mol. Cell. Biol.* **22**, 6336–6343.
73. Morrison, C. et al. (1999). The essential functions of human Rad51 are independent of ATP hydrolysis. *Mol. Cell. Biol.* **19**, 6891–6897.
74. Gupta, R. C., Folta-Stogniew, E., O'Malley, S., Takahashi, M., and Radding, C. M. (1999). Rapid exchange of A:T base pairs is essential for recognition of DNA homology by human Rad51 recombination protein. *Mol. Cell* **4**, 705–714.
75. Song, B., and Sung, P. (2000). Functional interactions among yeast Rad51 recombinase, Rad52 mediator, and replication protein A in DNA strand exchange. *J. Biol. Chem.* **275**, 15895–15904.
76. Eggler, A. L., Inman, R. B., and Cox, M. M. (2002). The Rad51-dependent pairing of long DNA substrates is stabilized by replication protein A. *J. Biol. Chem.* **277**, 39280–39288.
77. Van Komen, S., Petukhova, G., Sigurdsson, S., and Sung, P. (2002). Functional crosstalk among Rad51, Rad54, and RPA in heteroduplex DNA joint formation. *J. Biol. Chem.* **277**, 43578–43587.
78. Petukhova, G., Stratton, S., and Sung, P. (1998). Catalysis of homologous DNA pairing by yeast Rad51 and Rad54 proteins. *Nature* **393**, 91–94.

DSB REPAIR MACHINES IN EUKARYOTES 195

79. Sung, P. (1997). Yeast Rad55 and Rad57 proteins form a heterodimer that functions with replication protein A to promote DNA strand exchange by Rad51 recombinase. *Genes Dev.* **11,** 1111–1121.
80. Shinohara, A., Shinohara, M., Ohta, T., Matsuda, S., and Ogawa, T. (1998). Rad52 forms ring structures and co-operates with RPA in single-strand DNA annealing. *Genes Cells* **3,** 145–156.
81. Kagawa, W. et al. (2002). Crystal structure of the homologous-pairing domain from the human Rad52 recombinase in the undecameric form. *Mol. Cell* **10,** 359–371.
82. Mortensen, U. H., Bendixen, C., Sunjevaric, I., and Rothstein, R. (1996). DNA strand annealing is promoted by the yeast Rad52 protein. *Proc. Natl. Acad. Sci. USA* **93,** 10729–10734.
83. Milne, G. T., and Weaver, D. T. (1993). Dominant negative alleles of RAD52 reveal a DNA repair/recombination complex including Rad51 and Rad52. *Genes Dev.* **7,** 1755–1765.
84. Sung, P. (1997). Function of yeast Rad52 protein as a mediator between replication protein A and the Rad51 recombinase. *J. Biol. Chem.* **272,** 28194–28197.
85. New, J. H., Sugiyama, T., Zaitseva, E., and Kowalczykowski, S. C. (1998). Rad52 protein stimulates DNA strand exchange by Rad51 and replication protein A. *Nature* **391,** 407–410.
86. Shinohara, A., and Ogawa, T. (1998). Stimulation by Rad52 of yeast Rad51-mediated recombination. *Nature* **391,** 404–407.
87. Sugiyama, T., New, J. H., and Kowalczykowski, S. C. (1998). DNA annealing by RAD52 protein is stimulated by specific interaction with the complex of replication protein A and single-stranded DNA. *Proc. Natl. Acad. Sci. USA* **95,** 6049–6054.
88. New, J. H., and Kowalczykowski, S. C. (2002). Rad52 protein has a second stimulatory role in DNA strand exchange that complements replication protein-A function. *J. Biol. Chem.* **277,** 26171–26176.
89. Krejci, L. et al. (2002). Interaction with Rad51 is indispensable for recombination mediator function of Rad52. *J. Biol. Chem.* **277,** 40132–40141.
90. Krejci, L., Damborsky, J., Thomsen, B., Duno, M., and Bendixen, C. (2001). Molecular dissection of interactions between Rad51 and members of the recombination-repair group. *Mol. Cell. Biol.* **21,** 966–976.
91. Kaytor, M. D., and Livingston, D. M. (1996). Allele-specific suppression of temperature-sensitive mutations of the *Saccharomyces cerevisiae* RAD52 gene. *Curr. Genet.* **29,** 203–210.
92. Boundy-Mills, K. L., and Livingston, D. M. (1993). A *Saccharomyces cerevisiae RAD52* allele expressing a C-terminal truncation protein: Activities and intragenic complementation of missense mutations. *Genetics* **133,** 39–49.
93. Lovett, S. T. (1994). Sequence of the *RAD55* gene of *Saccharomyces cerévisiae*: Similarity of RAD55 to prokaryotic RecA and other RecA-like proteins. *Gene* **142,** 103–106.
94. Johnson, R. D., and Symington, L. S. (1995). Functional differences and interactions among the putative RecA homologs Rad51, Rad55, and Rad57. *Mol. Cell. Biol.* **15,** 4843–4850.
95. Hays, S. L., Firmenich, A. A., and Berg, P. (1995). Complex formation in yeast double-strand break repair: Participation of Rad51, Rad52, Rad55, and Rad57 proteins. *Proc. Natl. Acad. Sci. USA* **92,** 6925–6929.
96. Mazin, A. V., Bornarth, C. J., Solinger, J. A., Heyer, W. D., and Kowalczykowski, S. C. (2000). Rad54 protein is targeted to pairing loci by the Rad51 nucleoprotein filament. *Mol. Cell* **6,** 583–592.
97. Sigurdsson, S., Van Komen, S., Petukhova, G., and Sung, P. (2002). Homologous DNA pairing by human recombination factors Rad51 and Rad54. *J. Biol. Chem.* **277,** 42790–42794.
98. Clever, B. et al. (1997). Recombinational repair in yeast: Functional interactions between Rad51 and Rad54 proteins. *EMBO J.* **16,** 2535–2544.

99. Jiang, H. *et al.* (1996). Direct association between the yeast Rad51 and Rad54 recombination proteins. *J. Biol. Chem.* **271**, 33181–33186.
100. Petukhova, G., Van Komen, S., Vergano, S., Klein, H., and Sung, P. (1999). Yeast Rad54 promotes Rad51-dependent homologous DNA pairing via ATP hydrolysis-driven change in DNA double helix conformation. *J. Biol. Chem.* **274**, 29453–29462.
101. Ristic, D., Wyman, C., Paulusma, C., and Kanaar, R. (2001). The architecture of the human Rad54-DNA complex provides evidence for protein translocation along DNA. *Proc. Natl. Acad. Sci. USA* **98**, 8454–8460.
102. Havas, K. *et al.* (2000). Generation of superhelical torsion by ATP-dependent chromatin remodeling activities. *Cell* **103**, 1133–1142.
103. Saha, A., Wittmeyer, J., and Cairns, B. R. (2002). Chromatin remodeling by RSC involves ATP-dependent DNA translocation. *Genes Dev.* **16**, 2120–2134.
104. Klein, H. L. (1997). RDH54, a RAD54 homologue in *Saccharomyces cerevisiae*, is required for mitotic diploid-specific recombination and repair and for meiosis. *Genetics* **147**, 1533–1543.
105. Shinohara, M. *et al.* (1997). Characterization of the roles of the Saccharomyces cerevisiae RAD54 gene and a homologue of RAD54, RDH54/TID1, in mitosis and meiosis. *Genetics* **147**, 1545–1556.
106. Dresser, M. E. *et al.* (1997). DMC1 functions in a *Saccharomyces cerevisiae* meiotic pathway that is largely independent of the RAD51 pathway. *Genetics* **147**, 533–544.
107. Kraus, E., Leung, W. Y., and Haber, J. E. (2001). Break-induced replication: A review and an example in budding yeast. *Proc. Natl. Acad. Sci. USA* **98**, 8255–8262.
108. Bai, Y., and Symington, L. S. (1996). A Rad52 homolog is required for RAD51-independent mitotic recombination in *Saccharomyces cerevisiae. Genes Dev.* **10**, 2025–2037.
109. Davis, A. P., and Symington, L. S. (2001). The yeast recombinational repair protein Rad59 interacts with Rad52 and stimulates single-strand annealing. *Genetics* **159**, 515–525.
110. Petukhova, G., Stratton, S. A., and Sung, P. (1999). Single strand DNA binding and annealing activities in the yeast recombination factor Rad59. *J. Biol. Chem.* **274**, 33839–33842.
111. Smith, J., and Rothstein, R. (1999). An allele of RFA1 suppresses RAD52-dependent double-strand break repair in Saccharomyces cerevisiae. *Genetics* **151**, 447–458.
112. Sigurdsson, S. *et al.* (2001). Mediator function of the human Rad51B-Rad51C complex in Rad51/RPA-catalyzed DNA strand exchange. *Genes Dev.* **15**, 3308–3318.
113. Masson, J. Y. *et al.* (2001). Identification and purification of two distinct complexes containing the five RAD51 paralogs. *Genes Dev.* **15**, 3296–3307.
114. Thompson, L. H., and Schild, D. (2001). Homologous recombinational repair of DNA ensures mammalian chromosome stability. *Mutat. Res.* **477**, 131–153.
115. Chen, P. L. *et al.* (1998). The BRC repeats in BRCA2 are critical for RAD51 binding and resistance to methyl methanesulfonate treatment. *Proc. Natl. Acad. Sci. USA* **95**, 5287–5292.
116. Wong, A. K., Pero, R., Ormonde, P. A., Tavtigian, S. V., and Bartel, P. L. (1997). RAD51 interacts with the evolutionarily conserved BRC motifs in the human breast cancer susceptibility gene brca2. *J. Biol. Chem.* **272**, 31941–31944.
117. Sharan, S. K. *et al.* (1997). Embryonic lethality and radiation hypersensitivity mediated by Rad51 in mice lacking Brca2. *Nature* **386**, 804–810.
118. Mizuta, R. *et al.* (1997). RAB22 and RAB163/mouse BRCA2: Proteins that specifically interact with the RAD51 protein. *Proc. Natl. Acad. Sci. USA* **94**, 6927–6932.
119. Yang, H. *et al.* (2002). BRCA2 function in DNA binding and recombination from a BRCA2-DSS1-ssDNA structure. *Science* **297**, 1837–1848.
120. Elliott, B., and Jasin, M. (2002). Double-strand breaks and translocations in cancer. *Cell. Mol. Life Sci.* **59**, 373–385.

121. Zhou, B. B., and Elledge, S. J. (2000). The DNA damage response: Putting checkpoints in perspective. *Nature* **408,** 433–439.
122. Critchlow, S. E., and Jackson, S. P. (1998). DNA end-joining: From yeast to man. *Trends Biochem. Sci.* **23,** 394–398.
123. Goffeau, A. *et al.* (1996). Life with 6000 genes. *Science* **274,** 546, 563–567.
124. Roth, D. B., and Wilson, J. H. (1986). Nonhomologous recombination in mammalian cells: Role for short sequence homologies in the joining reaction. *Mol. Cell. Biol.* **6,** 4295–4304.
125. Moore, J. K., and Haber, J. E. (1996). Cell cycle and genetic requirements of two pathways of nonhomologous end-joining repair of double-strand breaks in *Saccharomyces cerevisiae*. *Mol. Cell. Biol.* **16,** 2164–2173.
126. Kramer, K. M., Brock, J. A., Bloom, K., Moore, J. K., and Haber, J. E. (1994). Two different types of double-strand breaks in *Saccharomyces cerevisiae* are repaired by similar RAD52-independent, nonhomologous recombination events. *Mol. Cell. Biol.* **14,** 1293–1301.
127. Grawunder, U. *et al.* (1997). Activity of DNA ligase IV stimulated by complex formation with XRCC4 protein in mammalian cells. *Nature* **388,** 492–495.
128. Feldmann, H., and Winnacker, E. L. (1993). A putative homologue of the human autoantigen Ku from *Saccharomyces cerevisiae*. *J. Biol. Chem.* **268,** 12895–12900.
129. Feldmann, H. *et al.* (1996). HDF2, the second subunit of the Ku homologue from *Saccharomyces cerevisiae*. *J. Biol. Chem.* **271,** 27765–27769.
130. Boulton, S. J., and Jackson, S. P. (1996). Identification of a *Saccharomyces cerevisiae* Ku80 homologue: Roles in DNA double strand break rejoining and in telomeric maintenance. *Nucleic Acids Res.* **24,** 4639–4648.
131. Boulton, S. J., and Jackson, S. P. (1998). Components of the Ku-dependent non-homologous end-joining pathway are involved in telomeric length maintenance and telomeric silencing. *EMBO J.* **17,** 1819–1828.
132. Porter, S. E., Greenwell, P. W., Ritchie, K. B., and Petes, T. D. (1996). The DNA-binding protein Hdf1p (a putative Ku homologue) is required for maintaining normal telomere length in *Saccharomyces cerevisiae*. *Nucleic Acids Res.* **24,** 582–585.
133. Ramos, W., Liu, G., Giroux, C. N., and Tomkinson, A. E. (1998). Biochemical and genetic characterization of the DNA ligase encoded by *Saccharomyces cerevisiae* open reading frame YOR005c, a homolog of mammalian DNA ligase IV. *Nucleic Acids Res.* **26,** 5676–5683.
134. Schar, P., Herrmann, G., Daly, G., and Lindahl, T. (1997). A newly identified DNA ligase of *Saccharomyces cerevisiae* involved in RAD52-independent repair of DNA double-strand breaks. *Genes Dev.* **11,** 1912–1924.
135. Teo, S. H., and Jackson, S. P. (1997). Identification of *Saccharomyces cerevisiae* DNA ligase IV: Involvement in DNA double-strand break repair. *EMBO J.* **16,** 4788–4795.
136. Wilson, T. E., Grawunder, U., and Lieber, M. R. (1997). Yeast DNA ligase IV mediates non-homologous DNA end joining. *Nature* **388,** 495–498.
137. Herrmann, G., Lindahl, T., and Schar, P. (1998). *Saccharomyces cerevisiae* LIF1: A function involved in DNA double-strand break repair related to mammalian XRCC4. *EMBO J.* **17,** 4188–4198.
138. Boulton, S. J., and Jackson, S. P. (1996). *Saccharomyces cerevisiae* Ku70 potentiates illegitimate DNA double-strand break repair and serves as a barrier to error-prone DNA repair pathways. *EMBO J.* **15,** 5093–5103.
139. Milne, G. T., Jin, S., Shannon, K. B., and Weaver, D. T. (1996). Mutations in two Ku homologs define a DNA end-joining repair pathway in *Saccharomyces cerevisiae*. *Mol. Cell. Biol.* **16,** 4189–4198.
140. Schiestl, R. H., Zhu, J., and Petes, T. D. (1994). Effect of mutations in genes affecting homologous recombination on restriction enzyme-mediated and illegitimate recombination in *Saccharomyces cerevisiae*. *Mol. Cell. Biol.* **14,** 4493–4500.

141. Tsukamoto, Y., Kato, J., and Ikeda, H. (1996). Effects of mutations of RAD50, RAD51, RAD52, and related genes on illegitimate recombination in Saccharomyces cerevisiae. Genetics **142**, 383–391.
142. Teo, S. H., and Jackson, S. P. (2000). Lif1p targets the DNA ligase Lig4p to sites of DNA double-strand breaks. Curr. Biol. **10**, 165–168.
143. Wilson, T. E., and Lieber, M. R. (1999). Efficient processing of DNA ends during yeast nonhomologous end joining. Evidence for a DNA polymerase beta (Pol4)-dependent pathway. J. Biol. Chem. **274**, 23599–23609.
144. Wu, X., Wilson, T. E., and Lieber, M. R. (1999). A role for FEN-1 in nonhomologous DNA end joining: The order of strand annealing and nucleolytic processing events. Proc. Natl. Acad. Sci. USA **96**, 1303–1308.
145. Hegde, V., and Klein, H. (2000). Requirement for the SRS2 DNA helicase gene in nonhomologous end joining in yeast. Nucleic Acids Res. **28**, 2779–2783.
146. Ooi, S. L., Shoemaker, D. D., and Boeke, J. D. (2001). A DNA microarray-based genetic screen for nonhomologous end-joining mutants in Saccharomyces cerevisiae. Science **294**, 2552–2556.
147. Valencia, M. et al. (2001). NEJ1 controls non-homologous end joining in Saccharomyces cerevisiae. Nature **414**, 666–669.
148. Kegel, A., Sjostrand, J. O., and Astrom, S. U. (2001). Nej1p, a cell type-specific regulator of nonhomologous end joining in yeast. Curr. Biol. **11**, 1611–1617.
149. Frank-Vaillant, M., and Marcand, S. (2001). NHEJ regulation by mating type is exercised through a novel protein, Lif2p, essential to the ligase IV pathway. Genes Dev. **15**, 3005–3012.
150. Astrom, S. U., Okamura, S. M., and Rine, J. (1999). Yeast cell-type regulation of DNA repair. Nature **397**, 310.
151. Lee, S. E., Paques, F., Sylvan, J., and Haber, J. E. (1999). Role of yeast SIR genes and mating type in directing DNA double-strand breaks to homologous and non-homologous repair paths. Curr. Biol. **9**, 767–770.
152. Moreau, S., Ferguson, J. R., and Symington, L. S. (1999). The nuclease activity of Mre11 is required for meiosis but not for mating type switching, end joining, or telomere maintenance. Mol. Cell. Biol. **19**, 556–566.
153. Cary, R. B. et al. (1997). DNA looping by Ku and the DNA-dependent protein kinase. Proc. Natl. Acad. Sci. USA **94**, 4267–4272.
154. Tseng, H. M., and Tomkinson, A. E. (2002). A physical and functional interaction between yeast Pol4 and Dnl4/Lif1 links DNA synthesis and ligation in non-homologous end joining. J. Biol. Chem. **277**, 45630–45637.
155. Stamato, T. D., Weinstein, R., Giaccia, A., and Mackenzie, L. (1983). Isolation of cell cycle-dependent gamma ray-sensitive Chinese hamster ovary cell. Somat. Cell. Genet. **9**, 165–173.
156. Li, Z. et al. (1995). The XRCC4 gene encodes a novel protein involved in DNA double-strand break repair and V(D)J recombination. Cell **83**, 1079–1089.
157. Hendrickson, E. A. (1997). Cell-cycle regulation of mammalian DNA double-strand-break repair. Am. J. Hum. Genet. **61**, 795–800.
158. Moshous, D. et al. (2001). Artemis, a novel DNA double-strand break repair/V(D)J recombination protein, is mutated in human severe combined immune deficiency. Cell **105**, 177–186.
159. Dronkert, M. L. et al. (2000). Disruption of mouse SNM1 causes increased sensitivity to the DNA interstrand cross-linking agent mitomycin C. Mol. Cell. Biol. **20**, 4553–4561.
160. Brendel, M., and Henriques, J. A. (2001). The pso mutants of Saccharomyces cerevisiae comprise two groups: One deficient in DNA repair and another with altered mutagen metabolism. Mutat. Res. **489**, 79–96.

161. Stewart, G. S. et al. (1999). The DNA double-strand break repair gene hMRE11 is mutated in individuals with an ataxia-telangiectasia-like disorder. *Cell* **99**, 577–587.
162. Carney, J. P. et al. (1998). The hMre11/hRad50 protein complex and Nijmegen breakage syndrome: Linkage of double-strand break repair to the cellular DNA damage response. *Cell* **93**, 477–486.
163. Varon, R. et al. (1998). Nibrin, a novel DNA double-strand break repair protein, is mutated in Nijmegen breakage syndrome. *Cell* **93**, 467–476.
164. Kraakman-van der Zwet, M. et al. (1999). Immortalization and characterization of Nijmegen breakage syndrome fibroblasts. *Mutat. Res.* **434**, 17–27.
165. Luo, G. et al. (1999). Disruption of mRad50 causes embryonic stem cell lethality, abnormal embryonic development, and sensitivity to ionizing radiation. *Proc. Natl. Acad. Sci. USA* **96**, 7376–7381.
166. Xiao, Y., and Weaver, D. T. (1997). Conditional gene targeted deletion by Cre recombinase demonstrates the requirement for the double-strand break repair Mre11 protein in murine embryonic stem cells. *Nucleic Acids Res.* **25**, 2985–2991.
167. Zhu, J., Petersen, S., Tessarollo, L., and Nussenzweig, A. (2001). Targeted disruption of the Nijmegen breakage syndrome gene NBS1 leads to early embryonic lethality in mice. *Curr. Biol.* **11**, 105–109.
168. O'Driscoll, M. et al. (2001). DNA ligase IV mutations identified in patients exhibiting developmental delay and immunodeficiency. *Mol. Cell* **8**, 1175–1185.
169. Li, L. et al. (2001). Role of the non-homologous DNA end joining pathway in the early steps of retroviral infection. *EMBO J.* **20**, 3272–3281.
170. Daniel, R., Katz, R. A., and Skalka, A. M. (1999). A role for DNA-PK in retroviral DNA integration. *Science* **284**, 644–647.
171. Biedermann, K. A., Sun, J. R., Giaccia, A. J., Tosto, L. M., and Brown, J. M. (1991). Scid mutation in mice confers hypersensitivity to ionizing radiation and a deficiency in DNA double-strand break repair. *Proc. Natl. Acad. Sci. USA* **88**, 1394–1397.
172. Fulop, G. M., and Phillips, R. A. (1990). The scid mutation in mice causes a general defect in DNA repair. *Nature* **347**, 479–482.
173. Hendrickson, E. A. et al. (1991). A link between double-strand break-related repair and V(D)J recombination: The scid mutation. *Proc. Natl. Acad. Sci. USA* **88**, 4061–4065.
174. Blunt, T. et al. (1995). Defective DNA-dependent protein kinase activity is linked to V(D)J recombination and DNA repair defects associated with the murine scid mutation. *Cell* **80**, 813–823.
175. Kirchgessner, C. U. et al. (1995). DNA-dependent kinase (p350) as a candidate gene for the murine SCID defect. *Science* **267**, 1178–1183.
176. Peterson, S. R. et al. (1995). Loss of the catalytic subunit of the DNA-dependent protein kinase in DNA double-strand-break-repair mutant mammalian cells. *Proc. Natl. Acad. Sci. USA* **92**, 3171–3174.
177. Badie, C. et al. (1997). A DNA double-strand break defective fibroblast cell line (180BR) derived from a radiosensitive patient represents a new mutant phenotype. *Cancer Res.* **57**, 4600–4607.
178. Riballo, E. et al. (1999). Identification of a defect in DNA ligase IV in a radiosensitive leukaemia patient. *Curr. Biol.* **9**, 699–702.
179. Gao, Y. et al. (1998). A targeted DNA-PKcs-null mutation reveals DNA-PK-independent functions for KU in V(D)J recombination. *Immunity* **9**, 367–376.
180. Gu, Y. et al. (2000). Defective embryonic neurogenesis in Ku-deficient but not DNA-dependent protein kinase catalytic subunit-deficient mice. *Proc. Natl. Acad. Sci. USA* **97**, 2668–2673.

181. Taccioli, G. E. et al. (1998). Targeted disruption of the catalytic subunit of the DNA-PK gene in mice confers severe combined immunodeficiency and radiosensitivity. *Immunity* **9,** 355–366.
182. Gu, Y. et al. (1997). Growth retardation and leaky SCID phenotype of Ku70-deficient mice. *Immunity* **7,** 653–665.
183. Li, G. C. et al. (1998). Ku70: A candidate tumor suppressor gene for murine T cell lymphoma. *Mol. Cell* **2,** 1–8.
184. Vogel, H., Lim, D. S., Karsenty, G., Finegold, M., and Hasty, P. (1999). Deletion of Ku86 causes early onset of senescence in mice. *Proc. Natl. Acad. Sci. USA* **96,** 10770–10775.
185. Barnes, D. E., Stamp, G., Rosewell, I., Denzel, A., and Lindahl, T. (1998). Targeted disruption of the gene encoding DNA ligase IV leads to lethality in embryonic mice. *Curr. Biol.* **8,** 1395–1398.
186. Frank, K. M. et al. (1998). Late embryonic lethality and impaired V(D)J recombination in mice lacking DNA ligase IV. *Nature* **396,** 173–177.
187. Gao, Y. et al. (1998). A critical role for DNA end-joining proteins in both lymphogenesis and neurogenesis. *Cell* **95,** 891–902.
188. Frank, K. M. et al. (2000). DNA ligase IV deficiency in mice leads to defective neurogenesis and embryonic lethality via the p53 pathway. *Mol. Cell* **5,** 993–1002.
189. Lee, Y., Barnes, D. E., Lindahl, T., and McKinnon, P. J. (2000). Defective neurogenesis resulting from DNA ligase IV deficiency requires Atm. *Genes Dev.* **14,** 2576–2580.
190. Gao, Y. et al. (2000). Interplay of p53 and DNA-repair protein XRCC4 in tumorigenesis, genomic stability and development. *Nature* **404,** 897–900.
191. Sekiguchi, J. et al. (2001). Genetic interactions between ATM and the nonhomologous end-joining factors in genomic stability and development. *Proc. Natl. Acad. Sci. USA* **98,** 3243–3248.
192. Adachi, N., Ishino, T., Ishii, Y., Takeda, S., and Koyama, H. (2001). DNA ligase IV-deficient cells are more resistant to ionizing radiation in the absence of Ku70: Implications for DNA double-strand break repair. *Proc. Natl. Acad. Sci. USA* **98,** 12109–12113.
193. Finnie, N. J., Gottlieb, T. M., Blunt, T., Jeggo, P. A., and Jackson, S. P. (1995). DNA-dependent protein kinase activity is absent in xrs-6 cells: Implications for site-specific recombination and DNA double-strand break repair. *Proc. Natl. Acad. Sci. USA* **92,** 320–324.
194. Baumann, P., and West, S. C. (1998). DNA end-joining catalyzed by human cell-free extracts. *Proc. Natl. Acad. Sci. USA* **95,** 14066–14070.
195. Hanakahi, L. A., Bartlet-Jones, M., Chappell, C., Pappin, D., and West, S. C. (2000). Binding of inositol phosphate to DNA-PK and stimulation of double-strand break repair. *Cell* **102,** 721–729.
196. Hanakahi, L. A., and West, S. C. (2002). Specific interaction of IP6 with human Ku70/80, the DNA-binding subunit of DNA-PK. *EMBO J.* **21,** 2038–2044.
197. Ma, Y., and Lieber, M. R. (2002). Binding of inositol hexakisphosphate (IP6) to Ku but not to DNA-PKcs. *J. Biol. Chem.* **277,** 10756–10759.
198. Hanakahi, L. A., and West, S. C. (2002). Specific interaction of IP6 with human Ku70/80, the DNA-binding subunit of DNA-PK. *EMBO J.* **21,** 2038–2044.
199. Ma, Y., and Lieber, M. R. (2002). Binding of inositol hexakisphosphate (IP6) to Ku but not to DNA-PKcs. *J. Biol. Chem.* **277,** 10756–10759.
200. Walker, J. R., Corpina, R. A., and Goldberg, J. (2001). Structure of the Ku heterodimer bound to DNA and its implications for double-strand break repair. *Nature* **412,** 607–614.
201. Chen, L., Trujillo, K., Sung, P., and Tomkinson, A. E. (2000). Interactions of the DNA ligase IV-XRCC4 complex with DNA ends and the DNA-dependent protein kinase. *J. Biol. Chem.* **275,** 26196–26205.

202. Ramsden, D. A., and Gellert, M. (1998). Ku protein stimulates DNA end joining by mammalian DNA ligases: A direct role for Ku in repair of DNA double-strand breaks. *EMBO J.* **17,** 609–614.
203. Hammarsten, O., and Chu, G. (1998). DNA-dependent protein kinase: DNA binding and activation in the absence of Ku. *Proc. Natl. Acad. Sci. USA* **95,** 525–530.
204. Yoo, S., and Dynan, W. S. (1999). Geometry of a complex formed by double strand break repair proteins at a single DNA end: Recruitment of DNA-PKcs induces inward translocation of Ku protein. *Nucleic Acids Res.* **27,** 4679–4686.
205. Chan, D. W. *et al.* (2002). Autophosphorylation of the DNA-dependent protein kinase catalytic subunit is required for rejoining of DNA double-strand breaks. *Genes Dev.* **16,** 2333–2338.
206. Chan, D. W., and Lees-Miller, S. P. (1996). The DNA-dependent protein kinase is inactivated by autophosphorylation of the catalytic subunit. *J. Biol. Chem.* **271,** 8936–8941.
207. Leuther, K. K., Hammarsten, O., Kornberg, R. D., and Chu, G. (1999). Structure of DNA-dependent protein kinase: Implications for its regulation by DNA. *EMBO J.* **18,** 1114–1123.
208. Hammarsten, O., DeFazio, L. G., and Chu, G. (2000). Activation of DNA-dependent protein kinase by single-stranded DNA ends. *J. Biol. Chem.* **275,** 1541–1550.
209. Martensson, S., and Hammarsten, O. (2002). DNA-dependent protein kinase catalytic subunit. Structural requirements for kinase activation by DNA ends. *J. Biol. Chem.* **277,** 3020–3029.
210. DeFazio, L. G., Stansel, R. M., Griffith, J. D., and Chu, G. (2002). Synapsis of DNA ends by DNA-dependent protein kinase. *EMBO J.* **21,** 3192–3200.
211. Ma, Y., Pannicke, U., Schwarz, K., and Lieber, M. R. (2002). Hairpin opening and overhang processing by an Artemis/DNA-dependent protein kinase complex in nonhomologous end joining and V(D)J recombination. *Cell* **108,** 781–794.
212. Cooper, M. P. *et al.* (2000). Ku complex interacts with and stimulates the Werner protein. *Genes Dev* **14,** 907–912.
213. Li, B., and Comai, L. (2000). Functional interaction between Ku and the Werner syndrome protein in DNA end processing. *J. Biol. Chem.* **275,** 28349–28352.
214. Orren, D. K. *et al.* (2001). A functional interaction of Ku with Werner exonuclease facilitates digestion of damaged DNA. *Nucleic Acids Res.* **29,** 1926–1934.
215. Mahajan, K. N., Nick McElhinny, S. A., Mitchell, B. S., and Ramsden, D. A. (2002). Association of DNA polymerase mu (pol mu) with Ku and ligase IV: Role for pol mu in end-joining double-strand break repair. *Mol. Cell. Biol.* **22,** 5194–5202.
216. Huang, J., and Dynan, W. S. (2002). Reconstitution of the mammalian DNA double-strand break end-joining reaction reveals a requirement for an Mre11/Rad50/NBS1-containing fraction. *Nucleic Acids Res.* **30,** 667–674.
217. Goedecke, W., Eijpe, M., Offenberg, H. H., van Aalderen, M., and Heyting, C. (1999). Mre11 and Ku70 interact in somatic cells, but are differentially expressed in early meiosis. *Nat. Genet.* **23,** 194–198.
218. Paull, T. T., and Gellert, M. (2000). A mechanistic basis for Mre11-directed DNA joining at microhomologies. *Proc. Natl. Acad. Sci. USA* **97,** 6409–6414.
219. DiBiase, S. J. *et al.* (2000). DNA-dependent protein kinase stimulates an independently active, nonhomologous, end-joining apparatus. *Cancer Res.* **60,** 1245–1253.

The Yeast and Plant Plasma Membrane H^+ Pump ATPase: Divergent Regulation for the Same Function

BENOIT LEFEBVRE,
MARC BOUTRY, AND
PIERRE MORSOMME

Unité de biochimie physiologique,
Institut des Sciences de la Vie,
University of Louvain,
B-1348 Louvain-la-Neuve,
Belgium

I. Introduction	204
II. PM H^+-ATPase Gene Organization and Expression	205
III. Structure and Activity	206
A. Topology and Structure	206
B. Enzymology	207
C. Oligomerization	209
IV. Functions	209
A. Transport	209
B. Cell Elongation	211
C. Internal pH Homeostasis	211
D. Abiotic Stress Resistance	211
V. Translational Regulation	212
VI. Posttranslational Regulation	212
A. Regulation of the Yeast PM H^+-ATPase	213
B. Regulation of the Plant PM H^+-ATPase	214
C. Different PM H^+-ATPase Isoforms Differ in Their Regulatory Properties	217
D. Other Regulatory Mechanisms of the PM H^+-ATPases	217
VII. Trafficking	218
A. PM H^+-ATPase and the Secretory Pathway	218
B. PM H^+-ATPase and ER Quality Control	219
C. PM H^+-ATPase and the COPII Complex	219
D. ER Export and Diacidic Signal	220
E. Plasma Membrane Targeting Sequence	222
F. PM H^+-ATPase and Rafts	222
G. Oligomerization, Rafts, and Trafficking	224
H. Conclusion	225
VIII. General Conclusions	225
References	228

The plasma membrane H^+-ATPase from fungi and plants is a P-type ATPase proton pump which plays a key role in the physiology of these organisms, controlling essential functions, such as nutrient uptake and intracellular pH regulation. In yeast and plant cells, its activity is regulated by a large number of environmental factors. Although yeast and plant H^+-ATPases share the same topology and function, these two organisms have developed different regulatory mechanisms in order to tightly control H^+-ATPase activity. Recent advances in the field have been mainly made using three different approaches: crystallography, biochemistry, and molecular genetics. The elucidation of the three-dimensional structure of the sarcoplasmic reticulum Ca^{2+}-ATPase in two different states of the catalytic cycle has led to a greater understanding of the structure-function relationships for both yeast and plant H^+-ATPases. Biochemical approaches have demonstrated differences in regulation of the activity of the yeast and the plant H^+-ATPases. In particular, much effort has been applied to elucidating the mechanism of regulation of the plant H^+-ATPase by its regulatory carboxy-terminal domain which involves the phosphorylation-dependent binding of 14-3-3 proteins. Finally, the combination of yeast genetics and biochemistry has helped to unravel the specific requirements for the trafficking of the H^+-ATPase from the endoplasmic reticulum to its final destination, the plasma membrane.

I. Introduction

The plasma membrane (PM)[1] proton-pump ATPase (PM H^+-ATPase), a 100-kDa polypeptide, is one of the most abundant proteins in the PM and is found in plants, fungi, and some archaebacteria. Using the energy provided by hydrolysis of ATP, it pumps protons out of the cell, creating a membrane potential and pH gradient, which, in turn, activate many secondary transporters. At the same time, PM H^+-ATPase is a high consumer of cellular ATP [accounting for 25–50% of the total ATP consumption in root hairs (1)] and is therefore expected to be tightly regulated.

Our laboratory has carried out research on PM H^+-ATPase for more than two decades. In 1980, we purified the yeast PM H^+-ATPase from *Saccharomyces pombe* (SpPMA1) (2), then cloned the gene in 1987 (3), whereas the

---

[1] Abbreviations: At, *Arabidopsis thaliana*; DIG, detergent-insoluble glycolipid-enriched domain; ER, endoplasmic reticulum; Hs, *Homo sapiens*; LPC, lysophosphatidylcholine; Nc, *Neurospora crassa*; Np, *Nicotiana plumbaginifolia*; PM, plasma membrane; PMA, plasma membrane H^+-ATPase; Sc, *Saccharomyces cerevisiae*; Sp, *Schizosaccharomyces pombe*; TMD, transmembrane domain; ts, temperature sensitive. All gene and protein names are preceded by the species abbreviation. The nomenclature used is *ScPMA1*, gene; ScPMA1, protein; *Scpma1-7*, mutant gene; ScPMA1-7, mutant protein; *Scpma1* Δ, gene deletion.

Saccharomyces cerevisiae ScPMA1 gene was cloned in 1986 by Serrano *et al.* (*4*). In 1989, three plant PM H$^+$-ATPase genes were isolated, two in *Arabidopsis* (*5, 6*) and one in *Nicotiana plumbaginifolia* (*NpPMA2*) (*7*). Eight more genes have been identified in the latter species [reviewed in (*8*)]. Meanwhile, work on *S. pombe* was abandoned to carry out more extended studies on *S. cerevisiae* ScPMA1. From this time, analysis of the H$^+$-ATPase in the unicellular yeast and higher plants gave us an overview of differences and similarities between the same protein in the two homologous families. Research on these two groups of organisms has provided complementary information about the enzyme. However, it is also interesting to see how gene family organization and enzyme regulatory properties have evolved to adapt to very distinct physiological contexts in yeast and plants. Recently, PM H$^+$-ATPase has also become a paradigm for studying the trafficking of polytopic proteins to the PM.

In this review, we will first describe the classification and family organization of the PM H$^+$-ATPase genes as well as the structural properties and main physiological roles of the enzyme. We will then focus on the different regulation mechanisms and, finally, deal with the trafficking aspect.

II. PM H$^+$-ATPase Gene Organization and Expression

In contrast to the vacuolar H$^+$-ATPase (V-type) or the mitochondrial and chloroplast H$^+$-ATPases (more accurately referred to as ATP synthases), the PM H$^+$-ATPase shares structural and mechanical properties with the P-type ATPase superfamily, which includes the well-characterized Ca^{2+}-, Na$^+$/K$^+$-, and H$^+$/K$^+$-ATPases. More precisely, the PM H$^+$-ATPase belongs to the P$_{III}$ family [(*9*), see also http://biobase.dk/~axe/Patbase.html]. The P-type family is defined by the fact that its members hydrolyze ATP through a phosphorylated aspartate intermediate during the catalytic cycle. Sixteen P-type ATPases have been identified in *S. cerevisiae* and 47 in *Arabipodsis*. P-type ATPases belong to the hydrolase superfamily and have probably evolved by the fusion of a phosphatase with a nucleotide-binding protein. Structure and sequence comparisons have highlighted similarities between the catalytic site of a haloacid dehalogenase and that of a P-type ATPase [reviewed in (*10*)]. The yeast H$^+$-ATPases and the *Neurospora crassa* enzyme, NcPMA1, share about 29% amino acid identity with the plant H$^+$-ATPases, revealing a common backbone that is conserved between these organisms.

PM H$^+$-ATPases are encoded by a family of two genes (89% amino acid identity) in *S. cerevisiae* and of about 10 genes in plants (*8, 11*). In yeast, only one gene, *ScPMA1*, is highly expressed and essential. The other, *ScPMA2*, has a very low expression and can be deleted without any effect on yeast growth, but can functionally replace *ScPMA1* if its transcription promoter is replaced by that of *ScPMA1*. The existence of two genes is probably explained by ancestral duplication of the yeast genome, with subsequent predominance of one of the two genes.

The different plant isoforms are organized into five subfamilies, which appeared early during plant evolution, as the same organization is found in monocotyledons (rice) and dicotyledons (*Nicotiana, Arabidopsis*) (8). In contrast to this conserved organization, gene composition within a subfamily varies between species, suggesting that gene duplication and/or deletion occurred more recently during evolution.

Genes belonging to subfamilies III, IV, and V (8) seem to be expressed in particular cell types, and their roles in the plant have not yet been precisely identified. The expression of subfamilies I and II has been characterized in more detail. Individual members can be expressed in a restricted number of cell types or in many cell types. However, if we sum up the expression of all the genes within these two subfamilies, we can conclude that both subfamilies are expressed in almost all cell types. There are, however, quantitative differences. The most abundant expression is found in those cell types that undergo intensive transport, i.e., the root epidermis, phloem, guard cells, and meristematic tissues. In *N. plumbaginifolia*, *NpPMA2* (subfamily I) and *NpPMA4* (subfamily II) are each expressed in many cell types and were therefore chosen as representative of their subfamilies to understand their respective roles (see below).

III. Structure and Activity

A. Topology and Structure

PM H^+-ATPases are composed of a single polypeptide of approximately 100 kDa, organized into 10 transmembrane domains (TMD) and four cytosolic regions, the NH_2- and COOH-terminal regions, a small loop between TMD2 and TMD3, and a large loop (bearing the catalytic site) between TMD4 and TMD5 (Fig. 1).

Although two-dimensional (2D) crystallographic and electron microscopic analysis of *N. crassa* (*12, 13*) and *Arabidopsis* (*14*) PM H^+-ATPases has been performed, the highest resolution (2.6 Å) for a P-type ATPase has been obtained by X-ray diffraction of a three-dimensional (3D) crystal of rabbit sarcoplasmic reticulum Ca^{2+}-ATPase (*15, 16*) (Fig. 2).

The N-terminal region and the small loop form the A-domain, whereas the large loop is divided into two domains, the phosphorylation (P) domain, which bears the aspartate residue that is phosphorylated during catalysis and that interacts with the membrane domain involved in cation translocation, and

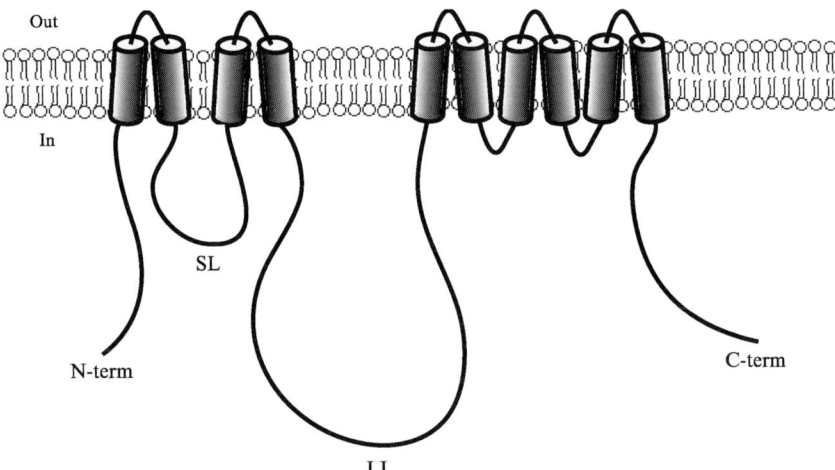

FIG. 1. PM H$^+$-ATPase topology. The PM H$^+$-ATPase is composed of 10 transmembrane domains and four exposed hydrophilic cytoplasmic domains, the N-terminal region (N-term); small loop (SL), large catalytic loop (LL), and C-terminal region (C-term).

the nucleotide-binding (N) domain, which binds ATP and that after structural rearrangement transfers the γ-phosphate to the P-domain. The C-terminal domain is almost absent in this Ca^{2+}-ATPase, but modeling of the *N. crassa* PM H$^+$-ATPases suggests that this region interacts with the N-domain (13).

B. Enzymology

The PM H$^+$-ATPases, which catalyze proton transport and ATP hydolysis, belong to the P-type ATPase family, so called because its members are characterized by a phosphorylated catalytic intermediate. During the catalytic cycle, the enzyme forms an aspartyl-phosphate intermediate and cycles between two different conformational states: E1 and E2 [reviewed in (17)]. The enzymatic properties of membrane-bound and purified detergent-solubilized PM H$^+$-ATPases have been extensively studied for several isoforms and mutants from yeast and plants [reviewed in (18)]. However, the main advance in the understanding of the enzymology of the P-ATPases came from the resolution of the 3D structure of the E1 and E2 states of the sarcoplasmic reticulum Ca^{2+}-ATPase (15, 16). Comparison of these two structures (Fig. 2) identified conformational changes that take place during catalysis. Basically, in the E2 state, the three cytoplasmic domains A, P, and N gather to form a single headpiece, and 6 of the 10 transmembrane helices undergo large-scale rearrangements. These modifications allow the subsequent release of calcium ions into the lumen of the sarcoplasmic reticulum (16).

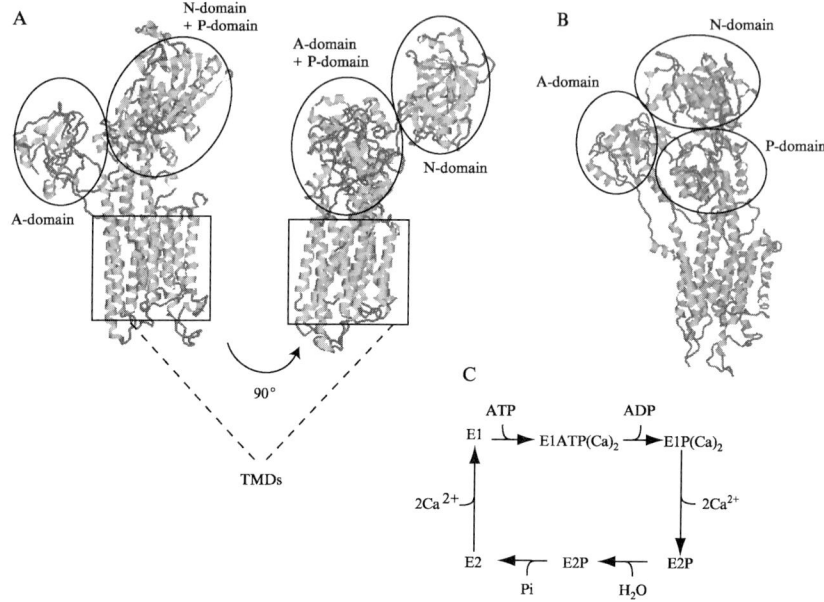

FIG. 2. 3D representation of the structure of the rabbit endoplasmic/sarcoplasmic reticulum Ca^{2+}-ATPase [reproduced with kind permission from Toyoshima *et al.* (*15*, *16*)]. (A) Lateral view of the Ca^{2+}-ATPase structure in the E1 state (*15*) The left picture shows the A-domain, composed of the N-terminal region and the small loop, the transmembrane domains (TMD), and the large loop. The two domains of the large loop, the phosphorylated domain (P-domain) and the nucleotide-binding domain (N-domain), are shown in the right picture. (B) Lateral view of the Ca^{2+}-ATPase structure in the E2 state (*16*). These structures were obtained using the Rasmol program (*166*) (E1 state: 1EUL, E2 state: 1IWO in PDB databank). (C) The scheme shows the Ca^{2+}-ATPase catalytic cycle based on two conformational states: E1 and E2. The same scheme applies to H^+-ATPase by replacing Ca^{2+} with H^+.

Such a resolution has not yet been achieved for the PM H^+-ATPases, as the best resolution obtained so far for the *Neurospora crassa* H^+-ATPase is 8 Å using electron microscopy of 2D crystals, followed by 3D map reconstruction (*12*). More recently, comparison of the ligand-free and Mg^{2+}/ADP-bound states (both in the E1 state) of this enzyme has shown rearrangements of the cytoplasmic domain following Mg^{2+}/ADP binding, accompanied by a modest conformational change in the transmembrane domain (*19*). Higher resolution is required to refine the mechanism of proton transport, although this will be complicated by the nature of the H^+ ion itself, which can bind to carboxyl and amino groups that are unrelated to the transport path. Another important unanswered question is the number of protons transported per ATP molecule hydrolyzed. This number is still unknown, but has been suggested to depend on the activation state of the PM H$^+$-ATPase in both yeasts (20) and plants (21, 22). Together with the stoichiometry, the determination of the mechanism of proton transport by the PM H$^+$-ATPase will be a major challenge.

C. Oligomerization

Although there is considerable evidence suggesting that PM H$^+$-ATPases exist as homooligomers, the degree of oligomerization is still uncertain, notably because of the difficulty in identifying the exact size of proteins interacting with detergents. In yeast, the *S. pombe* SpPMA1 is solubilized with detergent as octamers or decamers (2), whereas ScPMA1 is solubilized as trimers, dodecamers, or both, depending on the detergent used (23, 24). *N. crassa* PM H$^+$-ATPase is solubilized and purified as hexamers (25) and hexamers are also observed by electron microscopy of a 2D crystal (12); however, the oligomer size in the PM, calculated by radiation inactivation, corresponds to that of a dimer (26). When expressed in yeast, PM H$^+$-ATPases from the plant, *Arabidopsis* (14), and the archea, *Methanococcus jannaschii* (27), appear as dimers. After solubilization and purification, *N. plumbaginifolia* NpPMA2 was also found to be a dimer (O. Maudoux and J. Kanczewska, unpublished results).

Because most of these data were obtained using solubilized enzymes, it is possible that oligomerization is an artifact caused by the presence of detergents. The quaternary structure of PM H$^+$-ATPase has still to be determined within the PM, e.g., using cross-linking agents.

IV. Functions

A. Transport

In all organisms, the PM H$^+$-ATPase maintains the electrical and pH gradients across the PM and these, in turn, provide the energy used by channels and transporters for nutrient and ion transport across the membrane (Fig. 3). This role in plant and yeast organisms is thus comparable to the basic function of the Na$^+$/K$^+$-ATPase in animal cells. Rapid growth of yeast cells requires high transport of metabolites and thus relies heavily on PM H$^+$-ATPase-activated transport. As a consequence, the PM H$^+$-ATPase is very abundant, accounting for up to 10% of the PM proteins. As discussed later, this requires very strict control of its activity to avoid unnecessary ATP consumption.

In plants, the situation is more complex, as some cell types are more involved in transport than others and thus express larger amounts of PM H$^+$-ATPases. As shown by *in situ* immunolocalization or using the β-glucuronidase reporter, the PM H$^+$-ATPase is particularly abundant in certain cell types (10, 11, 28–31). This is the case in the root epidermis and to a lesser degree, the

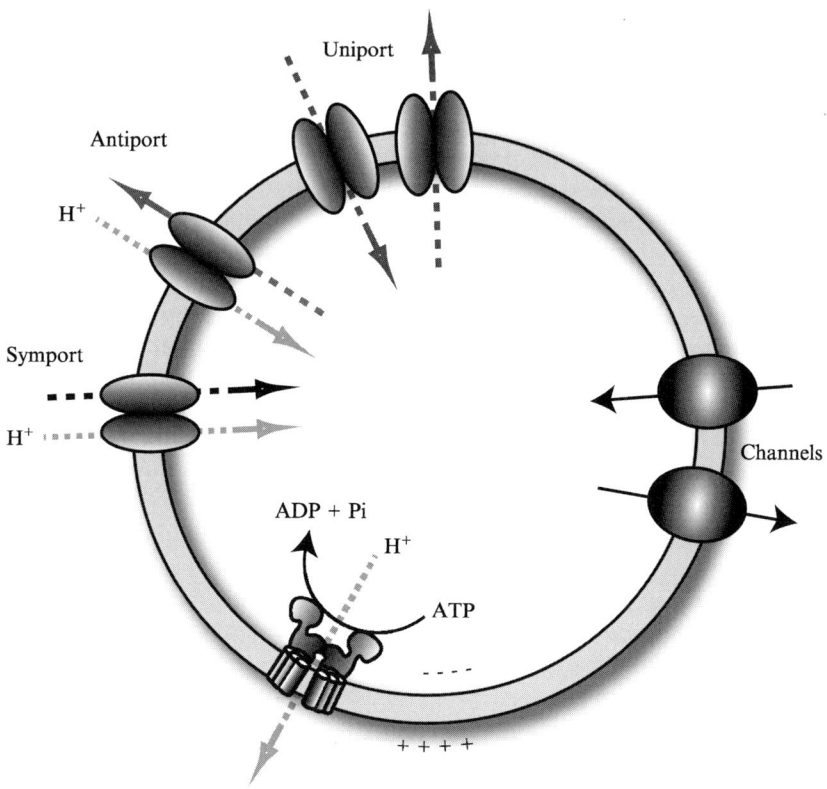

FIG. 3. Activation of the secondary transport systems by PM H^+-ATPase in plant and yeast cells. The PM H^+-ATPase transforms chemical energy (ATP) into electrical and osmotic energy (pH and H^+ gradients across the plasma membrane). This proton motive force is used by channels and secondary transporters (symporters, antiporters, or uniporters) to transport nutrients or ions across the plasma membrane.

subjacent cortical layer, which are involved in mineral nutrition (NO_3^-, NH_4^+, HPO_4^{2-}, K^+, Mg^{2+}, etc.). The phloem network is involved in the long distance transport of metabolites (sugars, amino acids, etc.), particularly between the source tissues (mature leaves, which generate organic molecules using photosynthetic energy) and sink tissues, which are net importers. The companion cells, which load the phloem vessels with sucrose released into the apoplasm (external medium) by the mesophyl cells, express high levels of PM H^+-ATPase. High PM H^+-ATPase expression is also found in the guard cells, the pairs of cells that control, in the epidermis, the size of the stomatal aperture, through which gas exchange (CO_2 and O_2) and water transpiration occur. Aperture size is controlled by osmotic effects that cause the guard cells to shrink or expand, and these osmotic modifications result from ion movement (e.g., K^+ and organic acids) through the PM, hence the importance of the PM H^+-ATPase. Finally, the PM H^+-ATPase is also abundant in meristematic regions, as these fast growing tissues require a large supply of organic resources.

B. Cell Elongation

An osmotic effect related to PM H^+-ATPase activity in plants has also been implicated in cell elongation, which is mainly triggered by an increase in cell volume. In addition to its indirect role in cell swelling, the PM H^+-ATPase is responsible for the acidification of the apoplasm that might be involved in cell wall loosening, possibly by triggering activation of endoglucanases (32) or a class of newly characterized enzymes, the expansins (33). Cell swelling and wall loosening result in cell elongation. Auxin, a phytohormone with a well-characterized role in cell elongation, activates the PM H^+-ATPase through auxin-binding proteins, including ABP_{57} (34). Moreover, fusicoccin, a fungal toxin, which irreversibly binds and activates PM H^+-ATPases (see below), partly mimics auxin by mediating cell elongation. In addition to activating PM H^+-ATPase, auxin has been reported to affect its expression (35, 36); however, contradictory results have been obtained in this regard (37, 38). The role of the PM H^+-ATPase proposed in the acid growth theory (39) is still debated.

C. Internal pH Homeostasis

Many metabolic reactions modify the cellular pH, either because they use or produce protons or because their substrates and products have weak acidic or basic functions with different pKs. It is tempting to hypothesize that the PM H^+-ATPase is involved in internal pH homeostasis when a low cytosolic pH requires the elimination of protons. There is no direct evidence for this; however, depending on the isoform, the optimal pH for the enzyme is between 6 and 7, suggesting that the pump might be activated when the cytosolic pH becomes too low.

D. Abiotic Stress Resistance

In contrast to animal cells, which are bathed in a saline external medium, plant and yeast cells are sensitive to NaCl. Although low Na^+ and high K^+ intracellular concentrations are conserved in these organisms, the main difference between the two groups is the presence, in animal cells, of the Na^+/K^+-ATPase pump (instead of the H^+-ATPase pump), which controls Na^+ homeostasis. In plants, PM H^+-ATPases are expected to be involved in salt tolerance, maintaining the transmembrane H^+ gradient used by PM Na^+/H^+ transporters to extrude Na^+ ions. Another means used by plants and yeasts to protect themselves from Na^+ toxicity is vacuole sequestration. Salt stress also results in osmotic stress taken in charge by more specific tolerance mechanisms, such as accumulation of osmoprotectants. PM H^+-ATPase mRNA accumulates in *Atriplex nummularia*, *N. tabacum*, or tomato in response to NaCl treatment (*40, 41*). Moreover, an *Arabidopsis* mutant with a knockout PM H^+-ATPase *AtAHA4* gene is more sensitive to salt stress (*42*).

In yeast, no increase in PM H^+-ATPase transcript levels is seen during salt stress (*43, 44*). In addition, it seems that PM H^+-ATPase down-regulation increases salt stress resistance. ScPMA1 mutants with reduced activity show increased salt tolerance (*45, 46*) and deletion of ScPTK2, a kinase involved in ScPMA1 activation, also results in increased salt tolerance and, more generally, cation tolerance (*47*). A decreased plasma membrane electrical potential could explain the decrease in Na^+ influx (*48*). The existence of a PM sodium pump in yeast (*49, 50*), not seen in plants, might also explain the different behavior of yeast and plant PM H^+-ATPases as regards salt tolerance.

V. Translational Regulation

Although there is no evidence for translation regulation of yeast PM H^+-ATPase, one feature of plant PM H^+-ATPase transcripts suggests that this does occur, this being that the 5' untranslated transcript region (leader) of several of them possesses a short (3-13 codon) upstream open reading frame (uORF) preceding the H^+-ATPase ORF [reviewed in (*8*)]. A large proportion of the ribosomes initiates translation at the uORF; however, some are able to reinitiate downstream translation of the main ORF (H^+-ATPase) (*51, 52*). This system might lead to a fine tuning; although only resulting in a maximal twofold difference in translation, this represents a lot of protein, considering the abundance of PM H^+-ATPase transcripts. The stimuli and factors involved in this regulation are unknown.

VI. Posttranslational Regulation

The catalytic activity of the yeast and plant PM H^+-ATPases is regulated by a large number of physiological signals. The best characterized mechanism of posttranslational regulation of the PM H^+-ATPase involves the autoinhibitory function of the C-terminal domain of the enzyme. It is now well established that this region inhibits the enzymatic activity and that the inhibition can be alleviated by different mechanisms, the most common being (de)phosphorylation of the enzyme.

A. Regulation of the Yeast PM H$^+$-ATPase

Initial data on posttranslational regulation of the yeast PM H$^+$-ATPase came from the observation that the enzyme is activated by addition of glucose to cells starved in a sugar-free medium (53). A similar phenotype is obtained by constitutive deletion of the 11 carboxy-terminal amino acids (54), suggesting that the PM H$^+$-ATPase contains a C-terminal inhibitory domain, the function of which can be switched off by glucose metabolism. Activation of the PM H$^+$-ATPase is characterized by higher ATPase hydrolytic activity, a higher affinity for MgATP, and an alkaline shift of the optimum pH (53). The C-terminal region of ScPMA1 contains putative phosphorylation sites, which are important for glucose-dependent regulation (55). In addition, *in vivo* phosphorylation of ScPMA1 is associated with increased ATPase activity during growth on glucose, and, upon glucose starvation, dephosphorylation occurs concomitantly with a decrease in enzymatic activity (56). Taken together, these observations led to the hypothesis that reversible, site-specific phosphorylation of the C-terminus of ScPMA1 serves to adjust ATPase activity in response to nutritional signals.

Several attempts have been made to identify the kinase(s) involved in this regulatory process, and different pathways have been proposed to transmit the signal leading from nutrient sensing to PM H$^+$-ATPase activation. For example, when glucose is added to starved cells, a transient increase in the cAMP concentration occurs that induces a protein phosphorylation cascade (57), and there is a concomitant increase in PM H$^+$-ATPase activity (53). The cAMP-dependent protein kinase might therefore be involved in regulation of ScPMA1 activity during growth in glucose. However, it is now clear that the link between these two effects is indirect (58). Other indirect evidence suggests that two different pathways are involved in glucose-dependent ScPMA1 regulation. In the first, glucose-induced activation of the PM H$^+$-ATPase would require the presence of ScSNF3 (a glucose sensor), the ScGPA2 protein (a G-protein), and the PKC signaling pathway (59), whereas, in the second, the kinase ScPTK2 would mediate ScPMA1 activation in response to glucose metabolism via Ser-899 in the C-terminal region (47). ScPTK2 has also been suggested to be a member of a protein kinase subfamily dedicated to the regulation of PM transporters (47). However, none of these proposed interactions has been shown to be directly involved in glucose-dependent PM H$^+$-ATPase phosphorylation.

To date, only one kinase activity has been shown to be directly responsible for PM H$^+$-ATPase phosphorylation in response to glucose metabolism, this being the casein kinases, ScYCK1 and ScYCK2, which phosphorylate ScPMA1 *in vitro* and regulate ScPMA1 activity, depending on the presence or absence of glucose (60). Surprisingly, the phosphorylation site identified in this study is located in the ATP-binding domain and not, as expected, in the C-terminal region and, second, the phosphorylation level decreases on glucose addition. Taken together with previous data, this strongly suggests that PM H^+-ATPase activity is modulated by the combined action of a down-regulating casein kinase I and an uncharacterized up-regulating kinase.

Finally, we still need to know which domains of the PM H^+-ATPase interact with the autoinhibitory C-terminal domain. One candidate is the large loop, which is the target for the casein kinase mentioned above. Alternatively, the stalk segment 5, which connects the catalytic domains to the membrane inserted region, contains residues mutation of which leads to constitutive activation of the PM H^+-ATPase even in the absence of glucose (61).

B. Regulation of the Plant PM H^+-ATPase

The C-terminus of plant PM H^+-ATPases differs from, and is longer than, that of their yeast homologues. Nevertheless, the plant C-terminus also acts as an inhibitory domain and can be removed by tryptic digestion (62) or genetic deletion (63). Removal of at least 51 residues from the C-terminal domain leads to a fully activated enzyme characterized by increased ATPase and H^+-pumping activities and a reduced K_m for ATP. Similar activation of the plant PM H^+-ATPase can be produced by single point mutations obtained in either the C-terminal region or other regions of the NpPMA2 isoform (21, 64, 65). These single point mutations render the C-terminus more sensitive to proteolytic degradation (64), suggesting that these substitutions promote the displacement of the C-terminus from the rest of the enzyme. Alanine-scanning mutagenesis through 87 residues of the C-terminal region of the AtAHA2 isoform identified 23 residues, concentrated in two stretches, mutation of which results in increased H^+-ATPase activity (65) (Fig. 4A).

Recent advances have been made in understanding the mechanism involved in regulation of plant PM H^+-ATPase activity via its autoinhibitory C-terminal domain by studying the action of the fungal toxin, fusicoccin. As mentioned above, fusicoccin treatment of plant cells induces physiological phenomena related to PM H^+-ATPase activation. *In vivo* treatment of intact plant tissues with the toxin stimulates PM H^+-ATPase activity (66–69). At the molecular level, fusicoccin activation of the plant PM H^+-ATPase has been shown to involve a fusicoccin-binding receptor, which consists of a complex of regulatory 14-3-3 proteins and the PM H^+-ATPase (70–72). Recent studies have shown that these proteins form a complex with the C-terminal regulatory domain of the PM H^+-ATPase, thus releasing its inhibitory action (73–75).

14-3-3 proteins, a family of highly conserved regulatory molecules that function as dimers, are present in all eukaryotic cells, in which they bind a

A

YALSGRAWDLVLEQRIAFTRKKDFGKEQRELQWAH
AQRTLHGLQVPDTKLFSEATNFNELNQLAEEAKRRA
EIARQRELHTLKGHVESVVKLKGLDIETIQQSYTV

B

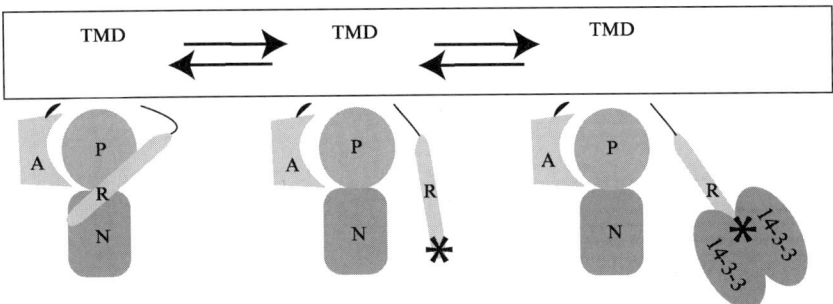

FIG. 4. Regulation of plant PM H$^+$-ATPase activity. (A) Sequence of the plant PM H$^+$-ATPase C-terminal autoinhibitory region (NpPMA2 isoform). Mutations in the NpPMA2 C-terminal region leading to activation of the pump are shown in the amino acid sequence in bold letters. The amino acid residues in NpPMA2 corresponding to those demonstrated, in AtAHA2, to belong to the inhibitory domain of the AtAHA2 C-terminal region are shown in the gray boxes. Two distinct inhibitory domains can be seen, although more clearly in AtAHA2 than in NpPMA2. (B) Regulatory mechanism of the PM H$^+$-ATPase. The C-terminal inhibitory region, also called the regulatory domain (R-domain), binds one or more other cytoplasmic domains (as yet unidentified) and inhibits the pump activity. Phosphorylation of the penultimate Thr and the subsequent binding of 14-3-3 proteins displace the C-terminal region and release pump inhibition.

variety of different substrates, thus regulating several different signaling pathways. In most cases, 14-3-3 proteins recognize their target by a specific phosphorylated sequence, the two most widespread motifs being RSXpSXP (76) and RXY/FXpSP (77). Although the plant PM H$^+$-ATPase C-terminal domain does not contain such a motif, attempts to identify a phosphorylated motif led to the discovery that phosphorylation of the penultimate Thr in the C-terminal region results in the formation of the binding site for 14-3-3 proteins (78–81).

The fusicoccin-binding site is generated after formation of the 14-3-3/PM H$^+$-ATPase complex (82–84), suggesting that fusicoccin stabilizes the preformed complex, rather than inducing its formation. However, fusicoccin treatment is far removed from normal physiological conditions, raising the question of whether this complex exists under normal conditions and how it is regulated. Heterologous expression of NpPMA2 in yeast allowed its purification as a complex with 14-3-3 proteins in the absence of fusicoccin treatment, showing that fusicoccin is not necessary for the formation of a stable complex (*81*) (Fig. 4B).

In plants, when guard cells from *Vicia faba* are exposed to blue light, the PM H^+-ATPase is activated, its C-terminus is phosphorylated on the penultimate Thr residue, and 14-3-3 proteins are coimmunoprecipitated with it (*85*). Four recently identified blue light receptors represent the first step in the signaling cascade leading from the blue light signal to stomatal opening as a result of PM H^+-ATPase phosphorylation and activation (*86*).

Other stresses are probably involved in PM H^+-ATPase activation. For example, it was recently proposed that osmotic stress might activate plant PM H^+-ATPase via the binding of 14-3-3 proteins (*87, 88*), and we have now shown this to be the case for both NpPMA2 and NpPMA4 in tobacco (M. Woloszynska, unpublished results). The physiological role of the C-terminus-mediated inhibition of PM H^+-ATPase activity has also been demonstrated in intact plants, with expression of a truncated version of the AtAHA3 isoform in transgenic plants leading to resistance to acid medium at the seedling stage (*89*).

Regulation of the binding of 14-3-3 proteins to the PM H^+-ATPase is probably mediated by a kinase/phosphatase cycle. Binding of 14-3-3 proteins to the PM H^+-ATPase C-terminal region requires a phosphorylated Thr at position C-2. The identity of the kinase involved is not known, but it is suggested to be localized in the PM (*80*). To dissociate the complex, the C-terminus would have to be dephosphorylated by a protein phosphatase, and a protein phosphatase (PP2A) that might do this has been purified from maize roots (*90*).

Phosphorylation of plant PM H^+-ATPases might be more complex than simply the phosphorylation of the penultimate Thr required for 14-3-3 binding. Several studies have reported some intriguing, and sometimes contradictory, effects of (de)phosphorylation on plant PM H^+-ATPase activity. It seems clear that phosphorylation of the penultimate Thr leads to 14-3-3 protein binding and PM H^+-ATPase activation. Syringomycin, a virulence factor produced by certain plant pathogens, stimulates *in vitro* Ca^{2+}-dependent protein phosphorylation of the red beet PM H^+-ATPase and activation of the pump (*91*). In addition, dephosphorylation of the PM H^+-ATPase correlates with a decrease in its activity (*92, 93*). However, a Ca^{2+}-dependent phosphorylation activity is reported to result in decreased PM H^+-ATPase activity (*94–96*). In a fashion similar to the yeast PM H^+-ATPase, the plant PM H^+-ATPase thus seems to be up- and/or down-regulated by phosphorylation of different sites, only one of which has been identified.

C. Different PM H$^+$-ATPase Isoforms Differ in Their Regulatory Properties

The yeast, *S. cerevisiae*, contains two isoforms of the PM H$^+$-ATPase, ScPMA1 and ScPMA2, which are differently regulated by posttranscriptional modification. Although ScPMA2 is expressed at a much lower level than ScPMA1 under normal conditions (97), it is possible to express ScPMA2 under the control of the *ScPMA1* promoter and to study its enzymatic and regulatory properties (98). This approach showed that the regulation of ScPMA2 and ScPMA1 by glucose metabolism is different, as, although addition of glucose increases the affinity of ScPMA2 for MgATP, it has no effect on the hydrolytic ATPase activity (98). Second, ethanol treatment of yeast cells activates PM H$^+$-ATPase activity in a strain expressing only ScPMA1, but not that in a strain expressing only ScPMA2 (99). The molecular basis of these differences is unknown.

In plants, different isoforms in *Arabidopsis* and *N. plumbaginifolia* have distinct enzymatic properties (100, 101). Recently, a study of the regulatory properties of NpPMA2 and NpPMA4, two representatives of the two main plant PM H$^+$-ATPase subfamilies, showed that the C-terminal regulatory regions of NpPMA2 and NpPMA4 cannot substitute for one other and in contrast to NpPMA2, NpPMA4 is poorly phosphorylated at its penultimate Thr and binds 14-3-3 proteins weakly (102). The molecular analysis of these differences identified a residue at position −4 that greatly influences the phosphorylation status of the penultimate Thr, binding of 14-3-3 proteins, and, subsequently, the ATPase activity of the enzyme (102). This difference at position C-4 (Ala in NpPMA2 and His in NpPMA4) can be generalized to all PM H$^+$-ATPases from various species belonging to the same subfamily as either NpPMA2 (Ala or Ser at C-4) or NpPMA4 (His or Asn at C-4). These data clearly show different regulatory properties of these two main PM H$^+$-ATPase subfamilies. However, whether this is related to different signal transducing systems in the plant remains to be determined.

D. Other Regulatory Mechanisms of the PM H$^+$-ATPases

The phosphorylation-mediated regulation of PM H$^+$-ATPase activity is probably not the only mechanism developed by organisms to control this essential activity. In yeast, two small hydrophobic proteins, ScPMP1 and ScPMP2, copurify with ScPMA1, and their deletion reduces PM H$^+$-ATPase activity by 50% (103, 104). The integral PM heat shock protein, ScHSP30, which is induced by either heat shock or weak organic acid stress, regulates the PM H$^+$-ATPase (105). Although phosphorylation of the C-terminus is not involved in stress induction of the PM H^+-ATPase, it appears to be involved in ScHSP30-mediated modulation (105).

In plants, a potential candidate recently isolated by the yeast two-hybrid technique is suggested to interact with the AtAHA1 PM H^+-ATPase and regulate its activity (106). As already mentioned, the auxin-binding protein, ABP_{57}, is also suggested to activate the plant PM H^+-ATPase via a direct interaction with indole acetic acid (IAA) (34). Finally, lysophosphatidylcholine (LPC) is the most powerful known phospholipid activator of the plant PM H^+-ATPase, and phospholipase A_2, which hydrolyzes phosphatidylcholine into LPC and free fatty acids, is present in the oat root PM, suggesting that LPC could be a natural regulator of PM H^+-ATPase activity (107). Although it has been suggested that LPC does not bind to the C-terminus of the PM H^+-ATPase (108), the activation mechanism is currently unknown.

VII. Trafficking
A. PM H^+-ATPase and the Secretory Pathway

In contrast to Ca^{2+}-ATPases, which are found in the PM, the vacuole, and the endoplasmic/sarcoplasmic reticulum in plants and yeasts, P-type H^+-ATPases have been identified only in the PM. However, PM H^+-ATPases are asymmetrically distributed around the cell in some plant cell types, such as the root epidermis and transfer cells (29, 31). This is reminiscent of the asymmetrical distribution of the Na^+/K^+- and H^+/K^+-ATPases in, for example, intestinal or stomach epithelium. Accumulation of plant PM H^+-ATPase in some kind of secretory vesicles responding to auxin has also been suggested (36) but not supported since then.

It has been known for several years that the yeast PM H^+-ATPase, ScPMA1, follows the secretory pathway to reach the PM. This was shown by blocking protein transport at different stages of the pathway using different secretory (*sec*) mutants blocked in the ER (*Scsec18*), the Golgi apparatus (*Scsec7*), or the secretory vesicles (*Scsec6* and *Scsec1*) (56, 109, 110). The specific sequences responsible for the targeting of the H^+-ATPase to the PM are unknown, and, more generally, little is known about the targeting information that drives polytopic proteins to the PM or segregates them from the endosome/vacuole pathway. Recently, some mechanistic aspects of ScPMA1 transport have been studied, highlighting new aspects of PM H^+-ATPase trafficking in the secretory pathway [reviewed in (111)]. Although an actin-binding protein has been suggested to modulate the ScPMA1 endocytosis process (112), the main efforts in deciphering ScPMA1 trafficking have concentrated on its biosynthetic pathway.

B. PM H$^+$-ATPase and ER Quality Control

ScPMA1 PM H$^+$-ATPase is synthesized on, and integrated into, the ER membrane, where it rapidly becomes fully folded (*113*). A number of ScPMA1 point mutations result in this enzyme being retained in the ER as a consequence of its misfolding, as shown by its higher sensitivity to trypsin [reviewed in (*114*)], and this correlates with abnormal ER proliferation. Most of the ER-retained ScPMA1 mutants are dominant lethal, consistent with oligomerization at the level of the ER (see below) and retention of the wild-type enzyme because of the presence of the mutant ATPase (*115*). Both the mutant and wild-type ATPases might be blocked in the ER by the quality control system. For example, the ScPMA1D378N mutant, in which the catalytically important Asp residue is replaced by Asn, exhibits a severe biosynthetic defect, leading to a dominant lethal phenotype. As judged by its extreme sensitivity to trypsin, this PM H$^+$-ATPase is very poorly folded. It becomes arrested in ER-derived membrane proliferations (*116*), fails to escape to the Golgi, and is translocated back into the cytosol for degradation by the proteasome (*117*). Insight into the nature of the ER quality control process has been obtained by screening an insertional genomic library for suppressors of the dominant lethal behavior of D378N (*117*). This approach has yielded a gene called *ScEPS1* (ER-retained *Scpma1* suppressing) that, when disrupted, prevents degradation of the D378N polypeptide and allows it to reach the cell surface. The product of the *ScEPS1* gene belongs to the protein disulfide isomerase family and may act as a membrane-bound chaperone. Its specificity for misfolded PM H$^+$-ATPase is not fully understood, but deleting *ScEPS1* has little or no effect on the biogenesis of wild-type PM H$^+$-ATPase or on the retention of other proteins that normally reside in the ER.

C. PM H$^+$-ATPase and the COPII Complex

Transport along the secretory pathway is mediated by vesicular carriers, which bud from one intracellular organelle and then are targeted to, and fuse with, the next compartment in the pathway. This mechanism has been characterized in yeast. Correctly folded secretory proteins travel from the ER to the Golgi apparatus in transport vesicles coated with coat protein complex II (COPII). COPII promotes ER-to-Golgi vesicle formation and separates biosynthetic cargo from ER-resident proteins. COPII coats are assembled on the surface of ER membranes by sequential binding of ScSAR1 GTPase, followed by the ScSEC23–ScSEC24 (ScSEC23/24) complex, then the ScSEC13–ScSEC31 (ScSEC13/31) complex (*118*). *In vivo*, ScSAR1 is activated by the ER-localized ScSEC12 guanine nucleotide exchange factor, which produces ScSAR1–GTP in transient association with cargo to be included in COPII vesicles. Activated ScSAR1 then recruits the ScSEC23/24 complex to ER membranes, forming a stabilized complex with cargo molecules (*119–121*). Finally, the ScSEC13/31 complex is recruited to these prebudding complexes, driving membrane curvature and the formation of typical 50-nm COPII vesicles (*122*).

Budding from ER membranes can be obtained *in vitro* in the presence of ScSAR1, ScSEC23/24, ScSEC13/31, and GTP (*118*). In *in vitro* assays using this minimal machinery, ScPMA1 is inefficiently packaged into ER-derived vesicles, but accumulates in vesicles when a yeast cytosol fraction is added to the assay (*123*). Efficient packaging of ScPMA1 therefore requires factors other than the minimal COPII complex. Indeed, deletion of *ScLST1*, an *ScSEC24* homologue, decreases the amount of ScPMA1 in the PM and results in its accumulation in the ER (*124*). Interestingly, in the absence of ScSEC24, ScLST1 is not sufficient for efficient packaging of ScPMA1, but a mixture of the ScSEC23–ScSEC24 and ScSEC23–ScLST1 complexes results in a 10-fold concentration of ScPMA1 10 in ER-derived vesicles (*123*).

Coimmunoprecipitation experiments have shown that ScPMA1 interacts with ScLST1 (*123*). Vesicles formed using a mixture of the ScSEC23–ScSEC24 and ScSEC23–ScLST1 complexes are 10–15% larger than those formed using ScSEC23–ScSEC24 alone. This was initially suggested to reflect the loading of large ScPMA1 oligomers into ER-derived vesicles (*123*). However, ScLST1 also appears to be necessary for loading ScPMA1 in its monomeric form, whereas overexpression of ScSEC24 results in packaging of the multimeric form of ScPMA1 into 50-nm sized vesicles (*123*). Two *ScSEC24* homologues are known in yeast, these being *ScLST1* and *ScISS1*; the latter has been shown to overlap in function with ScSEC24, as ScISS1 overexpression is able to complement *ScSEC24* lethal deletion. However, no specificity for cargo selection in COPII vesicles has been found for ScISS1 (*125, 126*).

COPII function seems to be highly conserved throughout eukaryotes. Plant *AtSEC12* and *AtSAR1* or *NtSAR1* complement the deletion of their yeast homologues (*127, 128*). Overexpression of either AtSEC12 (*129*) or a dominant mutant of NtSAR1 locked in its GTP-bound form (*130*) in tobacco blocks the ER-to-Golgi transport of reporters. Finally, AtSEC23 has been shown to be partly associated with ER membranes (*131*). Three SEC24 homologues are found in the *Arabidopsis* databank. As in yeast, two of these, AtSEC24B and AtSEC24C, share higher amino acid identity with each other (>50%) than with the third homologue, AtSEC24A (<30%), suggesting a common organization and, perhaps, specificity of the members of this family in yeast and plant organisms.

D. ER Export and Diacidic Signal

A subset of soluble and membrane proteins has been shown to be specifically loaded into COPII vesicles (*123, 132*) and several of these bear a motif that increases the ER exit rate. In animal cells, the vesicular stomatitis virus

G-protein (VSV-G) (*119, 133, 134*) and a PM potassium channel (*135*) have a common diacidic motif, D/EXD/E (in which X can be any amino acid), which promotes ER exit; in both cases, a more extended nonconserved motif seems to be necessary for a high rate of exit from the ER (*135, 136*). More recently, similar ER export signals have been found in the yeast ScSYS1 (Golgi protein) (*137*) and in ScYOR1 (PM ATP-binding cassette transporter) (*138*). Other known or potential diacidic motifs have been reviewed (*139*). The VSV-G cytoplasmic C-terminal region interacts directly with HsSAR1, but the extended diacidic motif is necessary for HsSEC23–HsSEC24 recruitment (*140*), while ScSYS1 binds ScSEC23–ScSEC24 directly through its C-terminal region (*137*). Other ER export motifs have been characterized; these are the FF motif, found in ScEMP24 and HsERGIC53 (*141, 142*), and the LV motif, found in ScEMP24 (*143*). The role of a C-terminal valine as an ER exit signal has also been highlighted by substitution of the ScERGIC53 FF motif and its addition to a reporter protein (*144*).

Because ScPMA1 interacts with ScLST1, we can expect to find ER export motifs in PM H^+-ATPases. However, no specific signal has yet been identified in ScPMA1. In our laboratory, we found that the plant NpPMA4 N-domain seems to be essential for the PM H^+-ATPase to leave the ER. A detailed analysis of this domain revealed a diacidic motif. Substitution of this motif with alanine leads to only partial PM localization, which is further decreased by substitution of the surrounding amino acids (B. Lefebvre, unpublished results). This motif is conserved in all *N. plumbaginifolia* PM H^+-ATPases and in 18 of 22 *Arabidopsis* and rice homologues. Our results suggest that it is not essential for PM targeting, but might influence the H^+-ATPase ER export rate.

Several other diacidic motifs have been found in the NpPMA4 sequence, including several in the C-terminal region in which all motifs known to be ER export signals are located (except for one in ScYOR1, which is in the N-terminal region). However, deletion of the NpPMA4 N-terminal region, small loop, or C-terminal region does not modify PM targeting, indicating that none of these regions is involved in ER exiting and PM targeting (B. Lefebvre, unpublished results).

Surprisingly, this plant PM H^+-ATPase diacidic motif is not conserved in yeast homologues, in which a Ser replaces the Asp residue. Phosphorylation of this residue may mimic the diacidic motif, which would be consistent with reported phosphorylation during trafficking (*56*). Alternatively, there are other diacidic motifs in ScPMA1, two of which are in the N-domain. Whether these predicted motifs in ScPMA1 are functional remains to be tested. Finally, we cannot exclude the possibility that the ScPMA1–ScLST1 interaction is mediated by another motif.

E. Plasma Membrane Targeting Sequence

Although no targeting-related sequence information has been clearly identified for vacuolar or PM polytopic proteins, single-membrane-span (monotopic) proteins are proposed to be localized in different secretory pathway compartments according to the length of their TMD. This might be related to differences in the thickness and composition of the lipid bilayer in different subcellular compartments, and fits with the observation that the PM is thicker than internal membranes. Increasing or decreasing the TMD length of different plant, yeast, or mammalian proteins results in their relocalization in different compartments of the secretory pathway (*145–149*). A longer PM TMD might also be connected with raft enrichment of the PM (see below), in which the sphingolipids have longer acyl chains.

Except for the ER export signal, which might accelerate exiting from the ER, no targeting-related sequence information for PM polytopic proteins has been identified in cytoplasmic domains, suggesting that it might be embedded in the transmembrane spans and may depend on their length and composition. Unfortunately, it is not really possible to predict the fitting of TMDs to a membrane from the primary sequence, as TMDs can be inserted into the lipid bilayer plane at angles other than 90°. The low number of membrane protein structures currently available does not allow the analysis of differences between proteins belonging to different compartments. However, although varying the TMD length modifies the localization of large amounts of several single-span proteins, complete relocalization of the modified protein is never achieved. Moreover, relocalization occurs with much longer kinetics than normal trafficking. This could mean that TMD length is not sufficient for correct targeting and that TMD composition is also important or that other targeting information is involved.

F. PM H^+-ATPase and Rafts

For a long time, membranes were considered as simple and homogeneous lipid bilayers, delimiting compartments and providing physical support for proteins, thus regulating cell "life." It is now clear that lipid bilayers are not homogeneous and that some lipids play roles in protein regulation and targeting or as secondary messengers.

One of the new insights into lipid heterogeneity is the existence of rafts (also called DIG for detergent-insoluble glycolipid-enriched domain, or DRMs for detergent-resistant membranes). These were first characterized in mammalian cells and consist of lipid and protein microdomains, enriched in cholesterol and sphingolipids, that have higher lateral cohesion than other membrane lipids (*150*). Sphingolipids are characterized by longer and highly saturated acyl chains. Even though it is clear that rafts exist, the main method used to determine the presence of a protein in rafts is rather indirect, as it is based on the protein being insoluble at 4 °C in Triton X-100.

The first evidence for a relation between ScPMA1 and sphingolipids was obtained in a yeast strain lacking sphingolipids and that, as a consequence, was sensitive to pH and salt, a phenotype similar to that displayed by strains with a defective PM H$^+$-ATPase (*151*). Then ScPMA1 was shown to be associated with the DIG in the PM (*23, 24, 152*). The fraction of DIG-associated ScPMA1 is about 50% (*24*). However, it is difficult to tell whether ScPMA1 equilibrates between the raft and nonraft lipid environment or whether only a part of the raft components is solubilized by detergent.

The presence of at least a part of the yeast PM H$^+$-ATPase in rafts was confirmed by a decreased ScPMA1-DIG association in a yeast strain deficient in the first step of sphingolipid synthesis and, consequently, in raft formation (*152*). Although, in mammalian cells, DIG seem to appear first in the Golgi, where sphingolipid synthesis takes place, in yeast, DIG are already seen in the ER (*23, 152*). It has therefore been postulated that ScPMA1 is incorporated into DIG at an early step of the secretory pathway (*24*), but the exact site of incorporation is still a matter of debate. Recently, an ScPMA1–DIG association was observed in COPII vesicles in an *in vitro* budding assay (*23*). However, attempts to detect DIG-associated ScPMA1 in the ER membranes have not been successful.

Very little is known about rafts in plant membranes. The amount of cholesterol is low, but similar sterols (mainly sitosterol and stigmasterol) and sphingolipids, for which some anabolic enzymes have been identified (*153, 154*), are present. Using liposomes, it has been shown that plant sterols promote the formation of liquid ordered domains with acyl chain-saturated lipids (*155*). Triton X-100-insoluble fractions have also been extracted from plants, and analysis of their protein content has revealed the presence of several glycosylphosphatidylinositol-anchored proteins (*156*), as in yeast and animal cells.

Immunoprecipitation of the microsomal fraction from carnation flowers and canola leaves using antibodies raised against plant PM H$^+$-ATPase revealed a subpopulation of vesicles with lipid contents that differed from that of the PM. One of the main differences was the higher degree of acyl chain saturation (*157*). In contrast to the situation in yeast, sterols and sphingolipids accumulate in plant vacuolar membranes at slightly, but significantly, lower concentrations than in the PM [reviewed in (*158*)], raising the question of whether lipid rafts occur in plant vacuolar membranes. However, no comprehensive description of the lipid composition of the vacuole and PM has been obtained in a single species. In addition, *in vitro*, different plant sterols show different abilities to form liquid-ordered phases (*155*) and the nature and length of the sphingolipids acyl chain might vary. At this stage, no conclusion can therefore be drawn about whether rafts exist in plant vacuoles.

G. Oligomerization, Rafts, and Trafficking

Recently, a correlation was found between PM H^+-ATPase oligomerization, DIG association, and targeting and/or stability. The first evidence came from a study of the yeast temperature-sensitive (ts) mutant, Scpma1–7, in which the newly synthesized PM H^+-ATPase is mistargeted to the vacuole without reaching the PM (159). The mutant protein is nevertheless transported through the Golgi apparatus, as it is rerouted to the PM in an Scvps1 Δ mutant, blocked at the level of Golgi-to-endosome/vacuole trafficking (160). DIG association and oligomerization of the mutant Scpma1–7 are also reduced (24). Overexpression of ScAST1, isolated as a multicopy suppressor of the Scpma1–7 mutation, restores oligomerization, DIG association, and PM localization of ScPMA1–7 (24, 159). Restoration seems to occur in the secretory pathway, as ScAST1 colocalizes with an early Golgi marker (24). Deletion of ScSOP4, which codes for an ER membrane protein, also restores the growth of a strain expressing Scpma1–7 (161). ScSOP4 deletion delays PM H^+-ATPase ER export, extending its residence time in contact with ER proteins and possibly increasing folding, oligomerization, and/or raft association. Moreover, an ER-to-Golgi blockage effect was confirmed by similar restoration of ScPMA1–7 PM targeting when the ER to Golgi transport step was slowed down in an Scsec13 ts mutant strain.

The ts phenotype of the ScPMA1–7 strain is probably due to mistargeting, as the protein is stabilized when it reaches the PM before the shift to restrictive temperature (159). This is also observed for another temperature-sensitive ScPMA1 mutant, ScPMA1–10, which is also degraded in the vacuole, but only after being endocytosed from the PM. Triton X-100 insolubility of ScPMA1–10 seems to be a prerequisite for its stabilization in the PM, which reaches a maximum 2 h after its synthesis at permissive temperature (162).

Bagnat and co-workers (24) found that Scvps1Δ restores ScPMA1–7 PM localization, but does not completely restore DIG association compared to an ScAST1 overexpressing strain. This suggests that the localization of ScPMA1–7 in the PM, in which sphingolipid and ergosterol concentrations are higher (163, 164), is not the only factor responsible for DIG association and oligomerization when ScAST1 is overexpressed. However, relocalization can still participate to the DIG association of the PM H^+-ATPase, but the process might be slow.

A second element suggesting a relationship between rafts and trafficking is that sphingolipid depletion leads to a decrease in ScPMA1 oligomerization and half live, due to degradation in the vacuole (23, 24, 115). Sphingolipids are depleted by blocking the first enzyme of the sphingolipid biosynthetic pathway (ScLCB1) using a specific inhibitor, myriocin, or using the ts strain Sclcb1-100 shifted to a restrictive temperature. In the Sclcb1-100 strain, depending on the conditions, ScPMA1 is either targeted to the PM, where it has a shorter half life and is degraded in the vacuole after ubiquitymation and endocytosis (115), a behavior similar to that of ScPMA1-10, or is directly degraded in the vacuole (23, 24, 115), a behavior similar to that of ScPMA1-7. Differences in sphingolipid depletion according to the conditions might thus affect ScPMA1 raft association and its targeting. A moderate decrease in sphingolipid synthesis might lead to PM targeting, but destabilize the enzyme, leading to its degradation, whereas a large decrease might inhibit PM targeting. Similarly, differences in the ability of ScPMA1-7 and ScPMA1-10 to oligomerize and/or associate with the raft might explain their behavior.

The functions of, and the relationship between, PM H^+-ATPase oligomerization and raft association are still unknown. However, even though they occur in COPII vesicles or earlier, raft association and/or oligomerization do not seem to be necessary for ER exiting, since a ScPMA1 monomeric form is degraded in the vacuole *in vivo* and efficiently packaged in COPII vesicle in an *in vitro* budding assay (23). However, they might be necessary for the Golgi-to-PM sorting step and for stabilization in the PM. Alternatively, they might prevent ScPMA1 from being recognized by the quality control system and entering the degradative pathway, either in the Golgi or in the PM. In this case, oligomerization and/or raft association would appear to be more linked to the folding and stability of the protein than to a sorting mechanism.

H. Conclusion

On the one hand, proton pumping has been observed in liposomes reconstituted with monomeric *N. crassa* PM H^+-ATPases (165), indicating that oligomerization is not a requirement for activity, whereas, on the other hand, a decrease in oligomerization and in Triton X-100 insolubility is correlated with vacuolar degradation and instability in the PM. Based on present data, we can propose a model for PM H^+-ATPase trafficking in which an ER export motif interacts with a specific SEC24 homologue to load the ATPase into a subpopulation of COPII vesicles. Oligomerization and raft association start in the ER and/or in COPII vesicles and might play a role in the segregation of proteins sent to the vacuole and the PM pathway and in the stabilization of the H^+-ATPase in the PM (Fig. 5).

VIII. General Conclusions

The plasma membrane H^+-ATPase appeared early during evolution, as it is found in some archaebacteria, in addition to plants and fungi. The basic function of this protein, coupling H^+ transport and ATP hydrolysis, has probably remained the same. Although the detailed 3D structure of the

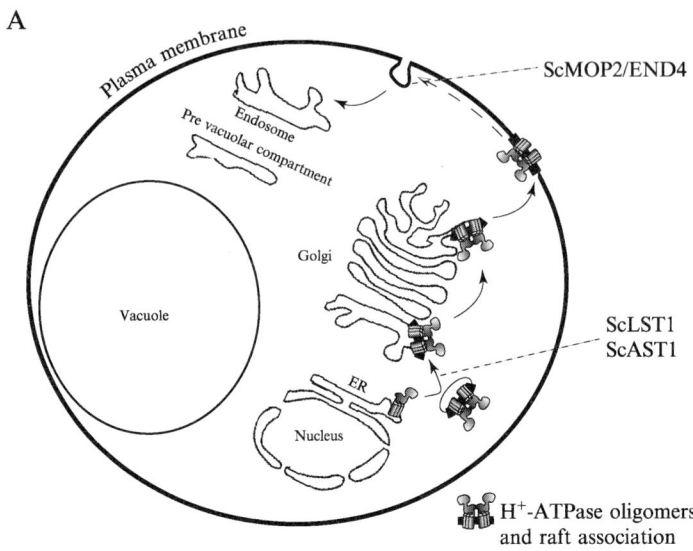

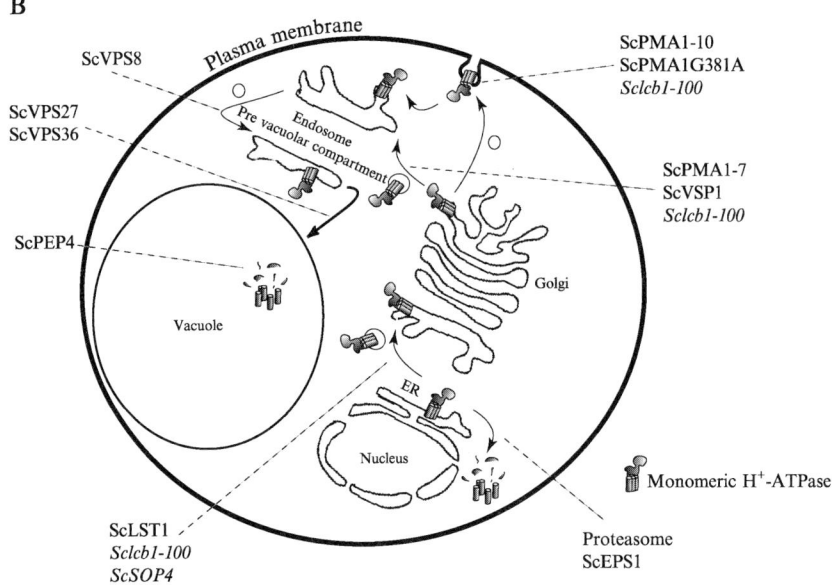

Ca^{2+}-ATPase has shed light on that of the entire P-type ATPase family, the resolution of the 3D structure of an H^+-ATPase is awaited, especially to identify the proton path through the H^+-ATPase membrane region. Another unresolved question is the number of protons transported per ATP molecule hydrolyzed. A major difficulty in solving these problems is the ubiquitous nature of the proton.

The H^+-ATPase is regulated in both yeast and plants; however, different regulatory systems have appeared during evolution, probably in response to different needs. In contrast to yeast, plants have a large H^+-ATPase gene family, the different products of which might play different physiological roles or respond to distinct regulatory pathways. Both yeast and plant H^+-ATPases are regulated by phosphorylation. However, the mechanisms of regulation (signaling pathway, kinases/phosphatases involved, target residues, and resulting H^+-ATPase structural modifications) have probably evolved differently. The most noticeable example is the involvement of 14-3-3-binding proteins in H^+-ATPase activation in plants, but not in yeast. These different regulatory features thus necessitate a detailed investigation of the regulatory aspects in both yeast and plants.

Finally, the yeast H^+-ATPase has been used as a paradigm to study how a polytopic protein is addressed through the secretory pathway to the plasma membrane. However, the structural determinants involved in this trafficking are not yet understood. In the plant, more work is needed to determine

---

FIG. 5. Trafficking of the yeast PM H^+-ATPase and related mutants. These schemes summarize data obtained in *S. cerevisae* on the trafficking and PM targeting of ScPMA1. (A) the pathway for the wild-type enzyme; (B) the pathways followed by abnormal H^+-ATPases or in yeast mutants affecting the transport of normal PM H^+-ATPase. PM H^+-ATPase is synthesized and translocated in the ER, where a first step of quality control involving ScEPS1 occurs. It is then transported to the Golgi via a subclass of COPII-coated vesicles containing ScLST1, an ScSEC24 homologue. A PM H^+-ATPase mutant that cannot correctly oligomerize and associate with rafts (**ScPMA1-7**) is rerouted to, and degraded in, the vacuole via the ScVPS1 pathway. It reaches the late endosome/multivesicular bodies/prevacuolar compartment via the ScVPS8 pathway, and the vacuole through the ScVPS27/VPS36 pathway (*160*). A prolonged stay in the ER (in ***Scsec13*** or ***Scsop4 Δ***) might improve oligomerization and raft association of ScPMA1-7, thus suppressing the PM targeting defect. Similarly, ScAST1, which is present in the early secretory pathway, was isolated as an *Scpma1-7* suppressor and restores oligomerization and/or raft association. The wild-type PM H^+-ATPase, which, under certain conditions, is not able to oligomerize and associate with rafts in the ***Sclcb1-100*** mutant is rerouted to, and degraded in, the vacuole or reaches the PM, where it is rapidly endocytosed and degraded. The **ScPMA1-10** mutant, which is not correctly associated with rafts, is rapidly endocytosed via the ScEND4 pathway and reaches the vacuole via the ScVPS27 pathway (*162*), where it is degraded. Similar endocytosis and vacuole degradation have been reported for **ScPMA1G381A** (*167*). (The names of machinery enzymes, which comprise the different trafficking pathways, are shown in normal type, whereas the names of mutant PM H^+-ATPases, which take these different pathways, are shown in bold.)

whether the trafficking system has been conserved during evolution. It will then be interesting to determine whether regulation occurs at the trafficking step.

Acknowledgments

The work performed in this laboratory was supported in part by grants from the Human Frontier Science Program Organization, the Interuniversity Poles of Attraction Program-Belgian State, Prime minister's Office for Scientific, Technical, and Cultural Affairs, and the Fonds National de la Recherche Scientifique (Belgium).

References

1. Sussman, M. R., and Harper, J. F. (1989). Molecular biology of the plasma membrane of higher plants. *Plant Cell* **1,** 953–960.
2. Dufour, J. P., and Goffeau, A. (1980). Molecular and kinetic properties of the purified plasma membrane ATPase of the yeast *Schizosaccharomyces pombe*. *Eur. J. Biochem.* **105,** 145–154.
3. Ghislain, M., Schlesser, A., and Goffeau, A. (1987). Mutation of a conserved glycine residue modifies the vanadate sensitivity of the plasma membrane H^+-ATPase from *Schizosaccharomyces pombe*. *J. Biol. Chem.* **262,** 17549–17555.
4. Serrano, R., Kielland-Brandt, M. C., and Fink, G. R. (1986). Yeast plasma membrane ATPase is essential for growth and has homology with (Na^+, K^+), K^+- and Ca^{2+}-ATPases. *Nature* **319,** 689–693.
5. Pardo, J. M., and Serrano, R. (1989). Structure of a plasma membrane H^+-ATPase gene from the plant *Arabidopsis thaliana*. *J. Biol. Chem.* **264,** 8557–8562.
6. Harper, J. F., Surowy, T. K., and Sussman, M. R. (1989). Molecular cloning and sequence of cDNA encoding the plasma membrane proton pump (H^+-ATPase) of *Arabidopsis thaliana*. *Proc. Natl. Acad. Sci. USA* **86,** 1234–1238.
7. Boutry, M., Michelet, B., and Goffeau, A. (1989). Molecular cloning of a family of plant genes encoding a protein homologous to plasma membrane H^+–translocating ATPases. *Biochem. Biophys. Res. Commun.* **162,** 567–574.
8. Arango, M., Gévaudant, F., Oufattole, M., and Boutry, M. (2002). The plasma membrane proton-ATPase: The significance of gene subfamilies. *Planta* **216,** 355–365.
9. Axelsen, K. B., and Palmgren, M. G. (1998). Evolution of substrate specificities in the P-type ATPase superfamily. *J. Mol. Evol.* **46,** 84–101.
10. Palmgren, M. G. (2001). Plant plasma membrane H^+-ATPases: Powerhouses for nutrient uptake. *Annu. Rev. Plant Physiol. Plant Mol. Biol.* **52,** 817–845.
11. Oufattole, M., Arango, M., and Boutry, M. (2000). Identification and expression of three new *Nicotiana plumbaginifolia* genes which encode isoforms of a plasma-membrane H^+-ATPase, and one of which is induced by mechanical stress. *Planta* **210,** 715–722.
12. Auer, M., Scarborough, G. A., and Kuhlbrandt, W. (1998). Three-dimensional map of the plasma membrane H^+-ATPase in the open conformation. *Nature* **392,** 840–843.
13. Kuhlbrandt, W., Zeelen, J., and Dietrich, J. (2002). Structure, mechanism, and regulation of the Neurospora plasma membrane H^+-ATPase. *Science* **297,** 1692–1696.

14. Jahn, T., Dietrich, J., Andersen, B., Leidvik, B., Otter, C., Briving, C., Kuhlbrandt, W., and Palmgren, M. G. (2001). Large scale expression, purification and 2D crystallization of recombinant plant plasma membrane H^+-ATPase. *J. Mol. Biol.* **309,** 465–476.
15. Toyoshima, C., Nakasako, M., Nomura, H., and Ogawa, H. (2000). Crystal structure of the calcium pump of sarcoplasmic reticulum at 2.6 Angstrom resolution. *Nature* **405,** 647–655.
16. Toyoshima, C., and Nomura, H. (2002). Structural changes in the calcium pump accompanying the dissociation of calcium. *Nature* **418,** 605–611.
17. Mac Lennan, D. H., Rice, W. J., and Green, N. M. (1997). The mechanism of Ca^{2+} transport by sarco(endo)plasmic reticulum Ca^{2+}-ATPases. *J. Biol. Chem.* **272,** 28815–28818.
18. Morsomme, P., and Boutry, M. (2000). The plant plasma membrane H^+-ATPase: Structure, function and regulation. *Biochim. Biophys. Acta.* **1465,** 1–16.
19. Rhee, K. H., Scarborough, G. A., and Henderson, R. (2002). Domain movements of plasma membrane H^+-ATPase: 3D structures of two states by electron cryo-microscopy. *EMBO J.* **21,** 3582–3589.
20. Venema, K., and Palmgren, M. G. (1995). Metabolic modulation of transport coupling ratio in yeast plasma-membrane H^+-ATPase. *J. Biol. Chem.* **270,** 19659–19667.
21. Morsomme, P., de Kerchove d'Exaerde, A., DeMeester, S., Thines, D., Goffeau, A., and Boutry, M. (1996). Single point mutations in various domains of a plant plasma membrane H^+-ATPase expressed in *Saccharomyces cerevisiae* increase H^+-pumping and permit yeast growth at low pH. *EMBO J.* **15,** 5513–5526.
22. Baunsgaard, L., Venema, K., Axelsen, K. B., Villalba, J. M., Welling, A., Wollenweber, B., and Palmgren, M. G. (1996). Modified plant plasma membrane H^+-ATPase with improved transport coupling efficiency identified by mutant selection in yeast. *Plant J.* **10,** 451–458.
23. Lee, M. C., Hamamoto, S., and Schekman, R. (2002). Ceramide biosynthesis is required for the formation of the oligomeric H^+-ATPase Pma1p in the yeast endoplasmic reticulum. *J. Biol. Chem.* **277,** 22395–22401.
24. Bagnat, M., Chang, A., and Simons, K. (2001). Plasma membrane proton ATPase Pma1p requires raft association for surface delivery in yeast. *Mol. Biol. Cell* **12,** 4129–4138.
25. Chadwick, C. C., Goormaghtigh, E., and Scarborough, G. A. (1987). A hexameric form of the *Neurospora crassa* plasma membrane H^+-ATPase. *Arch. Biochem. Biophys.* **252,** 348–356.
26. Bowman, B. J., Berenski, C. J., and Jung, C. Y. (1985). Size of the plasma membrane H^+-ATPase from *Neurospora crassa* determined by radiation inactivation and comparison with the sarcoplasmic reticulum Ca^{2+}-ATPase from skeletal muscle. *J. Biol. Chem.* **260,** 8726–8730.
27. Morsomme, P., Chami, M., Marco, S., Nader, J., Ketchum, K. A., Goffeau, A., and Rigaud, J. L. (2002). Characterization of a hyperthermophilic P-type ATPase from *Methanococcus jannashii* expressed in yeast. *J. Biol. Chem.* **277,** 29608–29616.
28. DeWitt, N. D., and Sussman, M. R. (1995). Immunocytological localization of an epitope-tagged plasma membrane proton pump (H^+-ATPase) in phloem companion cells. *Plant Cell* **7,** 2053–2067.
29. Bouchepillon, S., Fleurat Lessard, P., Fromont, J. C., Serrano, R., and Bonnemain, J. L. (1994). Immunolocalization of the plasma membrane H^+-ATPase in minor veins of *Vicia faba* in relation to phloem loading. *Plant Physiol.* **105,** 691–697.
30. Bouchepillon, S., Fleurat Lessard, P., Serrano, R., and Bonnemain, J. L. (1994). Asymmetric distribution of the plasma membrane H^+-ATPase in embryos of *Vicia faba* L with special reference to transfer cells. *Planta* **193,** 392–397.
31. Jahn, T., Baluska, F., Michalke, W., Harper, J. F., and Volkmann, D. (1998). Plasma membrane H^+-ATPase in the root apex: Evidence for strong expression in xylem parenchyma and asymmetric localization within cortical and epidermal cells. *Physiol. Plant.* **104,** 311–316.

32. Kotake, T., Nakagawa, N., Takeda, K., and Sakurai, N. (2000). Auxin-induced elongation growth and expressions of cell wall- bound exo- and endo-beta-glucanases in barley coleoptiles. *Plant Cell Physiol.* **41,** 1272–1278.
33. Cosgrove, D. J. (2000). Loosening of plant cell walls by expansins. *Nature* **407,** 321–326.
34. Kim, Y. S., Min, J. K., Kim, D., and Jung, J. (2001). A soluble auxin-binding protein, ABP57. Purification with anti-bovine serum albumin antibody and characterization of its mechanistic role in the auxin effect on plant plasma membrane H^+-ATPase. *J. Biol. Chem.* **276,** 10730–10736.
35. Frias, I., Caldeira, M. T., Perez Castineira, J. R., Navarro Avino, J. P., Culianez Macia, F. A., Kuppinger, O., Stransky, H., Pages, M., Hager, A., and Serrano, R. (1996). A major isoform of the maize plasma membrane H^+-ATPase: Characterization and induction by auxin in coleoptiles. *Plant Cell* **8,** 1533–1544.
36. Hager, A., Debus, H., Edel, H. G., Stransky, H., and Serrano, R. (1991). Auxin induces exocytosis and the rapid synthesis of a high-turnover pool of plasma-membrane H^+-ATPase. *Planta* **185,** 527–537.
37. Jahn, T., Johansson, F., Luthen, H., Volkmann, D., and Larsson, C. (1996). Reinvestigation of auxin and fusicoccin stimulation of the plasma-membrane H^+-ATPase activity. *Planta* **199,** 359–365.
38. Cho, H. T., and Hong, Y. N. (1995). Effect of IAA on synthesis and activity of the plasma-membrane H^+-ATPase of sunflower hypocotyls, in relation to IAA-induced cell elongation and H^+ excretion. *J. Plant Physiol.* **145,** 717–725.
39. Rayle, D. L., and Cleland, R. E. (1992). The acid growth theory of auxin-induced cell elongation is alive and well. *Plant Physiol.* **99,** 1271–1274.
40. Niu, X., Bressan, R. A., Hasegawa, P. M., and Pardo, J. M. (1995). Ion homeostasis in NaCl stress environments. *Plant Physiol.* **109,** 735–742.
41. Kalampanayil, B. D., and Wimmers, L. E. (2001). Identification and characterization of a salt-stress-induced plasma membrane H^+-ATPase in tomato. *Plant Cell Environ.* **24,** 999–1005.
42. Vitart, V., Baxter, I., Doerner, P., and Harper, J. F. (2001). Evidence for a role in growth and salt resistance of a plasma membrane H^+-ATPase in the root endodermis. *Plant J.* **27,** 191–201.
43. Serrano, R. (1996). Salt tolerance in plants and microorganisms: Toxicity targets and defense responses. *Int. Rev. Cytol.* **165,** 1–52.
44. Yale, J., and Bohnert, H. J. (2001). Transcript expression in *Saccharomyces cerevisiae* at high salinity. *J. Biol. Chem.* **276,** 15996–16007.
45. Withee, J. L., Sen, R., and Cyert, M. S. (1998). Ion tolerance of *Saccharomyces cerevisiae* lacking the Ca^{2+}/CaM-dependent phosphatase (calcineurin) is improved by mutations in *URE2* or *PMA1*. *Genetics* **149,** 865–878.
46. Nass, R., Cunningham, K. W., and Rao, R. (1997). Intracellular sequestration of sodium by a novel Na^+/H^+ exchanger in yeast is enhanced by mutations in the plasma membrane H^+-ATPase—Insights into mechanisms of sodium tolerance. *J. Biol. Chem.* **272,** 26145–26152.
47. Goossens, A., de La Fuente, N., Forment, J., Serrano, R., and Portillo, F. (2000). Regulation of yeast H^+-ATPase by protein kinases belonging to a family dedicated to activation of plasma membrane transporters. *Mol. Cell. Biol.* **20,** 7654–7661.
48. Serrano, R., Mulet, J. M., Rios, G., Marquez, J. A., de Larrinoa, I. F., Leube, M. P., Mendizabal, I., Pascual-Ahuir, A., Proft, M., Ros, R., and Montesinos, C. (1999). A glimpse of the mechanisms of ion homeostasis during salt stress. *J. Exp. Bot.* **50,** 1023–1036.
49. Haro, R., Garciadeblas, B., and Rodriguez-Navarro, A. (1991). A novel P-type ATPase from yeast involved in sodium transport. *FEBS Lett.* **291,** 189–191.
50. Wieland, J., Nitsche, A. M., Strayle, J., Steiner, H., and Rudolph, H. K. (1995). The Pmr2 gene-cluster encodes functionally distinct isoforms of a putative Na^+ pump in the yeast plasma-membrane. *EMBO J.* **14,** 3870–3882.

51. Michelet, B., Lukaszewicz, M., Dupriez, V., and Boutry, M. (1994). A plant plasma-membrane proton-ATPase gene is regulated by development and environment and shows signs of a translational regulation. *Plant Cell* **6**, 1375–1389.
52. Lukaszewicz, M., Jerouville, B., and Boutry, M. (1998). Signs of translational regulation within the transcript leader of a plant plasma membrane H^+-ATPase gene. *Plant J.* **14**, 413–423.
53. Serrano, R. (1983). In vivo glucose activation and of the yeast plasma membrane ATPase. *FEBS Lett.* **156**, 11–14.
54. Portillo, F., de Larrinoa, I. F., and Serrano, R. (1989). Deletion analysis of yeast plasma membrane H^+-ATPase and identification of a regulatory domain at the carboxyl-terminus. *FEBS Lett.* **247**, 381–385.
55. Eraso, P., and Portillo, F. (1994). Molecular mechanism of regulation of yeast plasma membrane H^+-ATPase by glucose. Interaction between domains and identification of new regulatory sites. *J. Biol. Chem.* **269**, 10393–10399.
56. Chang, A., and Slayman, C. W. (1991). Maturation of the yeast plasma membrane H^+-ATPase involves phosphorylation during intracellular transport. *J. Cell Biol.* **115**, 289–295.
57. Mazon, M. J., Gancedo, J. M., and Gancedo, C. (1982). Phosphorylation and inactivation of yeast fructose-bisphosphatase in vivo by glucose and by proton ionophores. A possible role for cAMP. *Eur. J. Biochem.* **127**, 605–608.
58. dos Passos, J. B., Vanhalewyn, M., Brandao, R. L., Castro, I. M., Nicoli, J. R., and Thevelein, J. M. (1992). Glucose-induced activation of plasma membrane H^+-ATPase in mutants of the yeast *Saccharomyces cerevisiae* affected in cAMP metabolism, cAMP-dependent protein phosphorylation and the initiation of glycolysis. *Biochim. Biophys. Acta.* **1136**, 57–67.
59. Souza, M. A., Tropia, M. J., and Brandao, R. L. (2001). New aspects of the glucose activation of the H^+-ATPase in the yeast *Saccharomyces cerevisiae*. *Microbiology* **147**, 2849–2855.
60. Estrada, E., Agostinis, P., Vandenheede, J. R., Goris, J., Merlevede, W., Francois, J., Goffeau, A., and Ghislain, M. (1996). Phosphorylation of yeast plasma membrane H^+-ATPase by casein kinase I. *J. Biol. Chem.* **271**, 32064–32072.
61. Miranda, M., Allen, K. E., Pardo, J. P., and Slayman, C. W. (2001). Stalk segment 5 of the yeast plasma membrane H^+-ATPase: Mutational evidence for a role in glucose regulation. *J. Biol. Chem.* **276**, 22485–22490.
62. Palmgren, M. G., Larsson, C., and Sommarin, M. (1990). Proteolytic activation of the plant plasma membrane H^+-ATPase by removal of a terminal segment. *J. Biol. Chem.* **265**, 13423–13426.
63. Regenberg, B., Villalba, J. M., Lanfermeijer, F. C., and Palmgren, M. G. (1995). C-terminal deletion analysis of plant plasma membrane H^+-ATPase: Yeast as a model system for solute transport across the plant plasma membrane. *Plant Cell* **7**, 1655–1666.
64. Morsomme, P., Dambly, S., Maudoux, O., and Boutry, M. (1998). Single point mutations distributed in 10 soluble and membrane regions of the *Nicotiana plumbaginifolia* plasma membrane PMA2 H^+-ATPase activate the enzyme and modify the structure of the C-terminal region. *J. Biol. Chem.* **273**, 34837–34842.
65. Axelsen, K. B., Venema, K., Jahn, T., Baunsgaard, L., and Palmgren, M. G. (1999). Molecular dissection of the C-terminal regulatory domain of the plant plasma membrane H^+-ATPase AHA2: Mapping of residues that when altered give rise to an activated enzyme. *Biochemistry* **38**, 7227–7234.
66. Rasi-Caldogno, F., and Pugliarello, M. C. (1985). Fusicoccin stimulates the H^+-ATPase of plasmalemma in isolated membrane vesicles from radish. *Biochem. Biophys. Res. Commun.* **133**, 280–285.
67. Aducci, P., Ballio, A., Blein, J. P., Fullone, M. R., Rossignol, M., and Scalla, R. (1988). Functional reconstitution of a proton-translocating system responsive to fusicoccin. *Proc. Natl. Acad. Sci. USA* **85**, 7849–7851.

68. Johansson, F., Sommarin, M., and Larsson, C. (1993). Fusicoccin activates the plasma-membrane H^+-ATPase by a mechanism involving the C-terminal inhibitory domain. *Plant Cell* **5**, 321–327.
69. Lanfermeijer, F. C., and Prins, H. B. A. (1994). Modulation of H^+-ATPase activity by fusicoccin in plasma-membrane vesicles from oat (*Avena sativa* L) roots—a comparison of modulation by fusicoccin, trypsin, and lysophosphatidylcholine. *Plant Physiol.* **104**, 1277–1285.
70. Oecking, C., Eckerskorn, C., and Weiler, E. W. (1994). The fusicoccin receptor of plants is a member of the 14-3-3 superfamily of eukaryotic regulatory proteins. *FEBS Lett.* **352**, 163–166.
71. Marra, M., Fullone, M. R., Fogliano, V., Pen, J., Mattei, M., and Aducci, P. (1994). The 30-kilodalton protein present in purified fusicoccin receptor preparations is a 14-3-3-like protein. *Plant Physiol.* **106**, 1497–1501.
72. Korthout, H. A., and de Boer, A. H. (1994). A fusicoccin binding protein belongs to the family of 14-3-3 brain protein homologs. *Plant Cell* **6**, 1681–1692.
73. Fullone, M. R., Visconti, S., Marra, M., Fogliano, V., and Aducci, P. (1998). Fusicoccin effect on the in vitro interaction between plant 14-3-3 proteins and plasma membrane H^+-ATPase. *J. Biol. Chem.* **273**, 7698–7702.
74. Jahn, T., Fuglsang, A. T., Olsson, A., Bruntrup, I. M., Collinge, D. B., Volkmann, D., Sommarin, M., Palmgren, M. G., and Larsson, C. (1997). The 14-3-3 protein interacts directly with the C-terminal region of the plant plasma membrane H^+-ATPase. *Plant Cell* **9**, 1805–1814.
75. Oecking, C., Piotrowski, M., Hagemeier, J., and Hagemann, K. (1997). Topology and target interaction of the fusicoccin-binding 14-3-3 homologs of *Commelina communis*. *Plant J.* **12**, 441–453.
76. Muslin, A. J., Tanner, J. W., Allen, P. M., and Shaw, A. S. (1996). Interaction of 14-3-3 with signaling proteins is mediated by the recognition of phosphoserine. *Cell* **84**, 889–897.
77. Yaffe, M. B., Rittinger, K., Volinia, S., Caron, P. R., Aitken, A., Leffers, H., Gamblin, S. J., Smerdon, S. J., and Cantley, L. C. (1997). The structural basis for 14-3-3: Phosphopeptide binding specificity. *Cell* **91**, 961–971.
78. Olsson, A., Svennelid, F., Ek, B., Sommarin, M., and Larsson, C. (1998). A phosphothreonine residue at the C-terminal end of the plasma membrane H^+-ATPase is protected by fusicoccin-induced 14-3-3 binding. *Plant Physiol.* **118**, 551–555.
79. Fuglsang, A. T., Visconti, S., Drumm, K., Jahn, T., Stensballe, A., Mattei, B., Jensen, O. N., Aducci, P., and Palmgren, M. G. (1999). Binding of 14-3-3 protein to the plasma membrane H^+-ATPase AHA2 involves the three C-terminal residues Tyr(946)-Thr-Val and requires phosphorylation of Thr(947). *J. Biol. Chem.* **274**, 36774–36780.
80. Svennelid, F., Olsson, A., Piotrowski, M., Rosenquist, M., Ottman, C., Larsson, C., Oecking, C., and Sommarin, M. (1999). Phosphorylation of Thr-948 at the C terminus of the plasma membrane H^+-ATPase creates a binding site for the regulatory 14-3-3 protein. *Plant Cell* **11**, 2379–2391.
81. Maudoux, O., Batoko, H., Oecking, C., Gevaert, K., Vandekerckhove, J., Boutry, M., and Morsomme, P. (2000). A plant plasma membrane H^+-ATPase expressed in yeast is activated by phosphorylation at its penultimate residue and binding of 14-3-3 regulatory proteins in the absence of fusicoccin. *J. Biol. Chem.* **275**, 17762–17770.
82. Baunsgaard, L., Fuglsang, A. T., Jahn, T., Korthout, H. A., de Boer, A. H., and Palmgren, M. G. (1998). The 14-3-3 proteins associate with the plant plasma membrane H^+-ATPase to generate a fusicoccin binding complex and a fusicoccin responsive system. *Plant J.* **13**, 661–671.
83. Piotrowski, M., Morsomme, P., Boutry, M., and Oecking, C. (1998). Complementation of the *Saccharomyces cerevisiae* plasma membrane H^+-ATPase by a plant H^+-ATPase generates a highly abundant fusicoccin binding site. *J. Biol. Chem.* **273**, 30018–30023.

84. Oecking, C., and Hagemann, K. (1999). Association of 14-3-3 proteins with the C-terminal antoinhibitory domain of the plant plasma-membrane H$^+$-ATPase generates a fusicoccin-binding complex. *Planta* **207**, 480–482.
85. Kinoshita, T., and Shimazaki, K. (1999). Blue light activates the plasma membrane H$^+$-ATPase by phosphorylation of the C-terminus in stomatal guard cells. *EMBO J.* **18**, 5548–5558.
86. Kinoshita, T., Doi, M., Suetsugu, N., Kagawa, T., Wada, M., and Shimazaki, K. (2001). *phot1* and *phot2* mediate blue light regulation of stomatal opening. *Nature* **414**, 656–660.
87. Kerkeb, L., Venema, K., Donaire, J. P., and Rodriguez-Rosales, M. P. (2002). Enhanced H$^+$/ATP coupling ratio of H$^+$-ATPase and increased 14-3-3 protein content in plasma membrane of tomato cells upon osmotic shock. *Physiol. Plant* **116**, 37–41.
88. Babakov, A. V., Chelysheva, V. V., Klychnikov, O. I., Zorinyanz, S. E., Trofimova, M. S., and De Boer, A. H. (2000). Involvement of 14-3-3 proteins in the osmotic regulation of H$^+$-ATPase in plant plasma membranes. *Planta* **211**, 446–448.
89. Young, J. C., DeWitt, N. D., and Sussman, M. R. (1998). A transgene encoding a plasma membrane H$^+$-ATPase that confers acid resistance in *Arabidopsis thaliana* seedlings. *Genetics* **149**, 501–507.
90. Camoni, L., Iori, V., Marra, M., and Aducci, P. (2000). Phosphorylation-dependent interaction between plant plasma membrane H$^+$-ATPase and 14-3-3 proteins. *J. Biol. Chem.* **275**, 9919–9923.
91. Suzuki, Y. S., Wang, Y. L., and Takemoto, J. Y. (1992). Syringomycin-stimulated phosphorylation of the plasma-membrane H$^+$-ATPase from red beet storage tissue. *Plant Physiol.* **99**, 1314–1320.
92. Vera-Estrella, R., Barkla, B. J., Higgins, V. J., and Blumwald, E. (1994). Plant defense response to fungal pathogens—Activation of host-plasma membrane H$^+$-ATPase by elicitor-induced enzyme dephosphorylation. *Plant Physiol.* **104**, 209–215.
93. Desbrosses, G., Stelling, J., and Renaudin, J. P. (1998). Dephosphorylation activates the purified plant plasma membrane H$^+$-ATPase—possible function of phosphothreonine residues in a mechanism not involving the regulatory C-terminal domain of the enzyme. *Eur. J. Biochem.* **251**, 496–503.
94. De Nisi, P., Dell'Orto, M., Pirovano, L., and Zocchi, G. (1999). Calcium-dependent phosphorylation regulates the plasma-membrane H$^+$-ATPase activity of maize (*Zea mays* L) roots. *Planta* **209**, 187–194.
95. Lino, B., Baizabal-Aguirre, V. M., and Gonzalez de la Vara, L. E. (1998). The plasma-membrane H$^+$-ATPase from beet root is inhibited by a calcium-dependent phosphorylation. *Planta* **204**, 352–359.
96. Schaller, G. E., and Sussman, M. R. (1988). Phosphorylation of plasma membrane H$^+$-ATPase of oat roots by a calcium-stimulated protein kinase. *Planta* **173**, 508–518.
97. Schlesser, A., Ulaszewski, S., Ghislain, M., and Goffeau, A. (1988). A second transport ATPase gene in *Saccharomyces cerevisiae*. *J. Biol. Chem.* **263**, 19480–19487.
98. Supply, P., Wach, A., and Goffeau, A. (1993). Enzymatic properties of the PMA2 plasma membrane-bound H$^+$-ATPase of *Saccharomyces cerevisiae*. *J. Biol. Chem.* **268**, 19753–19759.
99. Monteiro, G. A., Supply, P., Goffeau, A., and Sa-Correia, I. (1994). The in vivo activation of *Saccharomyces cerevisiae* plasma membrane H$^+$-ATPase by ethanol depends on the expression of the PMA1 gene, but not of the PMA2 gene. *Yeast* **10**, 1439–1446.
100. Palmgren, M. G., and Christensen, G. (1994). Functional comparisons between plant plasma membrane H$^+$-ATPase isoforms expressed in yeast. *J. Biol. Chem.* **269**, 3027–3033.
101. Luo, H., Morsomme, P., and Boutry, M. (1999). The two major types of plant plasma membrane H$^+$-ATPases show different enzymatic properties and confer differential pH sensitivity of yeast growth. *Plant Physiol* **119**, 627–634.

102. Dambly, S., and Boutry, M. (2001). The two major plant plasma membrane H^+-ATPase display different regulatory properties. *J. Biol. Chem.* **276**, 7017–7022.
103. Navarre, C., Ferroud, C., Ghislain, M., and Goffeau, A. (1992). A proteolipid associated with the plasma membrane H^+-ATPase of fungi. *Ann. N.Y. Acad. Sci.* **671**, 189–194.
104. Navarre, C., Catty, P., Leterme, S., Dietrich, F., and Goffeau, A. (1994). Two distinct genes encode small isoproteolipids affecting plasma membrane H^+-ATPase activity of *Saccharomyces cerevisiae*. *J. Biol. Chem.* **269**, 21262–21268.
105. Braley, R., and Piper, P. W. (1997). The C-terminus of yeast plasma membrane H^+-ATPase is essential for the regulation of this enzyme by heat shock protein Hsp30, but not for stress activation. *FEBS Lett.* **418**, 123–126.
106. Morandini, P., Valera, M., Albumi, C., Bonza, M. C., Giacometti, S., Ravera, G., Murgia, I., Soave, C., and De Michelis, M. I. (2002). A novel interaction partner for the C-terminus of *Arabidopsis thaliana* plasma membrane H^+-ATPase (AHA1 isoform): Site and mechanism of action on H^+-ATPase activity differ from those of 14-3-3 proteins. *Plant J.* **31**, 487–497.
107. Palmgren, M. G., Sommarin, M., Ulvskov, P., and Jorgensen, P. L. (1988). Modulation of plasma membrane H^+-ATPase by lysophosphatidylcholine, free fatty acids and phospholipase A2. *Physiol. Plant.* **74**, 11–19.
108. Gomes, E., Venema, K., Simon-Plas, F., Milat, M. L., Palmgren, M. G., and Blein, J. P. (1996). Activation of the plant plasma membrane H^+-ATPase. Is there a direct interaction between lysophosphatidylcholine and the C-terminal part of the enzyme?. *FEBS Lett.* **398**, 48–52.
109. Holcomb, C. L., Hansen, W. J., Etcheverry, T., and Schekman, R. (1988). Secretory vesicles externalize the major plasma membrane ATPase in yeast. *J. Cell Biol.* **106**, 641–648.
110. Brada, D., and Schekman, R. (1988). Coincident localization of secretory and plasma membrane proteins in organelles of the yeast secretory pathway. *J. Bacteriol.* **170**, 2775–2783.
111. Ferreira, T., Mason, A. B., and Slayman, C. W. (2001). The yeast Pma1 proton pump: A model for understanding the biogenesis of plasma membrane proteins. *J. Biol. Chem.* **276**, 29613–29616.
112. Na, S., Hincapie, M., McCusker, J. H., and Haber, J. E. (1995). MOP2 (SLA2) affects the abundance of the plasma membrane H^+-ATPase of *Saccharomyces cerevisiae*. *J. Biol. Chem.* **270**, 6815–6823.
113. Chang, A., Rose, M. D., and Slayman, C. W. (1993). Folding and intracellular transport of the yeast plasma-membrane H^+-ATPase: Effects of mutations in KAR2 and SEC65. *Proc. Natl. Acad. Sci. USA* **90**, 5808–5812.
114. Morsomme, P., Slayman, C. W., and Goffeau, A. (2000). Mutagenic study of the structure, function and biogenesis of the yeast plasma membrane H^+-ATPase. *Biochim. Biophys. Acta* **1469**, 133–157.
115. Wang, Q., and Chang, A. (2002). Sphingoid base synthesis is required for oligomerization and cell surface stability of the yeast plasma membrane ATPase, Pma1. *Proc. Natl. Acad. Sci. USA* **99**, 12853–12858.
116. DeWitt, N. D., dos Santos, C. F. T., Allen, K. E., and Slayman, C. W. (1998). Phosphorylation region of the yeast plasma-membrane H^+-ATPase—Role in protein folding and biogenesis. *J. Biol. Chem.* **273**, 21744–21751.
117. Wang, Q., and Chang, A. (1999). Eps1, a novel PDI-related protein involved in ER quality control in yeast. *EMBO J.* **18**, 5972–5982.
118. Barlowe, C., Orci, L., Yeung, T., Hosobuchi, M., Hamamoto, S., Salama, N., Rexach, M. F., Ravazzola, M., Amherdt, M., and Schekman, R. (1994). COPII—a membrane coat formed by Sec proteins that drive vesicle budding from the endoplasmic reticulum. *Cell* **77**, 895–907.
119. Aridor, M., Weissman, J., Bannykh, S., Nuoffer, C., and Balch, W. E. (1998). Cargo selection by the COPII budding machinery during export from the ER. *J. Cell Biol.* **141**, 61–70.

120. Kuehn, M. J., Herrmann, J. M., and Schekman, R. (1998). COPII-cargo interactions direct protein sorting into ER-derived transport vesicles. *Nature* **391**, 187–190.
121. Springer, S., and Schekman, R. (1998). Nucleation of COPII vesicular coat complex by endoplasmic reticulum to Golgi vesicle SNAREs. *Science* **281**, 698–700.
122. Antonny, B., Madden, D., Hamamoto, S., Orci, L., and Schekman, R. (2001). Dynamics of the COPII coat with GTP and stable analogues. *Nat. Cell Biol.* **3**, 531–537.
123. Shimoni, Y., Kurihara, T., Ravazzola, M., Amherdt, M., Orci, L., and Schekman, R. (2000). Lst1p and Sec24p cooperate in sorting of the plasma membrane ATPase into COPII vesicles in *Saccharomyces cerevisiae*. *J. Cell Biol.* **151**, 973–984.
124. Roberg, K. J., Crotwell, M., Espenshade, P., Gimeno, R., and Kaiser, C. A. (1999). LST1 is a SEC24 homologue used for selective export of the plasma membrane ATPase from the endoplasmic reticulum. *J. Cell Biol.* **145**, 659–672.
125. Kurihara, T., Hamamoto, S., Gimeno, R. E., Kaiser, C. A., Schekman, R., and Yoshihisa, T. (2000). Sec24p and Iss1p function interchangeably in transport vesicle formation from the endoplasmic reticulum in *Saccharomyces cerevisiae*. *Mol. Biol. Cell* **11**, 983–998.
126. Peng, R. W., De Antoni, A., and Gallwitz, D. (2000). Evidence for overlapping and distinct functions in protein transport of coat protein Sec24p family members. *J. Biol. Chem.* **275**, 11521–11528.
127. d'Enfert, C., Gensse, M., and Gaillardin, C. (1992). Fission yeast and a plant have functional homologues of the Sar1 and Sec12 proteins involved in ER to Golgi traffic in budding yeast. *EMBO J.* **11**, 4205–4211.
128. Takeuchi, M., Tada, M., Saito, C., Yashiroda, H., and Nakano, A. (1998). Isolation of a tobacco cDNA encoding Sar1 GTPase and analysis of its dominant mutations in vesicular traffic using a yeast complementation system. *Plant Cell Physiol.* **39**, 590–599.
129. Phillipson, B. A., Pimpl, P., daSilva, L. L., Crofts, A. J., Taylor, J. P., Movafeghi, A., Robinson, D. G., and Denecke, J. (2001). Secretory bulk flow of soluble proteins is efficient and COPII dependent. *Plant Cell* **13**, 2005–2020.
130. Takeuchi, M., Ueda, T., Sato, K., Abe, H., Nagata, T., and Nakano, A. (2000). A dominant negative mutant of sar1 GTPase inhibits protein transport from the endoplasmic reticulum to the Golgi apparatus in tobacco and Arabidopsis cultured cells. *Plant J.* **23**, 517–525.
131. Movafeghi, A., Happel, N., Pimpl, P., Tai, G. H., and Robinson, D. G. (1999). Arabidopsis Sec21p and Sec23p homologs. Probable coat proteins of plant COP-coated vesicles. *Plant Physiol.* **119**, 1437–1446.
132. Mizuno, M., and Singer, S. J. (1993). A soluble secretory protein is 1st concentrated in the endoplasmic-reticulum before transfer to the Golgi-apparatus. *Proc. Natl. Acad. Sci. USA* **90**, 5732–5736.
133. Nishimura, N., and Balch, W. E. (1997). A di-acidic signal required for selective export from the endoplasmic reticulum. *Science* **277**, 556–558.
134. Nishimura, N., Bannykh, S., Slabough, S., Matteson, J., Altschuler, Y., Hahn, K., and Balch, W. E. (1999). A di-acidic (DXE) code directs concentration of cargo during export from the endoplasmic reticulum. *J. Biol. Chem.* **274**, 15937–15946.
135. Ma, D. K., Zerangue, N., Lin, Y. F., Collins, A., Yu, M., Jan, Y. N., and Jan, L. Y. (2001). Role of ER export signals in controlling surface potassium channel numbers. *Science* **291**, 316–319.
136. Sevier, C. S., Weisz, O. A., Davis, M., and Machamer, C. E. (2000). Efficient export of the vesicular stomatitis virus G protein from the endoplasmic reticulum requires a signal in the cytoplasmic tail that includes both tyrosine-based and di-acidic motifs. *Mol. Biol. Cell* **11**, 13–22.
137. Votsmeier, C., and Gallwitz, D. (2001). An acidic sequence of a putative yeast Golgi membrane protein binds COPII and facilitates ER export. *EMBO J.* **20**, 6742–6750.

138. Epping, E. A., and Moye-Rowley, W. S. (2002). Identification of interdependent signals required for anterograde traffic of ATP-binding cassette transporter protein Yor1p. *J. Biol. Chem.* **277,** 34860–34869.
139. Bannykh, S. I., Nishimura, N., and Balch, W. E. (1998). Getting into the Golgi. *Trends Cell Biol.* **8,** 21–25.
140. Aridor, M., Fish, K. N., Bannykh, S., Weissman, J., Roberts, T. H., Lippincott-Schwartz, J., and Balch, W. E. (2001). The Sar1 GTPase coordinates biosynthetic cargo selection with endoplasmic reticulum export site assembly. *J. Cell Biol.* **152,** 213–229.
141. Dominguez, M., Dejgaard, K., Fullekrug, J., Dahan, S., Fazel, A., Paccaud, J. P., Thomas, D. Y., Bergeron, J. J. M., and Nilsson, T. (1998). gp25L/emp24/p24 protein family members of the cis-Golgi network bind both COP I and II coatomer. *J. Cell Biol.* **140,** 751–765.
142. Kappeler, F., Klopfenstein, D. R. C., Foguet, M., Paccaud, J. P., and Hauri, H. P. (1997). The recycling of ERGIC-53 in the early secretory pathway. *J. Biol. Chem.* **272,** 31801–31808.
143. Nakamura, N., Yamazaki, S., Sato, K., Nakano, A., Sakaguchi, M., and Mihara, K. (1998). Identification of potential regulatory elements for the transport of Emp24p. *Mol. Biol. Cell* **9,** 3493–3503.
144. Nufer, O., Guldbrandsen, S., Degen, M., Kappeler, F., Paccaud, J. P., Tani, K., and Hauri, H. P. (2002). Role of cytoplasmic C-terminal amino acids of membrane proteins in ER export. *J. Cell Sci.* **115,** 619–628.
145. Munro, S. (1995). An investigation of the role of transmembrane domains in Golgi protein retention. *EMBO J.* **14,** 4695–4704.
146. Pedrazzini, E., Villa, A., and Borgese, N. (1996). A mutant cytochrome b(5) with a lengthened membrane anchor escapes from the endoplasmic reticulum and reaches the plasma membrane. *Proc. Natl. Acad. Sci. USA* **93,** 4207–4212.
147. Watson, R. T., and Pessin, J. E. (2001). Transmembrane domain length determines intracellular membrane compartment localization of syntaxins 3, 4, and 5. *Am. J. Physiol. Cell Physiol.* **281,** 215–223.
148. Rayner, J. C., and Pelham, H. R. B. (1997). Transmembrane domain-dependent sorting of proteins to the ER and plasma membrane in yeast. *EMBO J.* **16,** 1832–1841.
149. Brandizzi, F., Frangne, N., Marc-Martin, S., Hawes, C., Neuhaus, J. M., and Paris, N. (2002). The destination for single-pass membrane proteins is influenced markedly by the length of the hydrophobic domain. *Plant Cell* **14,** 1077–1092.
150. Harder, T., and Simons, K. (1997). Caveolae, DIGs, and the dynamics of sphingolipid-cholesterol microdomains. *Curr. Opin. Cell Biol.* **9,** 534–542.
151. Patton, J. L., Srinivasan, B., Dickson, R. C., and Lester, R. L. (1992). Phenotypes of sphingolipid-dependent strains of *Saccharomyces cerevisiae*. *J. Bacteriol.* **174,** 7180–7184.
152. Bagnat, M., Keranen, S., Shevchenko, A., and Simons, K. (2000). Lipid rafts function in biosynthetic delivery of proteins to the cell surface in yeast. *Proc. Natl. Acad. Sci. USA* **97,** 3254–3259.
153. Lynch, D. V., and Fairfield, S. R. (1993). Sphingolipid long-chain base synthesis in plants—Characterization of serine palmitoyltransferase activity in squash fruit microsomes. *Plant Physiol.* **103,** 1421–1429.
154. Tamura, K., Mitsuhashi, N., Hara-Nishimura, I., and Imai, H. (2001). Characterization of an Arabidopsis cDNA encoding a subunit of serine palmitoyltransferase, the initial enzyme in sphingolipid biosynthesis. *Plant Cell Physiol.* **42,** 1274–1281.
155. Xu, X. L., Bittman, R., Duportail, G., Heissler, D., Vilcheze, C., and London, E. (2001). Effect of the structure of natural sterols and sphingolipids on the formation of ordered sphingolipid/sterol domains (rafts). *J. Biol. Chem.* **276,** 33540–33546.

156. Peskan, T., Westermann, M., and Oelmuller, R. (2000). Identification of low-density Triton X-100-insoluble plasma membrane microdomains in higher plants. *Eur. J. Biochem.* **267**, 6989–6995.
157. Madey, E., Nowack, L. M., Su, L., Hong, Y., Hudak, K. A., and Thompson, J. E. (2001). Characterization of plasma membrane domains enriched in lipid metabolites. *J. Exp. Bot.* **52**, 669–679.
158. Moreau, P., Bessoule, J. J., Mongrand, S., Testet, E., Vincent, P., and Cassagne, C. (1998). Lipid trafficking in plant cells. *Prog. Lipid Res.* **37**, 371–391.
159. Chang, A., and Fink, G. R. (1995). Targeting of the yeast plasma membrane H^+-ATPase: A novel gene *AST1* prevents mislocalization of mutant ATPase to the vacuole. *J. Cell Biol.* **128**, 39–49.
160. Luo, W., and Chang, A. (2000). An endosome-to-plasma membrane pathway involved in trafficking of a mutant plasma membrane ATPase in yeast. *Mol. Biol. Cell* **11**, 579–592.
161. Luo, W. J., Gong, X. H., and Chang, A. (2002). An ER membrane protein, Sop4, facilitates ER export of the yeast plasma membrane H^+-ATPase, Pma1. *Traffic* **3**, 730–739.
162. Gong, X., and Chang, A. (2001). A mutant plasma membrane ATPase, Pma1-10, is defective in stability at the yeast cell surface. *Proc. Natl. Acad. Sci. USA* **98**, 9104–9109.
163. Tuller, G., Nemec, T., Hrastnik, C., and Daum, G. (1999). Lipid composition of subcellular membranes of an FY1679-derived haploid yeast wild-type strain grown on different carbon sources. *Yeast* **15**, 1555–1564.
164. Zinser, E., and Daum, G. (1995). Isolation and biochemical-characterization of organelles from the yeast, *Saccharomyces cerevisiae*. *Yeast* **11**, 493–536.
165. Goormaghtigh, E., Chadwick, C., and Scarborough, G. A. (1986). Monomers of the Neurospora plasma membrane H^+-ATPase catalyze efficient proton translocation. *J. Biol. Chem.* **261**, 7466–7471.
166. Bernstein, H. J. (2000). Recent changes to RasMol, recombining the variants. *Trends Biochem. Sci.* **25**, 453–455.
167. Ferreira, T., Mason, A. B., Pypaert, M., Allen, K. E., and Slayman, C. W. (2002). Quality control in the yeast secretory pathway: A misfolded PMA1 H^+-ATPase reveals two checkpoints. *J. Biol. Chem.* **277**, 21027–21040.

The Genes Encoding Human Protein Kinase CK2 and Their Functional Links

WALTER PYERIN AND
KARIN ACKERMANN

Biochemische Zellphysiologie (B0200),
Deutsches Krebsforschungszentrum,
69120 Heidelberg, Germany

I. Introduction	240
II. Protein Kinase CK2	241
III. The Human Protein Kinase CK2 Genes	245
A. CK2 Loci in the Human Genome and Structural Organization of the CK2 Genes	245
B. CK2 Gene Structures Are Conserved	248
IV. Expression of Human CK2 Genes	250
A. Transcripts and Their Distribution	250
B. Transcript-Based Phylogeny of CK2	251
C. Regulation of Transcription	253
V. Functional Links of CK2 Genes	262
A. Genetic Perturbation of CK2 Affects Expression of Genes of All Cell-Cycle Phases and Often in a Subunit- and Isoform-Specific Manner	262
B. Links to Chromatin Remodeling Genes: A Global Role for CK2?	266
References	268

Protein kinase CK2 is a Ser/Thr phosphotransferase that occurs ubiquitously among eukaryotes. It is pleiotropic, vital, and highly conserved. CK2 is a tetramer composed of two catalytic (α) and two regulatory (β) subunits. Both subunits may occur in isoforms and both may play roles independent of the holoenzymes they form. Humans express α, α', and β subunits. The human genome contains four CK2 loci at different chromosomes, enclosing three active genes and a pseudogene. This chapter reviews the chromosomal location, structural organization, and expressional control of the genes. It shows that CK2's conservation can also be recognized at the nucleic acid level, that the three active genes have features in common, and that some of these are appropriate for a coordinate transcriptional regulation. In particular, an identical Ets1 double motif that cross-talks to multiple Sp1 (and other) sites is present in the α and β gene promoters, and CK2 holoenzyme but not CK2α phosphorylates Sp1, resulting in a loss of DNA binding. This is compatible with a negative feedback control according to which expression of α and β genes leads to an increased holoenzyme level and thus phosphorylation, which, in turn, decreases transcription. As a consequence, constant transcript levels of both genes are expected to adjust. In human cultured cells, this is indeed the case, independent of their respective proliferation or differentiation status. The chapter also provides an overview of functional links of the CK2 genes to cell-cycle-regulated genes. Based on comparative genome-wide transcript profiles of *Saccharomyces cerevisiae* wild-type and CK2 mutant strains, CK2 is shown to be involved in transcription regulation of various genes related to cell-cycle control, including genes encoding cyclins and components of spindle pole body formation and dynamics. Strikingly, most of the affected genes lack common elements in their promoters and expression of a large group of genes encoding chromatin remodeling factors is altered, compatible with the idea that CK2 plays a role in the global process of transcription-related chromatin remodeling. In addition, functional links of CK2 are seen to diverse metabolic and nutritional supply pathways, including MET genes responsible for methionine synthesis, and the PHO gene group responsible for phosphate maintenance, which, interestingly, is uncoupled from its central cyclin-Cdk control upon CK2 perturbation.

I. Introduction

Among the earliest protein phosphotransferases detected (in 1954) was an activity in liver homogenates that catalyzed the phosphoryl group transfer from ATP to serine and threonine residues of casein and that eluted in salt gradients from diethylaminoethanol (DEAE) cellulose columns as the second of two peaks. Consequently, it was operationally named casein kinase II (*1*). Later the kinase turned out to phosphorylate *in vivo* various proteins, but due to cellular compartmentation not casein, prompting an international expert meeting [International Symposium, "A Molecular and Cellular View of Casein Kinase II," Heidelberg, May 11–13, 1994; *Cell. Mol. Biol. Res.* **40** (5/6), 1994] to suggest replacing casein kinase II by the term "protein kinase CK2." This term is being used throughout this article.

CK2 had soon been found to occur in numerous tissues and organisms and to be quite different from other protein kinases, acting independently of the classic kinase-activating second messengers and rather affected by compounds such as heparin and polyamines inhibiting and stimulating activity, respectively (*2*). Before the 1980s, few endogenous substrate proteins were known and thus no information was available as to its biological role. The problem of a lack of substrates has been replaced by the problem of substrate numbers that are now over 300 (*3*), again hindering role definition. Nevertheless, CK2 may have a role in global processes related to transcription and transcription-directed signaling, because transcription factors, subunits of all three classes of RNA polymerases, and transcription-related signal transduction proteins make up a majority of substrates, and, further, CK2 is highest in level in the nuclear compartment of cells and expression-linked CK2-containing protein complexes have been found in *Saccharomyces cerevisiae* that have striking homologies to humans (4–6). A requirement of CK2 for expression of genes such as *fos* (7–9) is linked to cell-cycle entry.

CK2 is a tetramer composed of two catalytic and two regulatory subunits, α and β, respectively. The subunits may occur in isoforms; humans express α, α', and β subunits giving rise to respective holoenzyme variants. Coding sequences for these proteins in the human genome are found at four loci, enclosing three active genes and a pseudogene. The purpose of this review is to summarize the available information bearing on the genomic location, structure, and expressional control of these genes with a focus on the question of whether coordinate gene transcription may occur. Further, the transcripts, their tissue distribution, and CK2's phylogeny based on transcript sequences are considered. The review, moreover, deals with the functional links of CK2 genes. It surveys available data on the involvement of CK2 in gene control, focusing on the expression of cell-cycle genes whose link to CK2 is indicated by comparative genome-wide DNA array analysis of CK2-perturbed *S. cerevisiae* strains. Analysis of the cell-cycle reentry situation visualizes, among other relations, a global role for CK2: transcription-related chromatin remodeling.

II. Protein Kinase CK2

As far as has been investigated, each eukaryotic cell expresses more or less protein kinase CK2. This ubiquitous kinase was discovered in 1954 in liver homogenates and, therefore, was among the earliest protein phosphotransferases detected (1). CK2 was soon found to behave differently from other kinases, i.e., to act independently of second messengers such as cyclic nucleotides, phosphatidyl inositolphosphates, or calcium, to accept in addition to adenosine triphosphate (ATP), as most kinases do, other purine nucleoside triphosphates such as guanosine triphosphate (GTP) or inosine triphosphate (ITP) as cosubstrates, and to be responsive to compounds such as heparin and polyamines, inhibiting and stimulating activity, respectively (2). CK2 represents a pleiotropic, second messenger-independent, ubiquitous protein kinase with uncommon substrate and cosubstrate scope and activity control mode.

CK2 phosphorylates proteins at Ser or Thr located within sequence motifs $-(E/D/X)-S/T-(D/E/X')-(E/D/X)-(E/D)-(E/D/X)-$, where X and

X′ indicate any residue except basic residues and basic residues or Pro, respectively (3). Reports indicate that under certain conditions CK2 may also have Tyr phosphorylating (10, 11) and even nucleotidylating (12) activity. Physiologically important for cells is the fact that both Glu and Asp of the recognition motif may efficiently be replaced by phospho-Ser (or phospho-Tyr) as specific determinants, enabling CK2 to cross-talk to other kinases and participate in signaling networks.

CK2 has a highly conserved tetrameric structure composed of two catalytic subunits, CK2α, and two regulatory subunits, CK2β. The regulatory subunits determine whether CK2 phosphorylates a certain protein and, if so, to what extent (13). Biochemical evidence for the heterotetrameric holoenzyme has been available for many years (14–17) and has been further confirmed by recent crystallographic analyses (18, 19). According to these, the four CK2 subunits form a butterfly-like holoenzyme structure characterized by a zinc finger-mediated crescent-shaped β–β dimer that provides a stable basis for the complexation of two α subunits at distant positions excluding directly contacting one another (Fig. 1). In addition, the subunits may also play roles on their own. The catalytic subunits may phosphorylate proteins not accepted as substrates by holoenzymes (3, 13), and CK2β has been shown or suspected to have contact sites for other proteins, including diverse protein kinases such as c-Mos, A-Raf, or PKCζ, and components of signaling chains such as Dsh, FAF-1, p53, Topo II, or b-FGF (20), providing another junction for participation in signaling networks.

Both the catalytic and the regulatory subunits may occur in isoforms and, as a consequence, give rise to respective holoenzyme variants. Humans have one type of regulatory subunit, β, and two types of catalytic subunit isoforms, α and α′, thus forming $\alpha_2\beta_2$, $\alpha'_2\beta_2$, and $\alpha\alpha'\beta_2$ holoenzymes (13), the budding yeast S. cerevisiae expresses two catalytic (Cka1 and Cka2) and two regulatory (Ckb1 and Ckb2) subunit isoforms (21), and the worm Caenorhabditis elegans represents an example of only one catalytic and one regulatory subunit (22, 23). The human CK2β consists of 215 amino acid residues and the human CK2α and CK2α′ of 391 and 350 amino acid residues, respectively, of nearly identical sequence, except for the C-terminus (4, 24, 25). Generally, the primary structures of CK2 subunits are highly conserved. The dozens of CK2 subunit sequences of various organisms appearing on current computer-based searches (http://www.ncbi.nlm.nih.gov/; http://genius.dkfz-heidelberg.de/) are extremely similar.

The physiological role of CK2 has been the subject of continual debate and numerous corrections, and corrections of corrections (2–4, 20, 21, 26–30). The vast number of substrate proteins known today and their widely varying involvements in cellular processes at diverse levels of organization explain why this has necessarily been the case, and also why CK2 has so many physiological

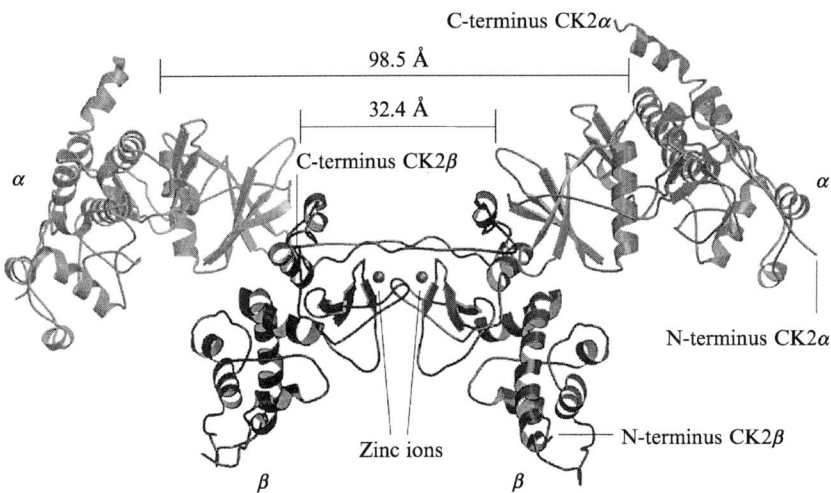

FIG. 1. Three-dimensional structure of the human CK2 holoenzyme $\alpha_2\beta_2$. The recombinant human protein kinase CK2 complex is shown in a view from above perpendicular to the molecular 2-fold axis with its approximate dimensions of $155 \times 90 \times 66$ Å. Indicated are the positions of the two α and the two β subunits (α; β) as well as the C- and N-termini of the individual subunits and of the central zinc ions. Picture created by K. Niefind (personal communication) using Bobscript [R. M. Esnouf, An extensively modified version of MolScript that includes greatly enhanced coloring capabilities. *J. Mol. Graph.* **15**, 132–134 (1997)] and Raster3D [E. A. Merritt and D. J. Bacon, Raster3D: Photorealistic molecular graphics. *Methods Enzymol.* **277**, 505–524 (1997)]. For further details see Chantalat et al. (18) and Niefind et al. (19).

and, as a consequence, pathophysiological implications. In any event, current knowledge indicates a strong relationship of CK2 to gene control. On the one hand, a majority of substrates comprise transcription factors and other proteins implicated in gene expression and transcription, including subunits of all three classes of RNA polymerases, and CK2 is predominantly nuclear and part of expression-linked protein complexes (3–6). On the other hand, many of the substrates are transcription-directed signal transduction proteins of growth signaling, stress signaling, and survival signaling (20), and CK2 is required for the expression of genes such as *fos* upon mitogenic cell stimulation (7–9), which is the key to subsequent waves of proliferation-linked gene expressions (31). Further, life is not compatible with a deletion of the catalytic CK2 subunits (32), and availability of regulatory CK2 subunits is essential for embryonic development and organogenesis (20). This, together with CK2's flexible signal-mediated cellular localization dynamics, has led to the hypothesis that CK2 represents a survival factor (20).

Central to cell growth, death, and differentiation is the cell cycle. CK2 has long been known to have various links to it (4, 21, 27). The cell cycle is governed by the highly conserved cell-cycle control system, primarily acting via cyclically activated protein kinases (CDKs) and their activating cyclins, CDK inhibitors and activators (acting either by complex formation or by changing phosphorylation status), and proteolytic degradation complexes. Additional regulatory devices allow the system to respond to various signals from both inside and outside the cell. CK2 is obviously not a component of the core control system. It rather relates to the system's signaling networks as indicated, for instance, by the inhibition of proliferation stimulation due to CK2 perturbation both at a nucleic acid level and protein level, which relates to reentry from G_0 phase [a nondividing—differentiation ("resting")— state], into the cell-cycle and the adjacent early G_1 phase (8, 9, 33–35). Because in G_1 cells decide whether to pass through the cell cycle or go to apoptosis, entry into the cycle has important implications for cell sociology, and thus many links to pathophysiological processes such as cancer development (see below). Further, CK2 may affect the expression of cell-cycle genes that encode cyclins responsible for the activation of cycle engine-driving CDKs or segregation apparatus components. This characterizes CK2 as a member of the group of factors that contributes to cell-cycle regulation at a level additional to the protein modification-based core component control. Evidence for this link of CK2 results from comparative genome-wide DNA array analyses of CK2-perturbed *S. cerevisiae* strains (35a). A further and particularly interesting result of the array studies is that CK2 affects the expression of various genes encoding chromatin remodeling proteins, which, together with CK2's known ability to phosphorylate chromatin structure proteins (36), is a strong indicator for a more global role for CK2: chromatin remodeling (see Section V).

The pathophysiological implications of CK2 are manyfold and often related to deviations in stoichiometry and in cellular levels of CK2 subunits. For instance, in theileriosis or East coast fever, a fatal lymphocytic disease of African cattle and Cape buffalos, *T. parva*-infected lymphoblastoid cells contain markedly increased amounts of CK2α mRNA and protein relative to noninfected cells, but because the parasite does not seem to contain a CK2β subunit, the host cell cytosol overflows with parasite CK2α, interfering with host control of cell-cycle regulation (37). In mating experiments with c-*myc* transgenic mice, the coexpression of CK2α markedly increased the rate of onset of fatal lymphoproliferative disease paralleled by a striking alteration in lymphocyte gene expression; elements of two excluding programs of lymphocyte differentiation became expressed (38). In mammary gland, overexpression of CK2α caused hyperplasia and dysplasia, which in 30% of cases developed into adenocarcinomas, often associated with Wnt pathway activation (39). Together with the *fas* mutation, CK2α overexpression dramatically exacerbated the lymphoproliferative process seen in a lupus-like autoimmune disease associated with the production of autoantibodies (40). CK2 might also play a role in processes underlying progressive disorders due to Alzheimer's disease, ischemia, chronic alcohol exposure, or human immunodeficiency virus (HIV) (41). High expression levels of CK2 are found in proliferative tissues and in tumors and causal relations are suspected (see Section IV). The underlying mechanisms, however, are not clear. Altering gene expression, overexpressed CK2α had been suspected to act as a transcription factor. Because CK2α overexpression induces aromatase, this had been tested with the human aromatase gene promoter and found not to be the case, indicating rather another, indirect, mechanism of action (42).

Despite the far-reaching physiological and pathophysiological implications of CK2, investigations of the human genes encoding the CK2 proteins and their control had been progressing comparatively slowly. A decade elapsed between unraveling the human CK2β gene structure, the very first eukaryotic CK2 gene solved (43), and understanding how it is regulated, which may include a pronounced coordinative element to the expression of the other CK2 genes (44, 45).

III. The Human Protein Kinase CK2 Genes

A. CK2 Loci in the Human Genome and Structural Organization of the CK2 Genes

1. CK2 LOCI

The human genome contains four CK2 loci. Two of these have been identified as subunit α loci, one as subunit α' locus, and one as subunit β locus. The loci are positioned at different chromosomes. The CK2α loci are at 11p15 and 20p13, the CK2α' locus is at 16q13, and the CK2β locus is at 6p21 (Fig. 2). Although our loci positioning at chromosomes 6, 11, and 20 by genomic fragment hybridization to elongated metaphase chromosomes (46, 47) match the data provided by the human genome project (*http://www.ncbi.nih.gov/genome/guide/human/*), the positioning of the CK2α' locus to the short arm of chromosome 16 (48) using *in situ* hybridization has been questioned; the locus of this gene seems rather positioned at the long arm, at 16q13 (K. Ackermann *et al.*, unpublished results).

2. CK2 GENES

Each of the CK2 loci accommodates one gene. Except for the gene at locus 11p15, all of the CK2 genes are composed of exons and introns. Exon numbers and sizes are not too far from each other (see below), but introns vary dramatically in lengths and, as a consequence, the CK2 genes differ significantly in

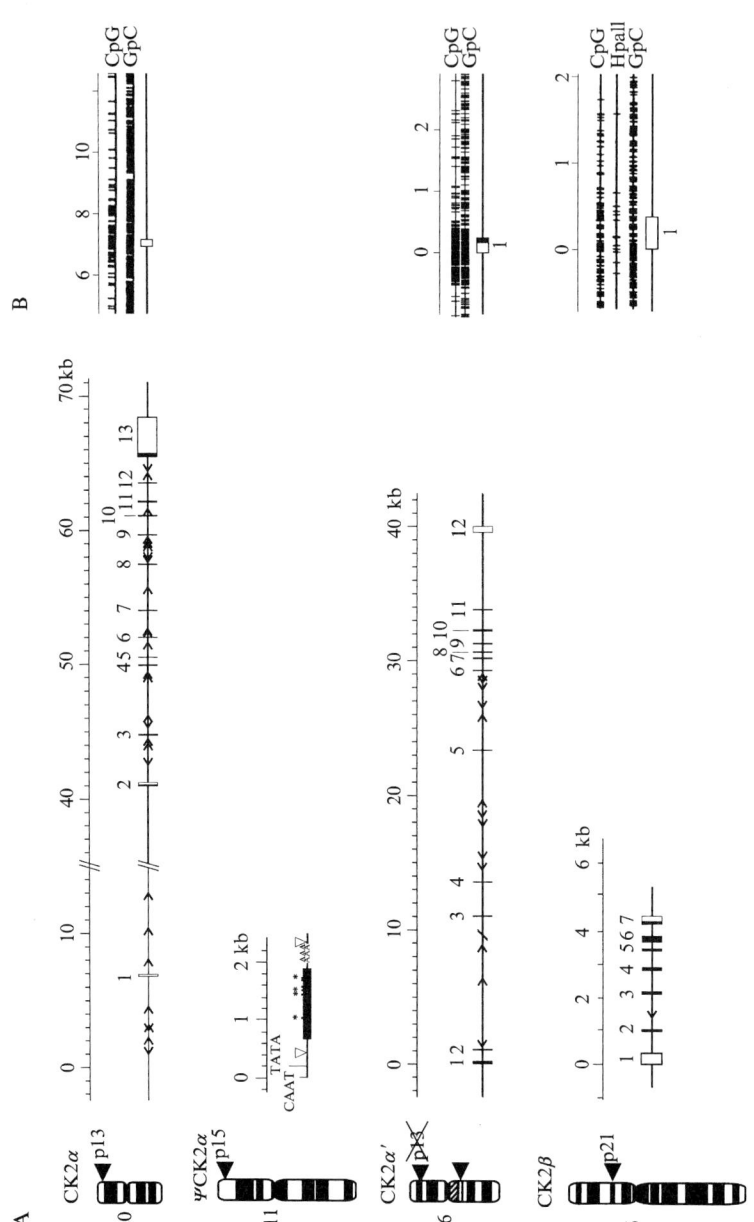

size: The CK2β gene spans roughly 4.2 kb, whereas the CK2α and CK2α' genes span 70 and 40 kb, respectively (Fig. 2).

The CK2α gene occurs in two versions, as a transcribed (locus 20p13) and a nontranscribed (locus 11p15) sequence. The active CK2α gene (*CSNK2A1*), originally isolated out of a genomic placental library, is composed of 13 exons, exon sizes ranging from 51 (exon 5) to 2960 bp (exon 13) and intron sizes from 527 (intron IV) to about 34,000 bp (intron I). Exon I is untranslated, the translation start site is located in the second exon, the stop codon in exon 13. The exon/intron boundaries conform to the canonical gt/ag rule (*47, 49*). Screening for repetitive elements revealed a considerable number of *Alu* repeats and several microsatellite sequences as well as a $(TA)_{10}$ and two CA repeats within introns. The inactive CK2α gene ($\Psi CK2\alpha$) spans 1.6 kb, is devoid of introns, has a short poly(A)$^+$ tail and direct flanking repeats, i.e., shows the typical features of a processed sequence, and is homologous to CK2α-cDNA, a typical pseudogene criterion. It has a complete open reading frame and several nucleotide exchanges, four of which would, if the sequence was expressed, result in amino acid exchanges. Interestingly, potential promoter elements (two TATA boxes and a CAAT box) are present in the adjacent 5' flanking region of the gene creating a theoretical possibility for transcription and thus the generation of a mutated but functionally active CK2α protein. So far, however, there is no hint that this occurs (see below) (*46, 50*).

The CK2α' gene (*CSNK2A2*) has very recently been solved by classic cloning and sequencing of a genomic fragment comprising roughly the 3' half of the gene (U. Wirkner and W. Pyerin, unpublished results) and *in silico* completion (segment NT-010406/genomic contig, BAC clone RPC1-11 159F6/accession number AC009107 containing the complete gene) (K. Ackermann *et al.*, unpublished results). To determine whether the originally proposed location at the p arm of chromosome 16 was possibly due to a pseudogene or related sequence, we compared all available relevant genomic contigs with CK2α' cDNA. We could not find any homology. The gene is composed of 12 exons, with exon sizes

FIG. 2. CK2 gene loci present in the human genome and structures of the CK2 genes. (A) Chromosomal locations and physical maps of the human genes coding for subunit α located at chromosomes 20p13 (CK2α; active gene) and 11p15 (ΨCK2α; processed pseudogene), for subunit α' (CK2α') and subunit β (CK2β) at 16p13 and 6p21, respectively. The exon/intron structure of the genes is schematically drawn; scaling is provided for each gene. Exons are numbered and given as filled boxes (coding sequences) and as open boxes (untranslated sequences). *Alu* repeats and their orientations are indicated by arrows. Triangles in ΨCK2α mark insertion sites and asterisks provide mutation sites causing amino acid exchanges; TATA and CAAT boxes are indicated. (B) CpG islands. Schematic drawing of the distribution of CpG and GpC dinucleotides given as vertical bars. Location of exon 1 is shown as filled or open box (translated or untranslated, respectively); scaling corresponds to scaling in (A). *Hpa*II, restriction sites of *Hpa*II given as vertical bars forming an HTF island (*Hpa*II tiny fragment fraction).

ranging from 51 (exon 4) to 444 bp (exon 12) and intron sizes from 375 (intron VII) to 9835 bp (intron II). The first exon contains the translation start site and exon 11 the translation stop codon; the last exon, exon 12, is therefore untranslated. The exon/intron boundaries conform to the gt/ag rule. It should be mentioned that the CK2α' sequences available in data bases need cautious consideration. Both incorrect sequences and sequence exchanges due to naming errors are stored. The CK2α' gene structure corresponds to a great extent to that of the active CK2α gene: 9 of the 12 exons (exons 2–10) have exactly the same size (exons 3–11 of the CK2α gene). The exon sequences, however, differ strongly, but the majority of nucleotide variations have no consequence for the amino acid sequence or cause conservative amino acid exchanges. Consequently, the amino acid sequence of CK2α' strongly resembles the CK2α sequence. Like in the CK2α gene, multiple *Alu* repeats are present within introns of the CK2α' gene. Also, several microsatellite sequences as well as poly(A) and poly(T) repeats within introns exist.

The CK2β gene (*CSNK2B*) was the first mammalian CK2 gene to be completely unraveled (*24, 43*). It was found by screening a leukocyte genomic library using a cDNA fragment representing 80% of the coding region. The gene is composed of seven exons, with exon sizes ranging from 76 (exon 5) to 329 bp (exon 1) and intron sizes from 145 (intron V) to 965 bp (intron II). The second exon contains the translation start site at position 12, i.e., the first exon remains untranslated, and exon 7 the stop codon. The exon/intron boundaries conform to the gt/ag rule. The second intron contains a single complete, reversely oriented *Alu* repeat.

B. CK2 Gene Structures Are Conserved

When the structures of the human CK2 genes are compared to those of evolutionarily distant organisms such as the nematode worm *C. elegans* (*22, 23*), three introns of both the CK2α and the CK2α' genes are seen located at corresponding positions (*47*; K. Ackermann *et al.*, unpublished results), and the same is true for the CK2β genes (*47*) (Fig. 3). This suggests that the catalytic subunit genes and the regulatory subunit genes both originate from one ancestral gene, and that both ancestors should have contained three introns at the conserved positions. This makes clear that not only the CK2 subunit proteins and the tetrameric holoenzyme structure they form are conserved (see above), but that conservation is even found for the genomic organization of CK2 loci, despite the remarkable size difference of the genes, 70 and 40 kb for the human CK2α and CK2α' genes but only 2.9 kb for the *C. elegans* CK2α gene. This characterizes CK2 as extremely conserved.

The differences in size may be a consequence of either stronger loss than insertion of intron sequences in the *C. elegans* gene during evolution, or stronger insertion than loss in the human genes. The accumulation of

GENES ENCODING HUMAN PROTEIN KINASE CK2 249

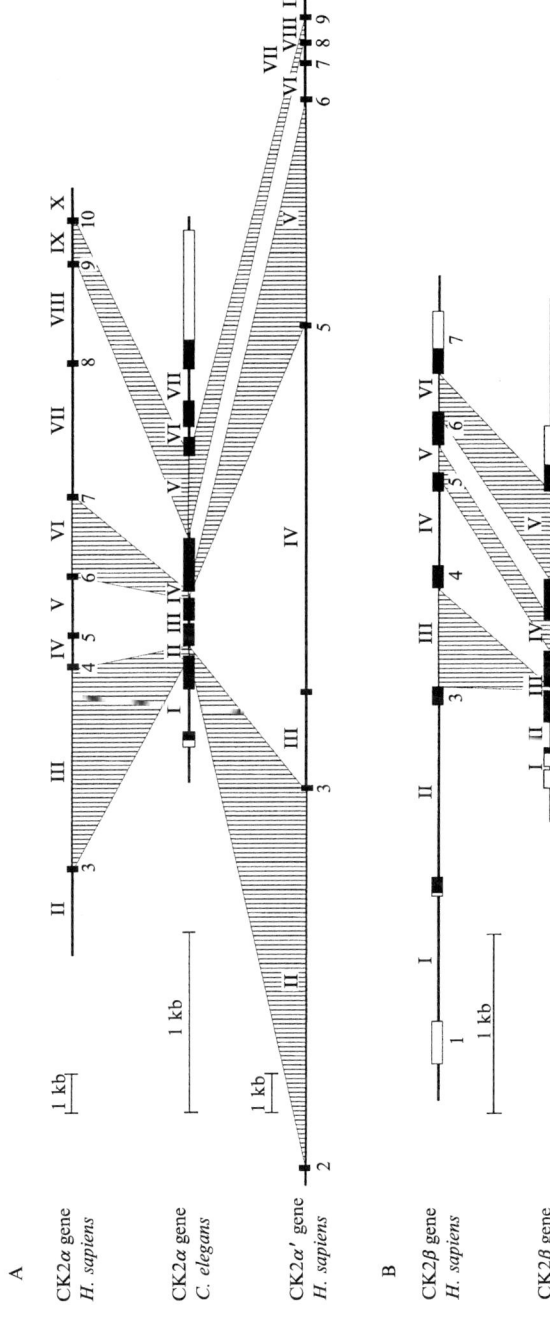

FIG. 3. Gene conservation. Structure comparisons of CK2 genes of humans and worm. For symbols see legend to Fig. 2. Introns are numbered in roman letters. Introns that are located at corresponding positions in the human CK2α, α', and β genes as well as in the corresponding C. elegans genes are connected by shaded areas. (A) CK2α and CK2α' genes of Homo sapiens compared to the CK2α gene of Caenorhabditis elegans. Note: the drawings of human and worm genes are not to the same scale. (B) CK2β gene of H. sapiens compared to C. elegans CK2β gene. Both genes are drawn to the same scale.

repetitive elements such as primate-specific *Alu* repeats in the human genes favors the second possibility. Statistically, *Alu* repeats occur every 4 kb, but in the human genes the frequency is compressed to one at every 1.5 kb. Further, microsatellites occur 3' of various repeats, including poly(A)$^+$ repeats, TA repeats, and CA repeats. It appears that the genes were targets for *Alu* insertion during different successive evolutionary periods, insertion of new repeats occurring into the poly(A)$^+$ region of already present repeats (*4, 47, 49, 51*; K. Ackermann *et al.*, unpublished results). The resulting conclusion that the human genes are located in relatively variable chromosomal regions is supported by observations such as a genetic polymorphism for CK2α (*52*), which, however, has not been sufficiently characterized for unambiguous assignment to one of the catalytic subunit genes. As a consequence of conservation, CK2 of humans differs very little from the CK2 of other mammalians. This is of considerable importance for any of the projections drawn from experimental results obtained with model organisms.

IV. Expression of Human CK2 Genes

A. Transcripts and Their Distribution

The transcription of each of the three active human CK2 genes results in more than one mRNA species (Fig. 4A). In cultured human cells such as the choriocarcinoma cell line JEG-3, for instance, three CK2α transcript species (4.5, 3.4, and 1.8 kb in length), up to five CK2α' transcript species (5.7, 4.5, 2.8, 2.0, and 1.0 kb), and several similar-sized CK2β transcript species (roughly 1.0 kb) are present (*24, 44, 45, 53*). Similar numbers and sizes of transcripts have been found in other human cells (*25*). However, Northern blot-assessed transcript sizes are rough estimates. Therefore, we compared the nucleotide sequences of the regions downstream of the termination sites in the CK2 genes [each containing several poly(A)$^+$ sites] with human expressed sequence tags (ESTs) of the dbEST database. Many of the ESTs end shortly after one of the potential poly(A)$^+$ signals. The *in silico* transcript lengths correlate well with those assessed experimentally.

Cells appear to keep the CK2 transcript levels constant, independent of their respective status. This has nicely been demonstrated in JEG-3 cells (Fig. 4A). When analyzed over time for availability of CK2 transcripts at seeding of very dilute cell suspensions and at various times postseeding until formation of confluent cell layers, a process running from a predominantly proliferating cell population over various stages of differentiation to a contact-inhibited fully differentiated cell population, the CK2 transcripts are found to remain practically unaltered throughout, both in number and level of mRNA species as compared to a standard GAPDH control. The increasing differentiation of cells with increasing cell density is clearly confirmed by the aromatase transcript levels; aromatase expression is a function of differentiation and transcript levels are low in predominantly proliferating cell populations and high in predominantly differentiated cell populations (53). Constant CK2 transcript situations are not restricted to human cells. It can, for instance, also be observed in yeast cells analyzed for transcript availability over several rounds of cell division and at various cell densities (54).

If investigated carefully enough, CK2 transcripts are found throughout all human cells and tissues, which is no surprise for a vital entity. However, both the level and the number of individual transcripts may vary considerably from tissue to tissue (K. Ackermann et al., unpublished results). Heart, placenta, skeletal muscle, and testis show a significantly higher level of transcripts than tissues such as lung, liver, kidney, or colon (Fig. 4B). The stoichiometry of subunit transcripts reflects the tetrameric stoichiometry of the gene products in several tissues but not in others. For instance, lung, liver, kidney, or leukocytes show in comparison to other tissues very weak transcript signals for CK2α and α', whereas their CK2β transcript levels resemble that of other tissues. In placenta, the situation is reversed; the α and α' transcripts exceed that of β. When investigated, the same was found by Bosc et al. (55). The varying stoichiometric situations either indicate that the tetrameric holoenzyme is not the exclusive form of CK2 in cells and individual subunit proteins may also complex to other proteins and thus participate in additional cellular processes, or a subunit-specific translational and/or posttranslational control may exist. Evidence is accumulating in favour of the former (3, 5, 6, 20, 35).

B. Transcript-Based Phylogeny of CK2

We have conducted phylogenetic analyses with all currently available mRNA sequences of both the catalytic and the regulatory subunits of protein kinase CK2 (K. Ackermann et al. unpublished results). Using the human CK2α and CK2α' (25) as well as CK2β (24) cDNAs as query DNAs, BLAST reveals a multitude of sequences that give significant alignments. Regions within multiple alignments chosen for α and α' cDNAs were nucleotides at positions 451–620 (corresponding to amino acid residues 152–207) and 501–650 (corresponding to amino acid residues 168–216). These regions contain the kinase domains VIb, VII, VIII, and partially, IX and are practically identical for α and α' in amino acid composition. The aligned region also contains the coding sequence for the activation segment whose conformation is conserved for isolated and complex-bound CK2α (19). For CK2β, the region enclosing nucleotide positions 351–561 (corresponding to amino acid residues 118–187) was chosen. As a result, the unrooted phylogenetic trees display for both the catalytic and the regulatory CK2 subunits prominent clades for mammalian

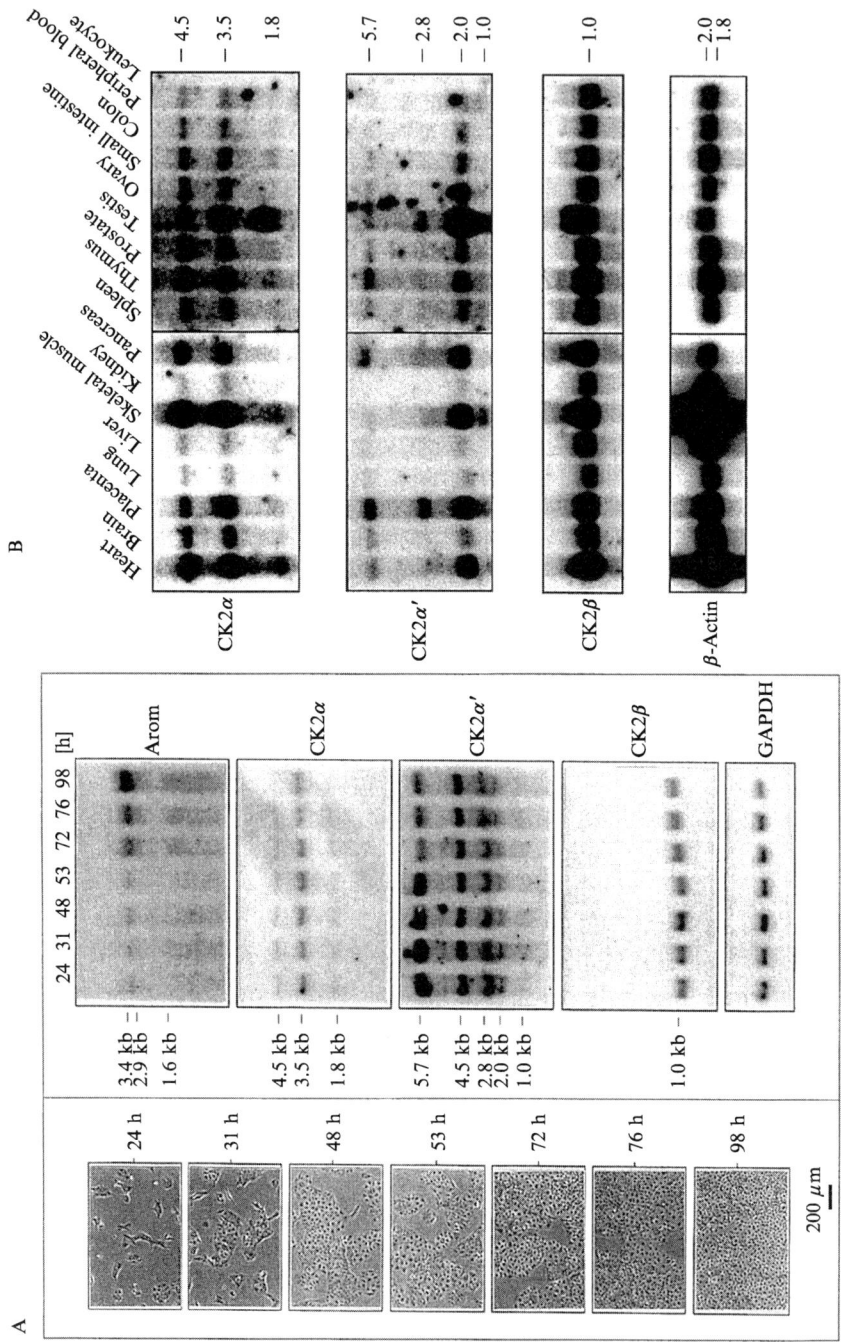

and plant sequences indicating their early diverging development (Fig. 5). In the tree based on human CK2α', zebrafish and bird branch off early as they do in the cluster formed within that tree by CK2α. Cow and mouse are both close to humans in CK2α', whereas rat and rabbit, which possess only a single catalytic subunit, are closer to human CK2α than mouse but less than cow. Thus, although branches considerably converge at the protein level [homologies 98 and 97% of human α and α' to bird, respectively (4, 56)], the principal relationships are the same, zebrafish remaining distantly related. The relationships are similarly indicated when the phylogenetic tree is based on CK2α (not shown). The CK2β tree also shows relatively large distances from humans to organisms such as zebrafish and bird. At the protein level, however, the distances are faded away completely.

C. Regulation of Transcription

1. Transcriptionally Active Regions and Their Expression Control Features

The active CK2α gene is characterized by two transcription start sites, the further 5'-located defining position +1, assigning position 50 to the second (stronger) start site (49). The sequence around exon 1 is characterized by a CpG island (see Fig. 2), and the adjacently upstream start site by a lack of a TATA box and the presence of CAAT boxes at nonstandard positions and the occurrence of various GC boxes (Table I). Together, these features assign a so-called housekeeping character to the CK2α gene. Systematic stepwise deletion from either end of an appropriately sized and positioned genomic fragment as well as specifically designed synthetic fragments, cloning into indicator gene vectors (luciferase), transfection of human cultured cells, and determination of luciferase activity in cell extracts (45, 49, 51, 57), localized promoter activity to positions −39 to 65 and an enhancer activity upstream of position −69. Maximum promoter activity is being developed at positions −9 to 46 (Fig. 6). The regions sourrounding TS1 and TS2 are functionally linked; all transcriptional activity is lost upon TS1 and TS2 elimination; at least one TS is required for transcriptional activity to develop. The transcriptionally

---

FIG. 4. CK2 transcripts. (A) Transcript situation in human cultured cells varying in differentiation/proliferation status. JEG-3 cells, harvested at indicated time points post-seeding, were analyzed by Northern blotting (10 μg of total RNA; specific probes for respective gene transcripts) for the presence of CK2α; CK2α' and CK2β gene transcripts as well as of aromatase (Arom) and GAPDH (GAPDH) gene transcripts. Transcript sizes are given on the left. Bar provides scale for cell culture. (B) Transcript situation in human tissues. Poly(A)$^+$ mRNA (2 μg) from indicated tissues was analyzed by Northern blotting as above for CK2 subunit transcripts. β-Actin and GAPDH are quantitation markers.

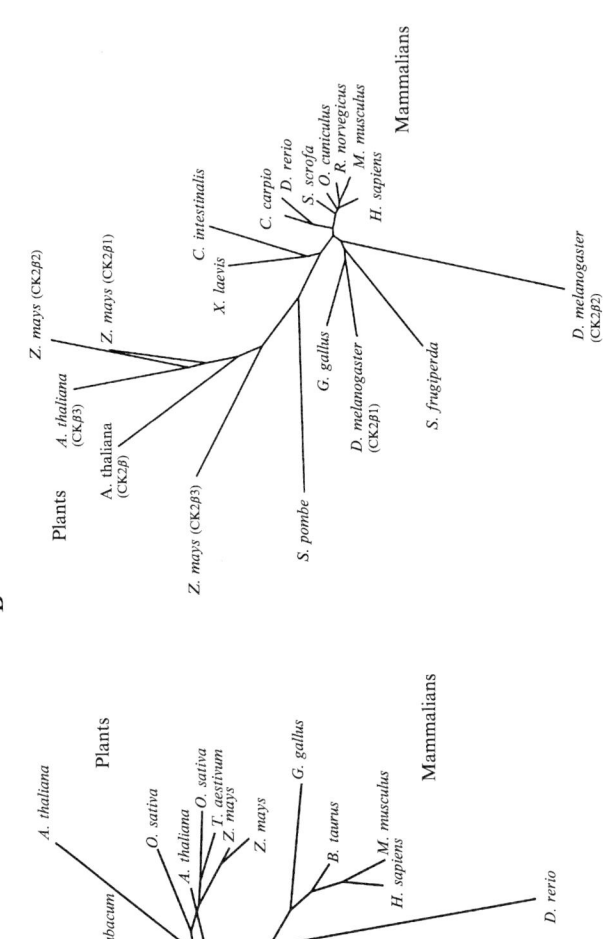

TABLE I
CHARACTERIZATION OF HUMAN CK2 GENES BY STRUCTURAL FEATURES OF IMPORTANCE FOR TRANSCRIPTION CONTROL[a]

Feature	Gene			
	CK2α	ΨCK2α	CK2α'	CK2β
Transcription start sites	2	1	(2)	3
TATA box	−	+	−	−
CAAT box				
At standard position	−	+	−	−
At nonstandard position	+	−	+	+
GC box	Multiple	−	Multiple	Multiple
CpG island around exon 1	+	−	+	+
Interrupted 5'-untranslated region	+	−	(−)	+

[a]The compared sequences were taken from Pyerin and colleagues (43, 46, 49; K. Ackermann et al., unpublished observations).

FIG. 5. Transcript-based evolutionary trees of CK2. Nucleotide sequences were retrieved by NCBI Entrez or EBI-SRS in FASTA format. Multiple alignments were performed with CLUSTAL, distance matrices from nucleotide sequences were calculated by DNADIST, and phylogenetic trees were constructed using the PHYLIP-package Version 3.5 [J. Felsenstein, PHYLIP—Phylogeny Inference Package (Version 3.2). *Cladistics* **5,** 164–166 (1989)]. (A) Unrooted phylogenetic tree of CK2 catalytic subunits. Note: Search conducted with human CK2α' cDNA as query sequence. Asterisk indicates organism with only one known catalytic subunit. Accession numbers for the sequences (Database: ALL EMBL): CK2α'/CK2α: M55268/M55265, *Homo sapiens*; X64692, pseudogene *Homo sapiens*; D90394/X54962, *Bos taurus*; AF012251/U51866, *Mus musculus*; M96173, αA pseudogene *Mus musculus*; X82232, αB pseudogene *Mus musculus*; X82233, αC pseudogene *Mus musculus*; M59457/M59456, *Gallus gallus*; X99964, S76875, *Danio rerio*; X62375, *Xenopus laevis*; M98451, *Oryctolagus cuniculus*; L15618, *Rattus norvegicus*; AB052133, *Triticum aestivum*; AF374474, *Nicotiana tabacum*; AF271237, X61387, AF239819, *Zea mays*; AB036787, *Oryza sativa* Nipponbare; AB036788, *Oryza sativa* Kasalath; AF370308, AY035088, AL132978, AC004401, *Arabidospis thaliana*. (B) Unrooted phylogenetic tree of CK2 regulatory subunits. Accession numbers for the sequences (Database: ALL EMBL): X57152, M30448, X16937, *Homo sapiens*; X56502, X80685, X52959, *Mus musculus*; L15619, *Rattus norvegicus*; M98450, *Oryctolagus cuniculus*; X56503, *Sus scrofa*; S76877, *Danio rerio*; AF133088, *Cyprinus carpio*; X62376, *Xenopus laevis*; AF360544, *Ciona intestinalis*; AF071211, *Spodoptera frugiperda*; U51209, M16535, *Drosophila melanogaster*; M59458, *Gallus gallus*; X74274, *Schizosaccharomyces pombe*; AF239816, AF239817, AF239818, *Zea mays*; U03984, AF068318, *Arabidopsis thaliana*.

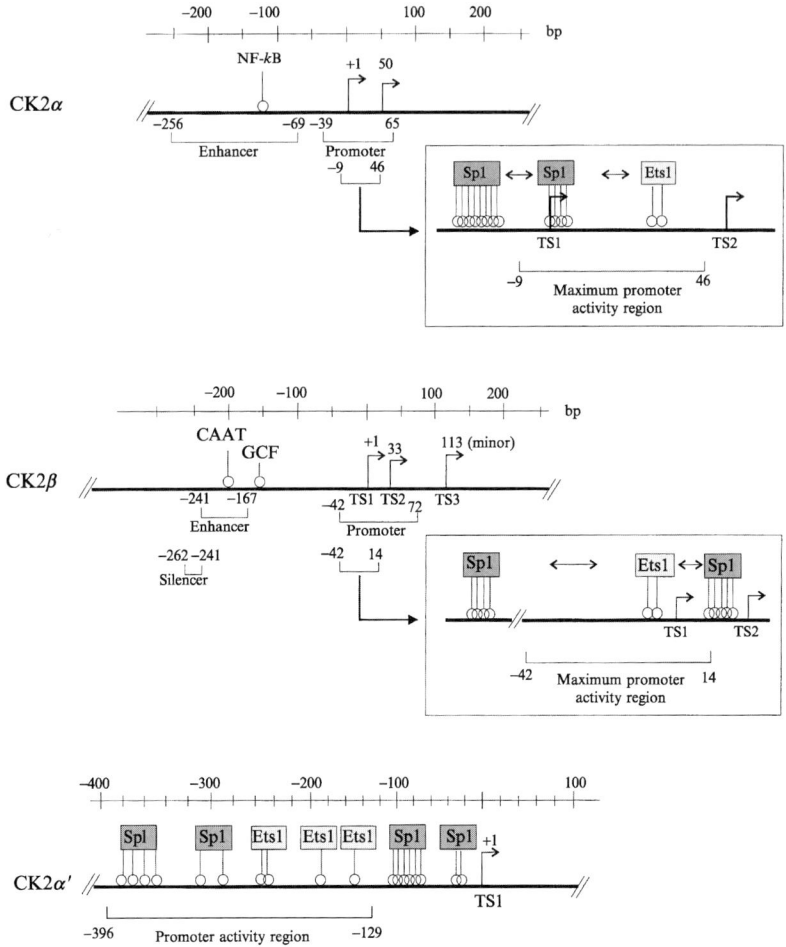

FIG. 6. Transcription control regions and features of the active human CK2 genes. Transcriptionally active regions were determined by luciferase reporter assays following transfection of 3′ and 5′ deletions of human genomic fragments (CK2α fragment, −256/144; CK2β fragment, −682/122; CK2α′ fragment, −1306/197) in cultured human cells. Provided are summarizing schemes of the promoter situation of the genes encoding CK2α, CK2β, and CK2α′. Numbers provide nucleotide positions; open circles give consensus sites for transcription factors as indicated; horizontal arrows indicate cross-talk between transcription factors; TS, transcription start site. Scale given above each gene.

active regions contain putative binding sites for various transcription factors, including GCF, Sp1, AP2, CTCF, Ets1, and NF-κB, and we have shown these factors to be present in the nuclear compartment of the cells employed in transcription control studies (44, 45, 49, 51, 57). Tested by site-directed mutagenesis of individual sites and of two or more sites simultaneously, evidence has been obtained that these sites and factors are strongly interlinked. The intensity of cross-talks between factors relates not only to the combination of sites but also to length of investigated DNA fragments and other circumstances. According to results obtained with a fragment likely to reflect the *in vivo* situation, Sp1, Ets1, and NF-κB contribute most significantly to transcriptional control. Aside from affecting transcription individually, these factors show strong cooperative effects: mutation of Sp1 sites amplifies Ets1 and NF-κB mutational effects, and Ets1 and NF-κB site mutations affect each other.

Different from all other CK2 genes, the CK2α pseudogene possesses a single transcription start site and upstream of it TATA and CAAT boxes at standard positions and other typical promoter features (Table I). Despite their presence, however, respective genomic fragments show no transcriptional activity in indicator gene assays, and no signals are obtained in Northern blots with gene-specific probes *(47, 49, 51)*. Therefore, the CK2α gene at locus 11p15 obviously represents a silent sequence, at least under normal conditions.

The CK2α′ gene structure and the transcriptionally relevant upstream sequence recently became known (K. Ackermann *et al.*, unpublished results). The data presented here have a preliminary character. There are unanswered questions concerning the presence of a putative *Alu* repeat in the 5′ region of the cDNA and resulting problems in defining transcription start sites that are worsened by an extremely high GC content and therefore complications in primer extension analyses. However, a transcription start site can be defined on the basis of ESTs, assuming position +1 is provided by the most 5′ situated ESTs. This start site is in agreement with the 5′ sequence of the CK2α′ transcripts.

There is no TATA box but there are CAAT boxes at nonstandard positions and multiple GC boxes can be found (Table I). Further, a CpG island is present around exon 1 (Fig. 2B). Despite the unanswered questions, we have identified promoter activity within the adjacently upstream sequence of exon 1. This is in accordance with *in silico* promoter prediction. Using indicator gene assays of systematically deleted genomic fragments the region with highest promoter activity comprises positions −396 to −129, and the activity is at least as strong as that of the CK2α and CK2β gene promoters (Fig. 6). This region contains various putative binding motifs for transcription factors, including multiple Sp1 and several Ets1 sites. Additional putative transcription factor binding sites are present in the region such as AP1, AP2, c-Myb, NF-1, and C/EBP.

The CK2β gene has three prominent transcription start sites, at positions +1 (TS1), 33 (TS2), and 113 (TS3) *(43)*. The upstream sequence lacks a TATA box, but has several GC boxes and CAAT boxes at nonstandard positions as well as the presence of a CpG island around the untranslated exon 1 (Table I).

This indicates that the CK2β gene has, like the CK2α gene, a housekeeping character. Systematic stepwise deletions from either end of a genomic fragment comprising exon 1 and adjacently upstream and downstream sequences (*44, 45*) indicate that the promoter spans roughly positions −42 to 72, and that the gene has enhancer and silencer elements upstream at positions −241 to −167 and −262 to −241, respectively. Maximum promoter activity is positioned at −42 to 14, and, consequently, transcription predominantly starts at TS1 and TS2 (Fig. 6). The upstream region of the CK2β gene possesses binding sites for transcription factors Sp1 and Ets1, but not NF-κB. In addition, AP2, GCF, CP1/CP2, and other CAAT-related motifs are present (*43–45*). Deleting motifs or mutating at several positions within affects transcriptional activity. Considering the results obtained with the longest investigated fragment comprising positions −241 to 122, the Ets1 and CAAT-related motifs contribute most significantly to activity, whereas weaker effects are seen with Sp1 sites. When simultaneously mutated, Sp1 enhances effects of Ets1 and CAAT-related motif mutations. The CAAT-related and other motifs, however, are not promoter located but rather are enhancer specific.

Transcriptional control of the CK2 genes should, therefore, decisively occur due to binding and cross-talking of these factors. Because each of the three active CK2 genes has its specific pattern of binding sites, this should allow for individual gene control, even in the same cell. As a consequence, transcript availability of individual CK2 subunits may be stoichiometric in one cell type but far from it in another. Cell-specific transcript compositions then should determine the CK2 transcript patterns of tissues, explaining the above observed variations in human tissues (see Fig. 4).

2. Is There Transcriptional Coordination of CK2 Subunit Expression?

Aside from differences due to the gene-specific patterns of transcription factor binding sites, the transcriptionally active regions of the three active CK2 genes share a number of features. Some of these are appropriate for a coordinate transcriptional regulation, primarily Ets1 and Sp1 sites. Ets1 belongs to a family of transcriptionally active proteins functioning frequently in combination with other factors, including Sp1 (*58, 59*). Ets1 and Sp1 motifs are present in the control region of all three genes (Fig. 6).

Most strikingly, however, is the presence of an Ets1 double motif that occurs in the promoters of the CK2α and CK2β genes. In both promoters, this 12-bp-long motif is identical in all its nucleotide positions, and the motif has similar relative distances to transcription start sites. The pairwise occurrence of Ets1 motifs has frequently been observed and seems to relate to a relatively low Ets1 binding affinity of individual sites depicted, aside from others, by DNase I footprints exceeding significantly the Ets-1 motif sequence (*60, 61*).

In TATA-less promoters, Ets motifs are usually located close to transcription initiation sites, and may participate in the formation of transcription initiation complexes (62–64). Further, some of the Ets family binding sites have been demonstrated to function as initiator elements for transcription in TATA-less genes (63–65). Both CK2 gene promoters have a number of Sp1 binding sites. The Sp1 motifs are present predominantly in the form of clusters of some three to eight motifs, overlapping each other or arranged side by side. Sp1 is a reported GC box binding activator of transcription found in many promoters and enhancers (42, 59, 66) that is capable of recruiting factors such as TFIID, which is also required by TATA-less promoters to activate RNA polymerase II (67) and which has the known ability to link distant transcription control elements, i.e., to mediate cooperation between transcriptionally active proteins (62, 66). Further, Sp1 elements have been shown to compensate for each other, requiring complete deletions to abolish effects provoked by ectopically expressed Sp1 (58). Ets1 and Sp1 may function separately to initiate CK2 gene transcription. More likely, however, seems to be concerted actions as reported for various genes, including virus-mediated Ets1–Sp1 complex formation and transactivation (58, 59, 68).

Concerted actions of Ets1 and Sp1 sites are in particular indicated by site-directed mutagenesis (44, 45, 57). Mutating the Ets1 site in the CK2α gene promoter, transcriptional activity is decreased to a significant extent, and mutation of nearby Sp1 clusters individually or simultaneously also affects more or less significantly activity. But when Ets1 and Sp1 motifs are mutated simultaneously, the effect is strongly intensified or the gene promoter is practically inactivated. A similar tendency has been observed for the CK2β gene promoter. The respective mediator proteins, Ets1 and Sp1, have been shown to be available in the nucleus of cells such as JEG-3 (see above) by a spectrum of different methods, including Western blotting, electrophoretic mobility shift assays (including up-shifts with specific antibodies), UV cross-linking, and affinity chromatography (44, 57). At least Sp1 is a CK2 substrate, and Ets1 possesses several minimal CK2 phosphorylation consensus sites (there is no direct evidence yet for phosphorylation by CK2). Two features of Sp1 phosphorylation are of importance. First, when tested under conditions that lead to a phosphorylation of other transcription factors such as UBF, CREB, and c-Jun (13), Sp1 phosphorylation occurs with the CK2 holoenzyme but not with individual CK2α. Second, phosphorylation alters the behavior of Sp1 measured as a significant decrease of its DNA binding capacity (44, 45, 57), a result supported by earlier data that identified phosphorylation at positions such as T579 (69).

The phosphorylation and its effect provide a basis for a working hypothesis of a coordinate transcription regulation of the the CK2α and CK2β genes, explaining why quasi–constant cellular CK2 transcript levels may be adjusted. As schematically outlined in Fig. 7, it is tempting to assume that activation of the

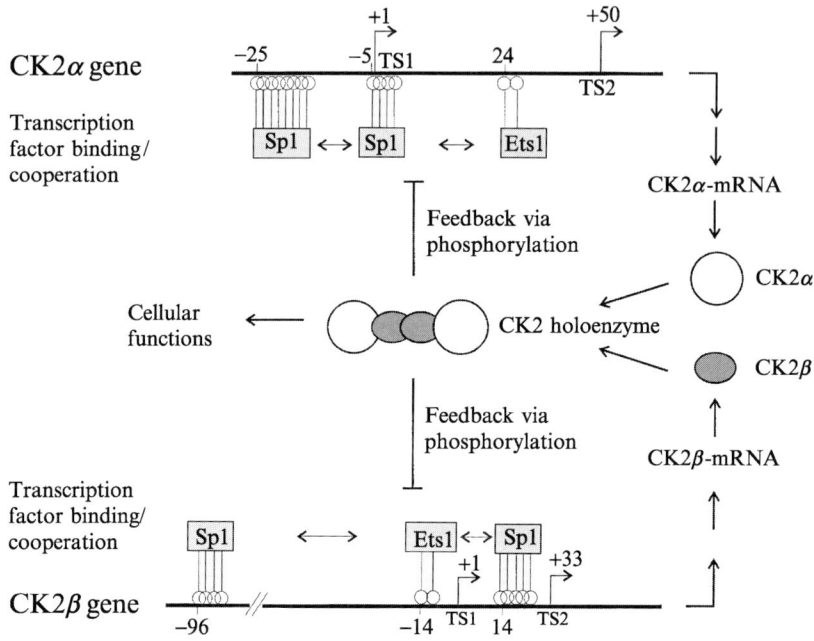

FIG. 7. Schematic representation of transcriptional coordination hypothesis of the human CK2α and CK2β genes. Ets1 response element (double motif; circles symbolizing individual Ets1 elements; not to scale) common to transcriptionally active regions of both the human CK2α (region −9 to 46) and CK2β (region −42 to 14) gene promoters might activate in cooperation with Sp1 response elements (overlapping motif clusters, circles symbolizing individual Sp1 elements; not to scale) the transcription of genes. The generated mRNAs (CK2α-mRNA; CK2β-mRNA) could be translated into CK2 subunit proteins (CK2α, open circle; CK2β, shaded circle), which may readily complex into a tetrameric CK2 holoenzyme. The holoenzyme might, aside from catalyzing various cellular functions, feed back to gene transcription via phosphorylation of Sp1 (and Ets1?) resulting in a down-regulation of transcription. For symbols see legend to Fig. 6.

CK2α and CK2β genes occurs under participation of Ets1 and Sp1 due to their promoter binding and cross-talk. Activation results in CK2α and CK2β gene transcription followed by translation into CK2α and CK2β protein. This process would continue until a certain cellular level of newly generated CK2α and CK2β is reached, which, considering their high mutual affinity (15), would readily complex to CK2 holoenzyme tetramers. Aside from serving cellular functions, the holoenzyme, but not the individual CK2α, would phosphorylate Sp1 (and Ets1?), loosening promoter binding and/or affecting cross-talks between Sp1 and Ets1. As a consequence, Ets1/Sp1-mediated transcriptional activation would, due to the increase in holoenzyme level and thus phosphorylation, decrease and the expression of both the CK2α and the CK2β gene would decline. The expected outcome would be constant transcript levels of both genes meeting exactly the situation observed in human cultured cells. Therefore, the scenario of a negative feedback control seems to apply nicely, coordinating CK2α and CK2β gene expressions.

The proposed negative feedback control applies only if both CK2α and CK2β are expressed. Expression of the CK2α gene alone would, if CK2β expression was disturbed and CK2β not available by any other form of intracellular storage, steadily increase. This may create harmful scenarios. For instance, theileriosis, a deadly African cattle disease with far-reaching economic problems, is characterized by high levels of CK2α and a lack of CK2β (37), and various links of asymmetric expression of CK2 subunits exist for tumorigenesis (37–39). In fact, various diseased states with deviating CK2 situations seem to link to Ets1. For instance, the Ets1 function is modulated by mitogenic signals transmitted via the Ras/Raf kinase pathway (70), and activated Raf kinase has been described as phosphorylating a conserved threonine (T38) causing activation of Ets1 and enhanced transcription of genes related to early mitogenic events (71), including CK2 genes (70). This may possibly help explain earlier reports on mitogen-induced increases in CK2 expression levels in cell cultures (28, 72), or on high expression levels of CK2 in proliferative tissues and in tumors (73, 74), matching the fact that Ras and Ets1 are both protooncogene products, and that Ets1 represents in addition a well-documented regulator of other protooncogenes and genes correlated with metastasis (75–77). The Ets1 double motif is highly conserved and is also found in all the promoters of CK2 genes, including human, mouse, nematode worm, and frog (45). Alignment of the human and mouse gene upstream sequences indicates 100% identity of the Ets1 motifs at only a 55% identity of the surrounding sequences. Appropriately, overexpression of Ets1, whose presence in the employed cells and motif-dependent binding to DNA has been documented [see also (44, 57)], causes moderate but significant increases of transcriptional activity. The impact of Ets1 on CK2 gene expression has also been supported by the antisense approach (57). Therefore, it is tempting to assume that Ets1 may represent a feature common to and important for the transcription machineries of both the CK2α gene and the CK2β gene, and thus may play a role in coordinating their expression.

Ets1 and Sp1 sites are also present in the promoter of the CK2α' gene. The promoter region contains two adjoining Ets1 motifs, although not completely identical to the Ets1 double motif in the two other CK2 genes, and also Sp1 multiple sites, that might be controlled in a like manner. This would explain the constant CK2α' transcript levels found in addition to that of the CK2α and CK2β transcript levels in JEG-3 cells. Such experiments are presently under investigation in our laboratory and preliminary results indicate a reduction in CK2α' promotor activity of about 50% upon mutation of either of the Ets-1 motifs both in JEG-3 and in HeLa cells.

V. Functional Links of CK2 Genes

Based on data from many different laboratories, it is suspected that CK2 plays a biological role in some global processes related to transcription and transcription-directed signaling (3, 20). The reentry into the cell cycle represents an excellent example in favor of this. If CK2 is perturbed at the nucleic acid level or protein level, cell-cycle entry stimulation of human cultured cells is inhibited, paralleled by repression of genes such as *fos* (7–9, 33–35, 78), whose expression is a prerequisite for subsequent waves of gene expressions essential for cell-cycle progression (31). To define CK2's role in a global context more precisely, we investigated the cell-cycle reentry situation by DNA array technology, expecting that due to the character of the data generated we could figure out the sites at which CK2 may act within the large and complicated network of gene expression control processes. Because such an analysis should be as broad as possible, the investigated portions of a genome should be as large as possible. Ideally, a complete genome is analyzed. The high conservation of both the CK2 complex and the cell cycle makes it easier to choose a suitable model genome and to project array data to the human situation.

We have been choosing the budding yeast *Saccharomyces cerevisiae* as a model system. Not only is the entire genome available, 6200 genes (79), but this organism has a number of advantageous features. It is genetically easily tractable, passes through the cell cycle in a haploid state (excluding compensation of genetic manipulations by respective healthy counterparts as in diploid states), and basic knowledge in cell-cycle genetics originates from it. Moreover, CK2 has been shown to be involved in yeast cell-cycle control (21). We have systematically investigated the effect of perturbed CKZ genes (21) on all genes of the genome for expressional alterations, i.e., for their functional links. Although not providing mechanistic explanations, functional links are suited to establish a map of action sites of CK2 within the knotty network of gene expression control processes.

A. Genetic Perturbation of CK2 Affects Expression of Genes of All Cell-Cycle Phases and Often in a Subunit- and Isoform-Specific Manner

S. cerevisiae has four CK2 genes, *CKA1* and *CKA2*, equivalents of human CK2α and CK2α' genes, and *CKB1* and *CKB2*, equivalents of the human CK2β gene. We have comparatively investigated in genome-wide screens (35a, 54) the effect of individual CK2 gene deletions or deletion combinations on the expression of genes in randomly growing cell populations (permanently cycling state) and in synchronized populations specifically at reentry into the cell cycle. The synchronization was by α pheromone treatment, arresting cells

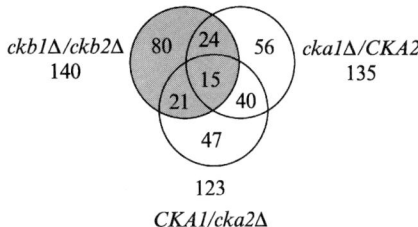

FIG. 8. Cell-cycle-regulated genes affected in expression by genetic perturbation of CK2 subunits. Venn diagram showing numbers of deviating genes that are at least one time point in the different CK2 subunit deletion strains. Significantly deviating expression was defined as an at least two-fold difference in transcript levels compared to wild type.

in a G_0-like differentiation state, and the investigated reentry time points corresponded to M/G_1 phase transition and early G_1 progression [0, 7, and 14 min postpheromone release; see (80)].

Significant alterations have been seen in the transcript profiles of both the regulatory ($ckb1\Delta$ $ckb2\Delta$) and catalytic ($cka1\Delta$ or $cka2\Delta$) CK2 subunit deletion strains. Collectively, nearly one-third of the roughly 900 known cell-cycle-regulated genes (80, 81) are significantly altered in expression in at least one of the CK2 mutants and at one of the time points. The expression alterations are frequently subunit specific, i.e., occur either upon regulatory or catalytic subunit deletions, and surprisingly often are also CK2 isoform specific, i.e., seen only upon deletion of the one or the other catalytic subunits (Fig. 8).

The CK2-linked genes relate to various cell-cycle phases. This is true for both the cells in a permanently cycling state and the cells about to reenter the cell cycle. The latter indicates somewhat surprisingly, compared to the expression alterations of immediate early genes at cell-cycle entry of human cells in culture (see above), that genetic perturbation of CK2 at cell-cycle entry may also have consequences for genes that peak in expression in a distant cycle phase (see below). The CK2-linked genes are either part of a defined process such as a metabolic pathway or a cell-cycle control element, or are involved in a superior process that might interfere with the expression control of many genes and thus may represent a global role for CK2, chromatin remodeling.

1. METABOLIC PATHWAY AND NUTRITION SUPPLY GENES

A group of genes, amounting to roughly one-fifth of the CK2-linked cell-cycle genes, is altered in cells reentering the cell cycle and also in permanently cycling cells. This group, therefore, is not specific for cell-cycle entry per se.

With few exceptions, the genes encode proteins involved in metabolic pathways and nutritional supply. The resulting deficiencies are therefore obviously either compensated for by available components present in the culture medium or by genetic mechanisms (21). An example of metabolic pathway genes linked to CK2 is the MET cluster responsible for the synthesis of methionine. This cluster is composed of 20 genes that are coordinately expressed, and the link to CK2 is highly subunit and isoform specific: at least eight MET genes are CK2 dependently expressed, the genes are strongly suppressed in the absence of the regulatory subunits ($ckb1\Delta$ $ckb2\Delta$ strain), and although the absence of one of the catalytic subunits ($cka1\Delta$ strain) elevates expression, the absence of the other ($cka2\Delta$ strain) has no effect (35a).

Examples of nutritional supply genes are the genes responsible for phosphate maintenance, the PHO genes. The PHO regulon is composed of genes that encode phosphate-supplying phosphatases (Pho5, 8, 11, 12) and membrane-bridging phosphate transporters (Pho84, 89). Together, they represent the executive part of the PHO regulon and are collectively expressed upon phosphate starvation. The expression is transcription regulated and dominated by a central transcription factor, Pho4. A cell-cycle-linked regulatory complex consisting of a cyclin (Pho80), a cyclin-dependent kinase (CDK; Pho85), and a CDK inhibitor (Pho81) controls this factor (82–84). When the regulatory CK2 subunit genes are deleted, all of the PHO genes encoding executive components are significantly repressed at any of the times analyzed as well as the central transcription factor gene *PHO4*. By contrast, the genes encoding the cell-cycle-linked control complex are unaltered (Fig. 9). This indicates an absolute requirement of CK2 for the expression of executive PHO genes. As a consequence, a deficiency in the regulatory CK2 subunits would permanently keep these genes repressed, independent of whether phosphate was available and what the phosphate sensor might signal, and would, therefore, uncouple the executive part of the PHO regulon from its cyclin-Cdk-regulatory part. The regulon is also negatively affected, although less dramatically and only temporarily at cell-cycle entry by deletion of catalytic subunit Cka1. By contrast, deletion of the other catalytic subunit, Cka2, has little effect, indicating that the PHO regulon is affected not only CK2 subunit specifically but also isoform specifically (84a). Because phosphate supply is, without exception, vital for all cells, the CK2 effect on the PHO regulon supports the survival factor hypothesis (20).

2. Cell-Cycle Progression and Exit Genes

In support of a survival factor function of CK2 (20) are the various genes altered in expression and involved in cell-cycle progression control and exit from it. These involve predominantly genes that relate to cell-cycle engine control and cell division and apoptosis machineries (35a).

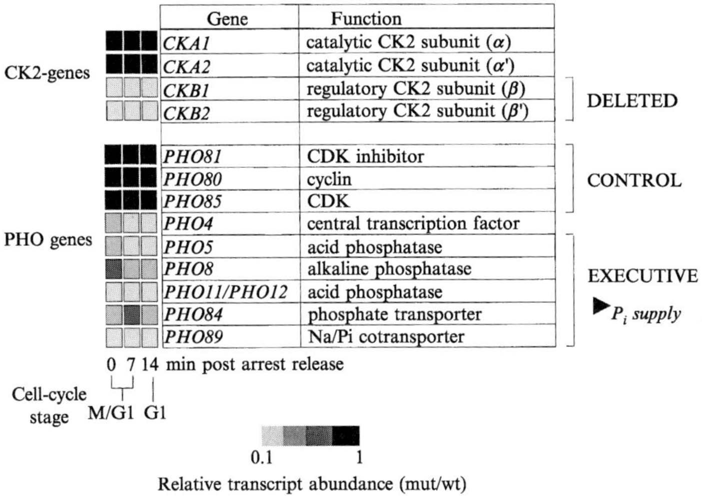

FIG. 9. Deletion of CK2 regulatory subunits uncouples phosphate supply genes (PHO genes) from their cyclin-Cdk control. S. cerevisiae wild-type and $ckb1\Delta\ ckb2\Delta$ strains were synchronized by pheromone treatment and comparatively analyzed for gene transcription at indicated times postpheromone arrest using oligonucleotide arrays. Relative transcript abundances are given according to scale (shades of gray) at the bottom. Transcription deviations are considered significant when greater than two-fold. Note: Ckb1 and Ckb2 transcripts are absent but Cka1 and Cka2 transcripts are unchanged as expected. CONTROL, cyclin-CDK regulator genes of the PHO pathway; EXECUTIVE, phosphate supply gene group.

The genes relating to cell-cycle engine control comprise several cyclin-encoding genes. A particularly important example for the entry into the cell cycle is the gene encoding Cln3, which is repressed by catalytic CK2 subunit deletion. This is a G_1 phase cyclin that complexes to Cdk1 (also named Cdc28), the only Cdk in yeast, which upon activation by Cln3 triggers the first step of a new cell-cycle round. Another cyclin-encoding gene, *CLN2*, whose product activates Cdk1 at a later stage in G_1, is repressed by regulatory CK2 subunit deficiency. Further, elevation of expression of *CDC20* is observed, which represents an example of a CK2-linked gene whose product acts at the other end of a cyclin's life, the proteolytic destruction. It encodes an anaphase-promoting complex subunit.

Genes relating to cell-cycle exit and apoptosis include, for instance, *SPO12*, encoding a putative positive regulator of exit from M phase. It is repressed by CK2 regulatory subunit deficiency. This deficiency also represses *BAT1*. This gene encodes a branched-chain amino acid transaminase and its human homolog *ECA39* is involved in apoptosis (85).

Various CK2-linked genes relate to the cell division machinery. Most prominent are genes encoding spindle pole body (SPB) components and SPB interaction partners, including factors of their control. For instance, *HCM1*, encoding a transcription factor involved in regulation of SPB assembly, is repressed by deficiencies in catalytic CK2 subunits; *NUF1*, a gene encoding an SPB component, is repressed by Cka1 deficiency but elevated by Cka2 deficiency; *KIP3*, encoding a kinesin-related protein, is enhanced by regulatory subunit deficiency; and *CIK1*, encoding an SPB-associated protein, is enhanced by Cka1 deficiency but repressed by regulatory CK2 subunits deficiency.

B. Links to Chromatin Remodeling Genes: A Global Role for CK2?

Members of gene groups such as the above considered MET or PHO genes have more or less common transcription regulation features. The majority of the CK2-affected genes, however, exhibit no common characteristics. Consequently, CK2 perturbation should alter their transcription via more global mechanisms. Indeed, we find strong hints for such mechanisms. The genes requiring CK2 for proper expression at cell-cycle entry include genes encoding proteins with a striking relation to chromatin remodeling and modification (Table II). Remodeling is of high global significance for the expression of genes, because the first steps in activating gene expression is altering the accessibility of chromatin to the transcription machinery. The expression of these genes, moreover, is not restricted to a certain cell-cycle phase; their expression peaks are rather distributed over all cycle phases, indicating roles at defined cell-cycle states. The gene products range from chromatin assembly proteins (Cac2), proteins involved in silencing and antisilencing of chromatin (Esc4; Asf1), chromatin remodeling transcription factors and cofactors (Swi1; Sri1), histone acetylase complex factors (Ahc1), histone deacetylases (Hos3; Hst4), to helicases (Mcm6). Not only does the expression of respective genes argue for a role of CK2 in chromatin remodeling, a number of observations with nuclear protein complexes do as well. All CK2 subunits, as individual proteins or diverse combinations, have been found complexed to general chromosomal remodeling factors in yeast (5, 6, 86), and CK2-mediated phosphorylation of nucleosome assembly proteins such as NAP1 and 2 has been shown to occur and affect function (36, 87). Further, CK2 has been reported to phosphorylate and possibly regulate human histone deacetylase 2 (88).

The involvement of CK2 in nucleosomal remodeling processes would explain divergent deviation patterns within certain gene groups. For instance, executive PHO genes have a Pho4 control in common and, consequently, their repression is paralleled by *PHO4* repression in *ckb1Δ ckb2Δ* mutants (see

TABLE II
GENES OF VARIOUS CELL-CYCLE PHASES INVOLVED IN CHROMATIN REMODELING AND
MODIFICATION ARE AFFECTED IN EXPRESSION BY CK2 PERTURBATION[a]

Cell-cycle phase	Gene	Gene product
G_1	CAC2	p60 subunit of chromatin assembly factor I
	ESC4	Establishes silent chromatin
	ASF1	Antisilencing protein
S	SWI1	Chromatin remodeling zinc-finger transcription factor
	TEL2	Telomere-binding protein
S/G_2	HOS3	Histone deacetylase
	CSE4	Similar to histone H3 and to centromere protein CENP-A
G_2/M	SRI1	Swi/SNF and RSC interacting protein 1
	AHC1	Component of Ada histone acetyltransferase complex
	MCM6	ATP-dependent DNA helicase
M/G_1	HST4	Histone deacetylase

[a]Global DNA array-based transcript profiles of S. cerevisiae wild-type and CK2 mutant strains (ckb1Δ ckb2Δ; cka1Δ; cka2Δ) were compared at cell-cycle entry and genes deviating in transcription by more than two-fold were classified as significantly affected. Genes were assigned to cell-cycle phases according to Spellman et al. (80). For details see text.

Fig. 9). But this is not so in cka1Δ or cka2Δ mutants. PHO4 expression is unaltered in these strains, whereas expression of PHO genes varies transiently and isoform specifically (84a). To explain this phenomenon, Pho4-mediated promoter activation is obviously not sufficient, and additional control devices are required. Targeting the transcription machinery properly to PHO gene promoters has repeatedly been related to chromatin remodeling (89). This could, therefore, be the CK2-affected control device.

Collectively, our observations are compatible with the assumption that CK2 plays a global role in transcription (and transcription-directed signaling) through a chromatin-remodeling capacity. This could occur both by affecting expression of genes encoding remodeling and modulating factors and by directly interacting (with and without phosphorylation) with remodeling and modifying proteins. Because of the global character of chromatin remodeling, such a role for CK2 could have consequences for diverse cellular processes and help explain many of the puzzling physiological and pathophysiological observations made in connection with CK2.

Acknowledgments

We would like to express our sincere appreciation to all members of our laboratory, past and present, for their commitment and dedication. In particular, we would like to acknowledge the impressive activity by Thomas Barz in establishing functional links of CK2 in the yeast model system, the help of Benedikt Brors in bioinformatics, and the excellent technical assistance of Andrea Waxmann. Research from our laboratory was supported by EU (CT 96-0047/GR-767.705/07), DFG (Py2/2-2), HGF (Strategiefonds III A1/H161; B5/H166), and DKFZ/MOS (Ca77).

References

1. Burnett, G., and Kennedy, E. P. (1954). The enzymatic phosphorylation of proteins. *J. Biol. Chem.* **211**, 969–980.
2. Tuazon, P. T., and Traugh, J. A. (1991). Casein kinase I and II—multipotential serine protein kinases: Structure, function, and regulation. *Adv. Sec. Mess. Phosphoprot. Res.* **23**, 123–164.
3. Pinna, L. A. (2002). Protein kinase CK2: A challenge to canons. *J. Cell Sci.* **115**, 3873–3878.
4. Pyerin, W., Ackermann, K., and Lorenz, P. (1996). Casein kinases. *In* "Protein Phosphorylation" (F. Marks, Ed.), pp. 117–174. Verlag Chemie, Weinheim.
5. Gavin, A. C., Bosche, M., Krause, R., Grandi, P., Marzioch, M., Bauer, A., Schultz, J., Rick, J. M., Michon, A. M. *et al.* (2002). Functional organization of the yeast proteome by systematic analysis of protein complexes. *Nature* **415**, 141–147.
6. Ho, Y., Gruhler, A., Heilbut, A., Bader, G. D., Moore, L., Adams, S.-L., Millar, A., Taylor, P., Bennett, K. *et al.* (2002). Systematic identification of protein complexes in *Saccharomyces cerevisiae* by mass spectrometry. *Nature* **415**, 180–183.
7. Gauthier-Rouvière, C., Basset, M., Blanchard, J.-M., Cavadore, J.-C., Fernandez, A., and Lamb, N. J. C. (1991). Casein kinase II induces *c-fos* expression via the serum response element pathway and p67SRF phosphorylation in living fibroblasts. *EMBO J.* **10**, 2921–2930.
8. Pepperkok, R., Lorenz, P., Ansorge, W., and Pyerin, W. (1994). Casein kinase II is required for transition of G0/G1, early G1, and G1/S phases of the cell cycle. *J. Biol. Chem.* **269**, 6986–6991.
9. Pepperkok, R., Herr, S., Lorenz, P., Pyerin, W., and Ansorge, W. (1993). System for quantitation of gene expression in single cells by computerized microimaging: Application to c-fos expression after microinjection of anti-casein kinase II antibody. *Exp. Cell Res.* **204**, 278–285.
10. Chardot, T., Shen, H., and Meunier, J. C. (1995). Dual specificity of casein kinase II from the yeast *Yarrowia lipolytica*. *C. R. Acad. Sci. III* **318**, 937–942.
11. Wilson, L. K., Dhillon, N., Thorner, J., and Martin, G. S. (1997). Casein kinase II catalyzes tyrosine phosphorylation of the yeast nucleolar immunophilin Fpr3. *J. Biol. Chem.* **272**, 12961–12967.
12. Mitchell, C., Plaho, J. A., and Roizman, B. (1994). Casein kinase II specifically nucleotidylylates *in vitro* the amino acid sequence of the protein encoded by the alpha 22 gene of herpes simplex virus 1. *Proc. Natl. Acad. Sci. USA* **91**, 11864–11868.
13. Bodenbach, L., Fauss, J., Robitzki, A., Krehan, A., Lorenz, P., Lozeman, F. J., and Pyerin, W. (1994). Recombinant human casein kinase II. A study with the complete set of subunits (alpha, alpha' and beta), site-directed autophosphorylation mutants and a bicistronically expressed holoenzyme. *Eur. J. Biochem.* **220**, 263–273.
14. Thornburg, W., and Lindell, T. J. (1977). Purification of rat liver nuclear protein kinase NII. *J. Biol. Chem.* **252**, 6660–6665.

15. Cochet, C., and Chambaz, E. M. (1983). Oligomeric structure and catalytic activity of G type casein kinase. Isolation of the two subunits and renaturation experiments. *J. Biol. Chem.* **258,** 1403–1406.
16. Pyerin, W., Burow, E., Michaely, K., Kübler, D., and Kinzel, V. (1987). Catalytic and molecular properties of highly purified phosvitin/casein kinase type II from human epithelial cells in culture (HeLa) and relation to ecto protein kinase. *Biol. Chem. Hoppe-Seyler* **368,** 215–227.
17. Krehan, A., Lorenz, P., Plana-Coll, M., and Pyerin, W. (1996). Interaction sites between catalytic and regulatory subunits in human protein kinase CK2 holoenzymes as indicated by chemical cross-linking and immunological investigations. *Biochemistry* **35,** 4966–4975.
18. Chantalat, L., Leroy, D., Filhol, O., Nueda, A., Benitez, M. J., Chambaz, E. M., Cochet, C., and Dideberg, O. (1999). Crystal structure of the human protein kinase CK2 regulatory subunit reveals its zinc finger-mediated dimerization. *EMBO J.* **18,** 2930–2940.
19. Niefind, K., Guerra, B., Ermakowa, I., and Issinger, O.-G. (2001). Crystal structure of human protein kinase CK2: Insights into basic properties of the CK2 holoenzyme. *EMBO J.* **20,** 5320–5331.
20. Ahmed, K., Gerber, D. A., and Cochet, C. (2002). Joining the cell survival squad: An emerging role for protein kinase CK2. *Trends Cell Biol.* **12,** 226–230.
21. Glover, C. V. C. (1998). On the physiological role of casein kinase II in *Saccharomyces cerevisiae*. *Prog. Nucleic Acids Res. Mol. Biol.* **59,** 95–133.
22. Hu, E., and Rubin, C. S. (1990). Casein kinase II from *Caenorhabditis elegans*. Properties and developmental regulation of the enzyme; cloning and sequence analyses of cDNA and the gene for the catalytic subunit. *J. Biol. Chem.* **265,** 5072–5080.
23. Hu, E., and Rubin, C. S. (1991). Casein kinase II from *Caenorhabditis elegans*. Cloning, characterization, and developmental regulation of the gene encoding the beta subunit. *J. Biol. Chem.* **266,** 19796–19802.
24. Jakobi, R., Voss, H., and Pyerin, W. (1989). Human phosvitin/casein kinase type II. Molecular cloning and sequencing of full-length cDNA encoding subunit beta. *Eur. J. Biochem.* **183,** 227–233.
25. Lozeman, F. J., Litchfield, D. W., Piening, C., Takio, A., Walsh, K. A., and Krebs, E. G. (1990). Isolation and characterization of human cDNA clones encoding the alpha and the alpha' subunits of casein kinase II. *Biochemistry* **29,** 8436–8447.
26. Pinna, L. A. (1990). Casein kinase 2: An 'eminence grise' in cellular regulation? *Biochim. Biophys. Acta* **1054,** 267–284.
27. Litchfield, D. W., and Lüscher, B. (1993). Casein kinase II in signal transduction and cell cycle regulation. *Mol. Cell. Biochem.* **127/128,** 187–199.
28. Litchfield, D. W., Dobrowolska, G., and Krebs, E. G. (1994). Regulation of casein kinase II by growth factors: A reevaluation. *Cell. Mol. Biol. Res.* **40,** 373–381.
29. Allende, J. E., and Allende, C. C. (1995). Protein kinases. 4. Protein kinase CK2: An enzyme with multiple substrates and a puzzling regulation. *FASEB J.* **9,** 313–323.
30. Ahmed, K. (1999). Nuclear matrix and protein kinase CK2 signaling. *Crit. Rev. Eukaryot. Gene Expr.* **9,** 329–336.
31. Iyer, V. R., Eisen, M. B., Ross, D. T., Schuler, G., Moore, T., Lee, J. C., Trent, J. M., Staudt, L. M., Hudson, J., Jr., Boguski, M. S., Lashkari, D., Shalon, D., Botstein, D., and Brown, P. O. (1999). The transcriptional program in the response of human fibroblasts to serum. *Science* **283,** 83–87.
32. Padmanabha, R., Chen-Wu, J. L., Hanna, D. E., and Glover, C. V. C. (1990). Isolation, sequencing, and disruption of the yeast CKA2 gene: Casein kinase II is essential for viability in *Saccharomyces cerevisiae*. *Mol. Cell. Biol.* **10,** 4089–4099.

33. Pepperkok, R., Lorenz, P., Jokobi, R., Ansorge, W., and Pyerin, W. (1991). Cell growth stimulation by EGF: Inhibition through antisense-oligodeoxynucleotides demonstrates important role of casein kinase II. *Exp. Cell Res.* **197,** 245–253.
34. Lorenz, P., Pepperkok, R., Ansorge, W., and Pyerin, W. (1993). Cell biological studies with monoclonal and polyclonal antibodies against human casein kinase II subunit beta demonstrate participation of the kinase in mitogenic signaling. *J. Biol. Chem.* **268,** 2733–2739.
35a. Barz, T., Ackermann, K., Dubois, G., Elis, R., and Pyerin, W. (2003). Genome-wide expression screens indicate a global role for protein kinase CK2 in chromatin remodelling. *J. Cell. Sci.* **116,** 1563–1577.
35b. Lorenz, P., Ackermann, K., Simoes-Wuest, P., and Pyerin, W. (1999). Serum-stimulated cell cycle entry of fibroblasts requires undisturbed phosphorylation and non-phosphorylation interactions of the catalytic subunits of protein kinase CK2. *FEBS Lett.* **448,** 283–288.
36. Li, M., Strand, D., Krehan, A., Pyerin, W., Heid, H., Neumann, B., and Mechler, M. (1999). Casein kinase 2 binds and phosphorylates the nucleosome assembly protein-1 (NAP1) in *Drosophila melanogaster*. *J. Mol. Biol.* **293,** 1067–1084.
37. Ole-MoiYoi, O. K. (1995). Casein kinase II in theileriosis. *Science* **267,** 834–836.
38. Seldin, D. C., and Leder, P. (1995). Casein kinase IIα transgene-induced murine lymphoma: Relation to theileriosis in cattle. *Science* **267,** 894–897.
39. Landesmann-Bollag, E., Romieu-Mourez, R., Song, D. H., Sonenshein, G. E., Cardiff, R. D., and Seldin, D. C. (2001). Protein kinase CK2 in mammary gland tumorigenesis. *Oncogene* **20,** 3247–3257.
40. Rifkin, R., Channavajhala, P. L., Kiefer, H. L. B., Carmack, A. J., Landesmann-Bollag, E., Beaudette, B. C., Jersky, B., Salant, D. J., Ju, S.-T., Marshak-Rothstein, A., and Seldin, D. C. (1998). Acceleration of *lpr* lymphoproliferative and autoimmune disease by transgenic protein kinase ck2α. *J. Immunol.* **161,** 5164–5170.
41. Blanquet, R. (2000). Casein kinase 2 as a potentially important enzyme in the nervous system. *Prog. Neurobiol.* **60,** 211–246.
42. Ackermann, K., and Pyerin, W. (1999). Protein kinase CK2α may induce gene expression but unlikely acts directly as a DNA-binding transcription-activating factor. *Mol. Cell. Biochem.* **191,** 129–134.
43. Voss, H., Wirkner, U., Jakobi, R., Hewitt, N. A., Schwager, C., Zimmermann, J., Ansorge, W., and Pyerin, W. (1991). Structure of the gene encoding human casein kinase II subunit beta. *J. Biol. Chem.* **266,** 13706–13711.
44. Krehan, A., Schmalzbauer, R., Böcher, O., Ackermann, K., Wirkner, U., Brouwers, S., and Pyerin, W. (2001). Ets1 is a common element in directing transcription of the α and the β genes of human protein kinase CK2. *Eur. J. Biochem.* **268,** 3243–3252.
45. Pyerin, W., and Ackermann, K. (2001). Transcriptional coordination of the genes encoding catalytic (CK2α) and regulatory (CK2β) subunits of human protein kinase CK2. *Mol. Cell. Biochem.* **227,** 45–57.
46. Wirkner, U., Voss, H., Lichter, P., Weitz, S., Ansorge, W., and Pyerin, W. (1992). Human casein kinase II subunit alpha: Sequence of a processed (pseudo)gene and its localization on chromosome 11. *Biochim. Biophys. Acta* **1131,** 220–222.
47. Wirkner, U., Voss, H., Lichter, P., Ansorge, W., and Pyerin, W. (1994). The human gene (*CSNK2A1*) coding for the casein kinase II subunit α is located on chromosome 20 and contains tandemly arranged *Alu* repeats. *Genomics* **19,** 257–265.
48. Yang-Feng, T. L., Naiman, T., Kopatz, I., Eli, D., Dafni, N., and Canaani, D. (1994). Assignment of the human casein kinase II alpha' subunit gene (*CSNK2A1*) to chromosome 16p13.2–p13.3. *Genomics* **19,** 173.

49. Wirkner, U., Voss, H., Ansorge, W., and Pyerin, W. (1998). Genomic organization and promoter identification of the human protein kinase CK2 catalytic subunit α (*CSNK2A1*). *Genomics* **48**, 71–78.
50. Wirkner, U., Voss, H., Lichter, P., and Pyerin, W. (1994). Human protein kinase CK2 genes. *Cell. Mol. Biol. Res.* **40**, 489–499.
51. Wirkner, U., and Pyerin, W. (1999). CK2alpha loci in the human genome: Structure and transcriptional activity. *Mol. Cell. Biochem.* **191**, 59–64.
52. Singh, S., Jantke, I., Simon, M., Meybohm, I., Boldyreff, B., Issinger, O.-G., and Goedde, H. W. (1994). *Pst*I identifies biallelic DNA polymorphism of the human casein kinase 2 alpha gene (CSNK2A1). *Hum. Genet.* **93**, 474.
53. Ackermann, K., Fauss, J., and Pyerin, W. (1994). Inhibition of cyclic AMP-triggered aromatase gene expression in human choriocarcinoma cells by antisense oligodeoxynucleotide. *Cancer Res.* **54**, 4940–4946.
54. Ackermann, K., Waxmann, A., Glover, C. V. C., and Pyerin, W. (2001). Genes targeted by protein kinase CK2: A genome-wide expression array analysis in yeast. *Mol. Cell. Biochem.* **227**, 59–66.
55. Bosc, D. G., Graham, K. C., Saulnier, R. B., Zhang, C., Prober, D., Gietz, R. D., and Litchfield, D. W. (2000). Identification and characterization of CKIP-1, a novel pleckstrin homology domain-containing protein that interacts with protein kinase CK2. *J. Biol. Chem.* **275**, 14295–14306.
56. Srinivasan, N., Antonelli, M., Jacob, G., Korn, I., Romero, F., Jedlicki, A., Dhanaraj, V., Sayed, M., Blundell, T. L., Allende, C. C., and Allende, J. E. (1999). Structural interpretation of site-directed mutagenesis and specificity of the catalytic subunit of protein kinase CK2 using comparative modelling. *Protein Eng.* **12**, 119–127.
57. Krehan, A., Ansuini, H., Böcher, O., Grein, S., Wirkner, U., and Pyerin, W. (2000). Transcription factors ets1, NF-kappa B, and Sp1 are major determinants of the promoter activity of the human protein kinase CK2alpha gene. *J. Biol. Chem.* **275**, 18327–18336.
58. Crepieux, P., Coll, J., and Stehelin, D. (1994). The ets family of proteins: Weak modulators of gene expression in quest for transcriptional partners. *Crit. Rev. Oncogen.* **5**, 615–638.
59. Dittmer, J., Pise-Masison, C. A., Clemens, K. E., Choi, K.-S., and Brady, J. N. (1997). Interaction of human T cell lymphotropic virus type I Tax, Ets1, and Sp1 in transactivation of the PTHrP P2 promoter. *J. Biol. Chem.* **272**, 4953–4958.
60. Ho, I.-C., Bhat, N. K., Gottschalk, L. R., Lindsten, T., Thompson, C. B., Papas, T. S., and Leiden, J. M. (1990). Sequence-specific binding of human Ets-1 to the T cell receptor alpha gene enhancer. *Science* **250**, 814–818.
61. Seth, A., Hodge, D. R., Thomson, D. M., Robinson, L., Panayiotakis, A., Watson, D. K., and Papas, T. S. (1993). Ets-family proteins activate transcription from HIV-1 long terminal repeat. *AIDS Res. Hum. Retroviruses* **9**, 1017–1023.
62. Gill, G., Pascale, E., Tseng, Z. H., and Tijan, R. (1994). A glutamine-rich hydrophobic patch in transcription factor Sp1 contacts the dTAFII110 component of the Drosophila TFIID complex and mediates transcriptional activation. *Proc. Natl. Acad. Sci. USA* **91**, 192–196.
63. Carter, R. S., Bhat, N. K., Basu, A., and Avadhani, N. G. (1992). The basal promoter elements of murine cytochrome c oxidase subunit IV gene consist of tandemly duplicated ets motifs that bind to GABP-related transcription factors. *J. Biol. Chem.* **267**, 23418–23426.
64. Carter, R. S., and Avadhani, N. G. (1994). Cooperative binding of GA-binding protein transcription factors to duplicated transcription initiation region repeats of the cytochrome c oxidase subunit IV gene. *J. Biol. Chem.* **269**, 4381–4387.
65. Sucharov, C., Basu, A., Carter, R. S., and Acadhani, N. H. (1995). A novel transcriptional initiator activity of the GABP factor binding ets sequence repeat from the murine cytochrome c oxidase Vb gene. *Gene Express.* **5**, 93–111.

66. Gunther, M., Frebourg, T., Laithier, M., Fossar, N., Bouziane-Ouartini, M., Lavialle, C., and Brison, O. (1995). An Sp1 binding site and the minimal promoter contribute to overexpression of the cytokeratin 18 gene in tumorigenic clones relative to that in nontumorigenic clones of a human carcinoma cell line. *Mol. Cell Biol.* **15**, 2490–2499.
67. Kollmar, R., Sukow, K. A., Spongale, S. K., and Farnham, P. J. (1994). Start site selection at the TATA-less carbamoyl-phosphate synthase (glutamine-hydrolyzing)/aspartate carbamoyl-transferase/dihydroorotase promoter. *J. Biol. Chem.* **269**, 2252–2257.
68. Block, K. L., Shou, Y., and Poncz, M. (1996). An Ets/Sp1 interaction in the 5'-flanking region of the megakaryocyte-specific alphaIIb gene appears to stabilize Sp1 binding and is essential for expression of this TATA-less gene. *Blood* **88**, 2071–2080.
69. Armstrong, S. A., Barry, D. A., Leggett, R. W., and Mueller, C. R. (1997). Casein kinase II-mediated phosphorylation of the C-terminus of Sp1 decreases its DNA binding activity. *J. Biol. Chem.* **272**, 13489–13495.
70. Orlandini, M., Semplici, F., Ferruzzi, R., Meggio, F., Pinna, L. A., and Oliviero, S. (1998). Protein kinase CK2alpha' is induced by serum as a delayed early gene and cooperates with Ha-ras in fibroblast transformation. *J. Biol. Chem.* **273**, 21291–21297.
71. Yang, B.-S., Hauser, C. A., Henkel, G., Colman, M. S., VanBeveren, C., Stacey, K. J., Hume, D. A., Maki, R. A., and Ostrowski, M. C. (1996). Ras-mediated phosphorylation of a conserved threonine residue enhances the transactivation activities of c-Ets1 and c-Ets2. *Mol. Cell. Biol.* **16**, 538–547.
72. Ackerman, P., and Osheroff, N. (1989). Regulation of casein kinase II activity by epidermal growth factor in human A-431 carcinoma cells. *J. Biol. Chem.* **264**, 11958–11963.
73. Faust, R. A., Gapany, M., Tristani, P., Davis, A., Adams, G. L., and Ahmed, K. (1996). Elevated protein kinase CK2 activity in chromatin of head and neck tumors: Association with malignant transformation. *Cancer Lett.* **101**, 31–35.
74. Daya-Makin, M., Sanghera, J. S., Mogentale, T. L., Lipp, M., Parchomchuk, J., Hogg, J. C., and Pelech, S. L. (1994). Activation of a tumor-associated protein kinase (p40TAK) and casein kinase 2 in human squamous cell carcinomas and adenocarcinomas of the lung. *Cancer Res.* **54**, 2262–2268.
75. Bhat, N. K., Fischinger, P. J., Seth, A., Watson, D. K., and Papas, T. (1996). Isolation and characterization of a novel gene expressed in multiple cancers. *Int. J. Oncol.* **8**, 841–846.
76. Sato, M., Morii, E., Komori, T., Kawahata, H., Sugimoto, M., Terai, K., Shimizu, H., Yasui, T., Ogihara, H., Yasui, N., Ochi, T., Kitamura, Y., Ito, Y., and Nomura, S. (1998). Transcriptional regulation of osteopontin gene in vivo by PEBP2alphaA/CBFA1 and ETS1 in the skeletal tissues. *Oncogene* **17**, 1517–1525.
77. Gambarotta, G., Boccaccio, C., Giordano, S., Ando, M., Stella, M. C., and Comoglio, P. M. (1996). Ets up-regulates MET transcription. *Oncogene* **13**, 1911–1917.
78. Pyerin, W., Pepperkok, R., Ansorge, W., and Lorenz, P. (1992). Early cell growth stimulation is inhibited by casein kinase II antisense oligodeoxynucleotides. *Ann. N.Y. Acad. Sci.* **660**, 295–297.
79. Goffeau, A., Barrell, B. G., Bussey, H., Davis, R. W., Dujon, B., Feldmann, H., Galibert, F., Hoheisel, J. D., Jacq, C., Johnston, M., Louis, E. J., Mewes, H. W., Murakami, Y., Philippsen, P., Tettelin, H., and Oliver, S. G. (1996). Life with 6000 genes. *Science* **274**, 563–567.
80. Spellman, P. T., Sherlock, G., Zhang, M. Q., Iyer, V. R., Anders, K., Eisen, M. B., Brown, P. O., Botstein, D., and Futcher, B. (1998). Comprehensive identification of cell cycle-regulated genes of the yeast Saccharomyces cerevisiae by microarray hybridization. *Mol. Biol. Cell* **9**, 3273–3297.
81. Cho, J. R., Campbell, M. J., Winzeler, E. A., Steinmetz, L., Conway, A., Wodicka, L., Wolfsberg, T. L., Gabrielian, A. T., Landsman, D., Lockhart, D. J., and Davis, R. W. (1998). A genome-wide transcriptional analysis of the mitotic cell cycle. *Mol. Cell* **2**, 65–73.

82. Lenburg, M. E., and O'Shea, E. K. (1996). Signaling phosphate starvation. *Trends Biol. Sci.* **21,** 383–387.
83. Wykhoff, D. D., and O'Shea, E. K. (2001). Phosphate transport and sensing in *Saccharomyces cerevisiae*. *Genetics* **159,** 1491–1499.
84a. Barz, T., Ackermann, K., and Pyerin, W. (2003). Perturbation of protein kinase CK2 uncouples executive part of phosphate maintenance pathway from cyclin-CDK control. *FEBS Lett.* **537,** 210–214.
84b. Moffat, J., Huang, D., and Andrews, B. (2000). Functions of Pho85 cyclin-dependent kinases in budding yeast. *Prog. Cell Cycle Res.* **4,** 97–106.
85. Eden, A., and Benvenisty, N. (1999). Involvement of branched-chain amino acid aminotransferase (Bcat1/Eca39) in apoptosis. *FEBS Lett.* **457,** 255–261.
86. Krogan, N. J., Kim, M., Ahn, S. H., Zhong, G., Kobor, M. S., Cagney, G., Emili, A., Shilatifard, A., Buratowski, S., and Greenblatt, J. F. (2002). RNA polymerase II elongation factors of *Saccharomyces cerevisiae*: A targeted proteomics approach. *Mol. Cell. Biol.* **22,** 6979–6992.
87. Rodriguez, P., Pelletier, J., Price, G. B., and Zannis-Hadjopoulos, M. (2000). NAP-2: Histone chaperone function and phosphorylation state through the cell cycle. *J. Mol. Biol.* **298,** 225–238.
88. Sun, J. M., Chen, H. Y., Moniwa, M., Litchfield, D. W., Seto, E., and Davie, J. R. (2002). The transcriptional repressor Sp3 is associated with CK2-phosphorylated histone deacetylase 2. *J. Biol. Chem.* **277,** 35783–35786.
89. McAndrew, P. C., Svaren, J., Martin, S. R., Hörz, W., and Goding, C. R. (1998). Requirements for chromatin modulation and transcription activation by the Pho4 acidic activation domain. *Mol. Cell. Biol.* **18,** 5818–5827.

Heterochromatin, Position Effects, and the Genetic Dissection of Chromatin

JOEL C. EISSENBERG AND
LORI L. WALLRATH

Department of Biochemistry and Molecular Biology, St. Louis School of Medicine, St. Louis, Missouri 63104

I.	Introduction	276
II.	Heterochromatin and Gene Silencing	278
III.	Heterochromatin and Gene Activation	278
IV.	Genetic Strategies for Dissection of Heterochromatin	279
V.	*cis*-Spreading, *trans*-Inactivation, and Heterochromatic Associations	281
VI.	Molecular Composition of Heterochromatin and the Regulation of Heterochromatin Silencing	283
	A. DNA Structure and Heterochromatin	283
	B. The Heterochromatic Nucleosome	283
	C. Histone Modifications That Distinguish Heterochromatin from Euchromatin	285
	D. Nucleosome–DNA Interactions	287
	E. Nonhistone Proteins and Heterochromatin	288
	F. RNA as a Component of Heterochromatin	290
VII.	Setting Up Heterochromatin and Euchromatin Domains	291
VIII.	Summary and Future Directions	292
	References	293

The partitioning of the eukaryotic chromosome regions into heterochromatin and euchromatin reflects fundamental differences in the packaging of DNA. The observation that chromosome rearrangments juxtaposing euchromatic sequences with heterochromatic regions usually result in silencing of nearby euchromatic genes (termed "position effect silencing") led to genetic screens for position effect modifiers. These screens uncovered key proteins involved in chromatin assembly, including structural components of heterochromatin and euchromatin, and enzymes that covalently modify histones. In addition, recent data implicates RNA in the targeting of chromosome regions for heterochromatin assembly. Here, we review the genetic, cytological, and biochemical properties of heterochromatin and summarize recent data suggesting mechanisms for heterochromatin assembly. © 2003, Elsevier (USA).

I. Introduction

The mechanism by which chromosomes in eukaryotic nuclei are packaged to accommodate regulated gene expression is a major problem in biology. Chromosomal organization is reflected in the cytological partitioning of the interphase nucleus into zones of condensed heterochromatin and dispersed euchromatin. Euchromatin is commonly thought to represent the transcriptionally active regions of chromosomes, whereas heterochromatin is thought to be transcriptionally inert, but this functional dichotomy is certainly overly simplistic. In this review, we will use the term "heterochromatin" in its original cytological definition (1) as the material in the interphase nucleus that remains condensed and densely staining after telophase in the cell cycle. For types of chromatin that share with heterochromatin the ability to silence transcription but for which the cytological definition is not fulfilled, we will use the term "silencing chromatin." Intensive genetic analysis and recent insights into the molecular composition of heterochromatin point to a distinctive form of nucleosome, the "heterochromatic nucleosome," discussed below. The composition of the heterochromatic nucleosome, including modified forms of histones and associations with DNA, and nonhistone chromosomal proteins defines the architectural foundation on which cytological and genetic heterochromatin is built.

The primary genetic assay for the dissection of heterochromatin is position-effect variegation (PEV). Heterochromatic PEV is the variegated silencing that occurs when a euchromatic gene is misplaced in or near heterochromatin by rearrangement or transposition (Fig. 1). The variegation of mutant and wild-type tissue observed first in *Drosophila* PEV was a puzzle to its earliest students. Mutations giving rise to such phenotypes were called "eversporting"—"sport" being an early word for mutant—suggesting a mutational basis to the loss of function. Subsequent genetic studies demonstrated the position-effect nature of the phenotype (2, 3). Although one study using Southern blot analysis suggested that somatic DNA elimination accompanies the variegation (4), a subsequent reinvestigation revealed that the missing DNA remained in the gel after transfer (5), and later studies found no evidence for DNA elimination in diploid tissue (6). Thus, the silencing of PEV is a purely epigenetic phenomenon, mediated by changes in the chromatin template that do not affect the DNA sequence.

PEV has been observed in fungi, plants, and animals, but has been exploited as a genetic assay primarily in yeast and *Drosophila*. The variegated nature of heterochromatic silencing in PEV makes it ideal for genetic identification and characterization of nuclear factors required for heterochromatin formation and maintenance. The purpose of this review is to summarize many of the key genetic strategies and observations that lead to our current understanding of how the nucleus is organized for transcription.

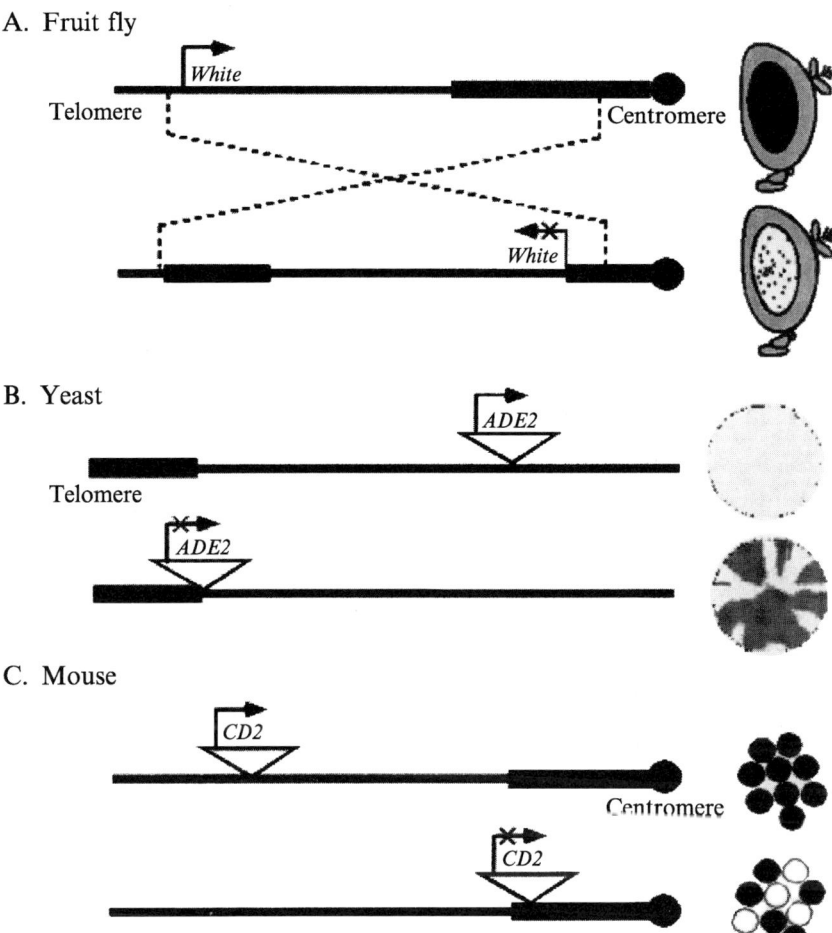

FIG. 1. The observation of PEV in multiple eukaryotes. (A) PEV observed in the fruit fly *Drosophila* when the *white* gene, required for red eye pigmentation, becomes juxtaposed to centric heterochromatin by a chromosomal inversion. In the absence of a rearrangement, the *white* gene resides in euchromatin and is expressed to give a red eye phenotype. When juxtaposed to heterochromatin, the *white* gene is silenced in a subset of the cells within the eye, giving rise to a red and white variegated phenotype. (B) PEV observed in the yeast *S. cerevisiae* when the *ADE2* gene is inserted near the telomere. At a euchromatic position, the *ADE2* transgene is expressed, giving rise to a white color yeast colony. At a telomeric position, the *ADE2* transgene exhibits PEV, giving rise to a variegated red and white sectored colony. (C) PEV observed in mouse T cells when the *CD2* gene is integrated near the centric region of a chromosome. At a euchromatic position, the *CD2* transgene is expressed and the receptor can be found on all T cells (indicated by black). At a centric position, the *CD2 trans*gene is expressed only in a subset of T cells, giving rise to a mosaic population of T cells expressing the receptor.

II. Heterochromatin and Gene Silencing

The most familiar and conspicuous examples of naturally occurring gene inactivation by heterochromatinization are the Barr body X chromosome of mammalian females (7) and the inactivation of the paternal chromosome in coccid beetles (8). In both cases, whole chromosomes remain condensed throughout interphase, and most or all of the resident genes are inactive. Well-characterized examples of heterochromatin-like silencing in yeast are the silent mating type cassettes (9, 10); copies of mating type information are maintained in silent form, and are used as templates for gene conversion at the active *MAT* locus, where the information is capable of being expressed.

Experimentally, the potent silencing activity of heterochromatin is also evidenced by PEV mediated by the pericentric heterochromatin of mammals, flies, and fission yeast and the telomeric heterochromatin of flies, fission yeast, and budding yeast (Fig. 1). In *Drosophila*, PEV silencing is initiated early in embryonic development, and is mitotically stable in proliferating cells as evidenced by the persistence of silencing through most of development (11). In mice, PEV silencing of an X-linked transgene is similarly initiated early in development and stably maintained (12). In humans, examples of PEV include females heterozygous for the X-linked disorder anhidrotic ectodermal dysplasia. Such females are mosaics of skin cells carrying the normal allele linked to the inactive X chromosome giving rise to patches of skin with no sweat glands, together with regions of cells carrying the normal allele linked to the active X, giving rise to patches of skin with normal sweat glands (13). PEV in humans has also been proposed to be the molecular basis for specific diseases (14). It should be emphasized, though, that examples of PEV are the result of juxtaposition of genes with heterochromatin. There is no evidence to date that the normal function of either pericentric or telomeric heterochromatin is to silence genes. These regions may function indirectly in gene regulation as sinks for silencing factors that could be recruited to specific euchromatic loci during development and differentiation.

III. Heterochromatin and Gene Activation

Despite the reputation of heterochromatin as a transcriptionally inert zone, there are examples of genes that function normally within heterochromatin, notably in *Drosophila* and *Arabidopsis* [reviewed in (15–17)]. Indeed, *Drosophila* genes that reside normally in heterochromatin exhibit a kind of reciprocal PEV, in that rearrangements displacing such genes away from heterochromatin result in reduced or variegated expression (18–20). In larvae lacking the heterochromatin-associated protein HP1 (see below), two heterochromatic genes were found to have reduced expression—and one of them variegated (*21*), suggesting that heterochromatic genes require structural features of heterochromatin to maintain normal expression. Thus, genes in heterochromatin are not merely escaping the silencing properties of heterochromatin, as if they reside within blocks of euchromatin intercalated between heterochromatin. Instead, heterochromatin functions as an "activator" of heterochromatic genes.

The paradox of how heterochromatin can be both a silencer of euchromatic genes and an activator of heterochromatic genes can be better understood in light of recent experiments in yeast. Transcriptional activation by an upstream activating sequence (UAS) in yeast normally occurs over distances of ca. 1 kb or less upstream of the target promoter. The placement of both a UAS and its target within a subtelomeric domain of silencing chromatin permits activation by a downstream UAS located over a distance of 1.4–1.9 kb 3′ to its target promoter (*22*). The same constructs placed elsewhere in the genome showed no activation. The activating effect of subtelomeric position effect has been interpreted as facilitated DNA looping by subtelomeric chromatin, which is believed to interact with the telomeric silencing chromatin though a DNA looping/foldback mechanism (*23*). The intrinsic compaction/looping at the telomere could act to bring the downstream UAS and promoter together. By analogy, the requirement of heterochromatic genes for heterochromatin packaging can be understood as heterochromatin-facilitated communication between enhancers and promoters of heterochromatic genes.

Chromosomal position effects thus clearly demonstrate a functional subdivision of the nucleus into domains with different potential to accommodate transcription. This functional subdivision reflects the cytological subdivision of the nucleus into heterochromatin and euchromatin. The underlying basis for this organization lies in the distinct molecular composition of heterochromatin and euchromatin. We now turn to the genetic strategies used to identify the molecular composition of heterochromatin, some of the key genes identified by these strategies, and current models of how the gene products act to organize heterochromatin.

IV. Genetic Strategies for Dissection of Heterochromatin

In an attempt to identify the functional components of heterochromatin, several *Drosophila* investigators exploited chromosome rearrangements that cause variegation of a visible marker to screen for mutations that modify the extent of variegation (*24–26*). Usually, such screens have begun with a rearrangement causing variegation of *white*, an eye color gene (Fig. 1). Mutations that dominantly suppress or enhance *white* variegation are easily visible as resulting in adult eyes that are nearly fully pigmented (suppressor of PEV) or nearly unpigmented (enhancer of PEV). To verify that such mutations affect a general property of heterochromatin and not merely the regulation of the *white* gene, they must be retested with different rearrangements involving variegation of different genes (e.g., *brown, Stubble, yellow*) and different regions of heterochromatin. Dominant suppressors of PEV are a priori candidates for genes encoding structural components of heterochromatin or their modifiers, whereas dominant enhancers of PEV are candidates for genes encoding factors promoting gene activation through chromatin effects.

Analogous strategies to those used in *Drosophila* have been used in yeast to identify genes that encode subunits of silencing chromatin. Initially, the phenotype screened was mating type switching and the ability to sporulate. All yeast cells carry silenced copies of both **a** and α mating type genes, but the mating type of an individual cell is determined by which set of mating type genes, either **a** or α, is copied into the *MAT* locus, where it is expressed. Mutations in haploid cells that cause loss of silencing of the silent mating type genes result in expression of both **a** and α genes, effectively making the mutant cells sterile. Screens for such mutations were initially done using wild-type cells. However, mating type silencing in yeast is quite robust, with redundant silencing elements, making this assay fairly insensitive. Subsequent screens were carried out using cells with genetically weakened mating type cassettes, making the screens more sensitive to mutations in genes with modest effects on silencing, yielding additional silencing factors. The subsequent recognition that yeast telomeres also can impose PEV (telomeric position effect; TPE, Fig. 1) provided a second assay; a reporter gene inserted in a subtelomeric position shows variegated silencing (27), and mutations that either enhance or suppress this silencing have uncovered components of telomeric silencing chromatin, as well as genes required for the maintenance of the telomeres themselves (28).

A particular virtue of using *Saccharomyces cerevisiae* for chromatin structure studies is that yeast has only two copies of each of the histone genes, in contrast to *Drosophila*, which has ca. 110 copies. Thus, in addition to identifying genes encoding nonhistone components of silencing chromatin, yeast screens have yielded mutations in histone genes, directly implicating the nucleosome as an active participant in the silencing mechanism. An important additional strategy available in yeast is the ability to transform variegating cells with recombinant libraries to identify factors that, when overexpressed, modify silencing chromatin.

Although genetic screens for modifiers of position-effect silencing cannot distinguish between direct and indirect effects on silencing, these screens have generally identified factors that act directly on the silencing chromatin or the silenced gene. These factors include structural proteins, as well as enzymes that modify those proteins. Molecular cloning and characterization of the mutated genes have given us many of the actors in the silencing mechanism and clues as to how the mechanism works.

V. cis-Spreading, trans-Inactivation, and Heterochromatic Associations

Two types of PEV have been described. Recessive PEV is characterized by cis-inactivation of genes placed adjacent to a block of heterochromatin on the same chromosome. In these cases, the neighboring heterochromatin behaves as though it were "spreading" into the flanking or inserted DNA. Indeed, in the giant polytene chromosomes of Drosophila, it is possible to visualize the cytological "heterochromatinization" of the variegating euchromatic material near the heterochromatin–euchromatin junction (24, 29, 30). Immunostaining such chromosomes with antibodies to the heterochromatin-associated protein HP1 shows that HP1 protein also appears to spread across the breakpoint (31).

Some instances of dominant PEV have been described in Drosophila. The best-characterized example is caused by the $brown^{Dominant}$ (bw^D) mutation. This mutation is caused by an insertional transposition of a ca. 1 Mb block of heterochromatic satellite DNA at the brown locus in Drosophila (32). The brown locus on the bw^D chromosome appears as an unusually dense band, and HP1 binds at this site on the mutant chromosome. The bw^D allele causes dominant PEV of the wild-type brown allele on the normal homolog. In contrast to recessive PEV, however, the inactivated allele of bw does not acquire the cytological appearance of heterochromatin, nor does it recruit HP1 (33). The dominant PEV caused by bw^D is pairing dependent, suggesting that intimate contact between the mutant and wild-type locus is required. These results suggest a mechanism by which some structural feature of silencing spreads, in cis or in trans across paired homologs, from heterochromatin to euchromatin.

An alternative mechanism for heterochromatic position effects proposes that euchromatic genes become silenced by being drawn into a region of the nucleus in which silencing factors are at high concentrations, increasing the probability that the chromatin of such genes will bind such factors and thereby be silenced (34). Such a model helps explain genetic data showing that as chromosomal distance from a block of heterochromatin (or a telomere in yeast) increases, the probability that a variegating marker will be silenced decreases (35, 36). A correlation between a physical association with constitutive heterochromatin and silencing during interphase has been observed by immunoFISH analysis in flies (37–39). Interestingly, trans-silencing by bw^D

(see above) is correlated with the cytological association of the $brown^+$ allele with a major block of heterochromatin in interphase nuclei (37). Similarly, the repression of several B cell genes was correlated with heterochromatin associations (39). The notion that physical association with a nuclear region rich in silencing factors is sufficient to impose silencing gains further support from experiments in yeast showing that artificial tethering of a locus to the nuclear membrane is sufficient to confer significant silencing (40). Because heterochromatin is usually located in close proximity with the nuclear membrane, this region of the nucleus is likely to be enriched in silencing factors.

Recent observations challenge a model whereby nuclear localization dictates gene expression. First, Sass and Henikoff (41) described a chromosomal transposition of the *brown* locus in *Drosophila* that results in the displacement of *brown* to the pericentric heterochromatin. Although *cis*-silencing by heterochromatin of the displaced *brown* locus was evident phenotypically, little or no silencing of a wild-type *brown* allele in *trans* was observed. Surprisingly, the *brown* homolog that escapes silencing nevertheless is found to be as frequently and intimately associated with the pericentric heterochromatin as with other examples where efficient silencing is observed. Second, the nuclear repositioning of the *Rag-1* and *TdT* genes next to heterochromatin in MHC-deficient immature thymocytes occurs significantly later than transcriptional silencing of these genes (42), suggesting that heterochromatic association does not cause the silencing. Furthermore, silencing of *Rag-1* and *TdT* in a thymic lymphoma cell line was not accompanied by association with the pericentric heterochromatin, suggesting a centric position is not required (42). One way to reconcile these observations is if protein complexes are shared between certain silenced loci and heterochromatin, causing an aggregation of the silenced locus with natural aggregates of heterochromatin. In other words, heterochromatin association may be a result, rather than a cause, of silencing.

Taking together the studies on the effects of nuclear organization on gene expression, a hypothesis has emerged that the location of a gene within the nucleus places it in a "position of potential" (43). In other words, certain regions of the nucleus favor one transcriptional state over another, yet the ultimate fate of gene expression relies on variables that are specific to a particular gene. Supporting this theory, it was shown that the expression of a *Drosophila* transgene depended not only upon its position in the nucleus relative to a centromere, but also the context of the genomic environment flanking the transgene (44). Whether a gene will be expressed or not depends ultimately on characteristics of the local chromatin structure.

VI. Molecular Composition of Heterochromatin and the Regulation of Heterochromatin Silencing

The big picture of position-effect silencing that emerges from genetics and cytological study is that heterochromatic regions of chromosomes have the ability to propagate silencing effects into neighboring euchromatin in *cis* and in *trans*. To understand the mechanism of silencing, it is first necessary to understand the molecular properties of heterochromatin that distinguish it from euchromatin. In the following sections, we will survey the primary molecular signatures of heterochromatin and the kinds of interactions that underlie the assembly and maintenance of silencing chromatin.

A. DNA Structure and Heterochromatin

The DNA sequence composition of constitutive heterochromatin differs from that of euchromatin primarily insofar as the heterochromatin is generally rich in repetitious DNA sequence, including satellite DNA arrays, relative to euchromatin. There is still considerable overlap between euchromatin and heterochromatin DNA sequences, as many transposable elements are found in both regions. Of course, for regions that are heterochromatic in some tissues and euchromatic in others, like the mammalian Barr body X-chromosome, DNA sequence content is unchanged when the sequences are packaged as heterochromatin or euchromatin. This observation suggests that DNA sequence content cannot be the sole determinant for heterochromatin formation. Indeed, genetic studies in *Drosophila* show that the repetitious nature of the DNA sequences, not the sequences themselves, can be a critical factor. The normally euchromatic *white* gene sequences of *Drosophila* become silenced when three or more copies of *white* transgenes are placed in close proximity (45), and this silencing requires certain heterochromatin-associated factors (46). On the other hand, multicopy arrays of other sequences do not display this property (47), suggesting that there are certain sequence requirements for heterochromatin formation.

B. The Heterochromatic Nucleosome

The smallest functional unit of chromatin is the nucleosome (48), and a clear distinction can be made between the nucleosomal structure of heterochromatin and euchromatin. Two characteristic aspects of heterochromatin structure are the covalent modifications marking heterochromatic nucleosomes and the organization of heterochromatic nucleosomal arrays.

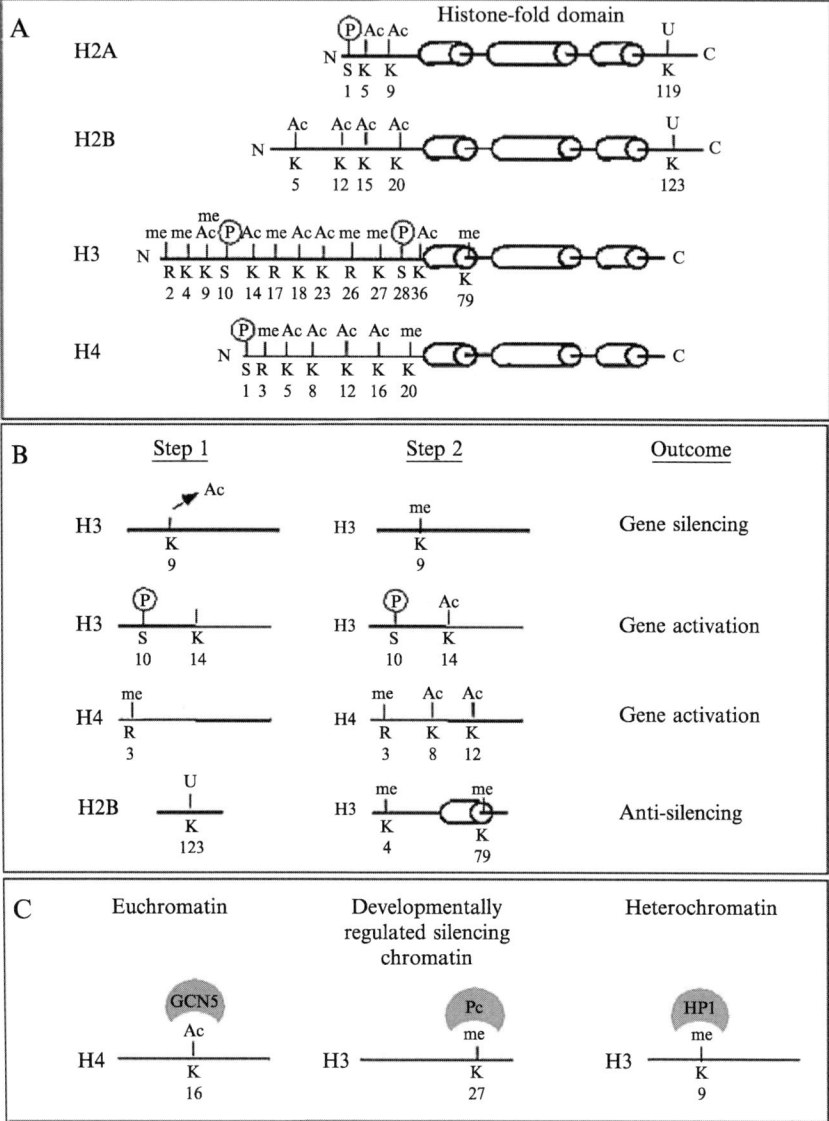

FIG. 2. Posttranslational modifications of the core histones and their implications. (A) Diagram of the four core histones (H2A, H2B, H3, and H4) with the α-helical regions within the histone fold domain shown as cylinders. Modifications include acetylation (Ac) at lysines (K), methylation (me) at K and arginine (R) residues, ubiquitination (U) at K residues, and phosphorylation (P) at serines (S). N and C represent the amino and carboxy terminal ends of each histone, respectively. (B) Interactions between histone modifications. The modification depicted in step 1 promotes the modification(s) depicted in step 2. The net outcome is the effect on gene

C. Histone Modifications That Distinguish Heterochromatin from Euchromatin

A large catalog of covalent histone modifications has been compiled (Fig. 2A). Although the physiological function of these modifications is diverse, several observations point to mechanisms by which specific histone modifications can promote heterochromatin formation. Specifically, both heterochromatin in higher eukaryotes and silencing chromatin in yeast are enriched in histones that are hypoacetylated (49–52). This is not to say that acetylated histones are entirely absent from heterochromatin: for example, histone H4 acetylated at lysine 12 is somewhat enriched in the heterochromatic chromocenter of *Drosophila* polytene chromosomes and the silencing chromatin of the yeast mating type (53, 54). Overall, both the variety and extent of acetylated histone isoforms are low in heterochromatin relative to euchromatin. The reduced acetylation of heterochromatin may be explained in part by the fact that certain transcription activators appear to recruit histone acetylases specifically to promoters (55, 56). Thus, the low levels of histone acetylation in heterochromatin can be explained by the relatively low gene density. Furthermore, histone deacetylases have been found in complexes containing nonhistone chromosomal proteins that preferentially associate with heterochromatin (57, 58). The hypoacetylation of the histones in heterochromatin, specifically at lysine 9 of histone H3, permits histone methylation, a second modification that is associated with heterochromatin (Fig. 2B).

Heterochromatin is enriched in histone H3 that is di- or trimethylated at lysine 9. Members of a conserved family of histone methyltransferases responsible for generating this epigenetic mark are SUV39H1 and SUV39H2 in mouse (59), SU(VAR)3-9 in *Drosophila* (57), and Clr4p in *Schizosaccharomyces pombe* (60). The connection between histone hypoacetylation and methylation in heterochromatin probably reflects a concerted mechanism, as SU(VAR)3-9 family methylases in flies and mammals are physically associated with histone deacetylase enzyme (57, 58).

In female mammals, the Barr body X-chromosome is preferentially enriched in lysine 9-methylated H3 (61, 62), and chromatin immunoprecipitation experiments show that this methylated histone is associated with the

---

expression listed at the right. (C) Evidence supporting the histone code hypothesis. The diagram shows how histone modifications serve as specific recognition motifs for nonhistone chromosomal proteins. The modifications and their associated proteins correlate with particular types of chromatin. The GCN5 bromo domain interacts with acetylated lysine 16 of H3, an event that signifies euchromatin. The chromo domain of POLYCOMB interacts with methylated lysine 27 of H3 to generate developmentally regulated silencing chromatin at homeotic loci and the chromo domain of HP1 interacts with methylated lysine 9 of H3, an event that signifies heterochromatin.

promoters of inactive genes on the Barr body X-chromosome (61). Interestingly, this methylation is unaffected in Suv39h1/Suv39h2 double knockout mouse cells (62), indicating that methylation of the inactive X is catalyzed by one or more separate histone methylases. There are more than 20 predicted protein sequences encoded in the mouse genome that could be histone methylases, at least four of which have been shown to methylate lysine 9 of histone H3 (61, 63–66). Evidence suggests that lysine 9 methylation in histone H3 is required to maintain silencing in PEV and for proper chromosome segregation (see below). Similarly, fission yeast silencing chromatin at the silent mating type cassettes is enriched in H3 methylated at lysine 9 (60). Thus, it appears that in yeast, flies, and mammals, methylation of lysine 9 of H3 is a defining feature of silencing chromatin.

What is the function of the histone modifications found in heterochromatin? Insight into this question came from the discovery that di- and tri-methylated lysine 9 of histone H3 serve as specific recognition motifs for heterochromatin protein 1 (HP1, discussed below, Fig. 2C) (67, 68). The importance of this discovery is exemplified by the fact that the relationship is conserved in diverse organisms such as S. pombe, Drosophila, and mammals. The interaction between the methylated residue and HP1 is direct support for the "histone code hypothesis" (69). This hypothesis states that modifications of histone tails serve as specific recognition motifs for nonhistone chromosomal proteins. A similar mechanism functions in the generation of developmentally regulated silencing chromatin where the SET domain containing protein EN-HANCER OF ZESTE methylates lysine 27 of H3, which, in turn, serves as a recognition motif for the silencing protein POLYCOMB (70) (Fig. 2C). The histone code is also utilized for generating active chromatin; transcriptional activators such as BRAHMA and Taf$_{250}$ specifically interact with acetylated histone H4 (71), a histone modification enriched in euchromatin (Fig. 2C). Thus, the code can be "translated" into a biological outcome such as transcriptional activation or silencing depending on the properties of the specific protein reading the code.

Clearly, the code is much more complex than a single modification on a histone tail. Certain modifications are frequently found in association with each other. For example, phosphorylation of serine 10 on H3 has been shown to enhance acetylation of lysine 14 on H3 (72, 73) (Fig. 2B). Similarily, methylation of arginine 3 of H4 assists the acetylation of lysine 8 and lysine 12 of H4 (74), These combinations of modifications are generated by modifying enzymes that require, or are inhibited by, tail modifications, other than the one they produce. Revealing further complexity in the code, it was recently discovered that modifications on one histone can affect the modification on a different histone, a process termed "*trans*-histone" regulation. Ubiquitation of lysine 123 of histone H2B regulates the methylation of lysine 4 and lysine 79 of histone H3 (75, 76) (Fig. 2B). These modifications are thought to prevent the formation of silencing chromatin and therefore have antisilencing capabilities. What is not yet clear is whether the modifications take place on histones within the same nucleosome, or on histones within different nucleosomes. It is tempting to speculate that interactions between modifications of histones within different nucleosomes might be one mechanism used to regulate the folding and unfolding of the chromatin fiber.

Although it is relatively easy to understand how the histone code can regulate gene expression on a gene-by-gene basis through local targeting of specific enzyme complexes, a more global role for histone modifications has also been proposed. The theory behind this hypothesis is that the modifications often change the overall charge of the histone tails, thereby generally influencing interactions with regulatory factors. Evidence for this hypothesis comes from studies in which transcription elongation was effected by the overall charge of the histone tail, rather than modification of specific residues (77). Global acetylation and deacetylation by enzymes that survey large genomic domains are thought to be responsible for controlling basal levels of gene expression and resetting the ground state after targeted modifications have been utilized (78). Thus, it is likely that the histone code attracts modifying enzymes that act at the local level to regulate gene expression, but these same enzymes might also act at the global level to define domains of similar transcriptional potential.

D. Nucleosome–DNA Interactions

Early footprinting studies in yeast suggested that a difference in nucleosomal positioning distinguished the chromatin of the *S. cerevisiae* silenced mating type cassettes from the same sequences in which silencing was lost in a Sir mutant background (79, 80). In *Drosophila*, PEV silencing is also accompanied by a more highly positioned, regular nucleosomal array at the silenced locus, as judged by the more regular cleavage patterns of heterochromatic nucleosomal linker DNA by micrococcal nuclease (81, 82). Higher occupancy of heterochromatic template DNA by nucleosomes could account for the dramatic reduction in DNA accessibility to restriction endonucleases in heterochromatic sequences, compared to the same sequences in euchromatin (81, 82). The fact that that GAGA factor, TFIID, and RNA polymerase footprints are reduced or missing at a PEV-silenced transgene (83) suggests that heterochromatic nucleosome positioning could also occlude the DNA template for access by the transcriptional machinery. The mechanisms by which nucleosomal positions are established and enforced in heterochromatin are unknown, but probably involve interactions with specific nonhistone proteins (see below).

E. Nonhistone Proteins and Heterochromatin

The identification and isolation of genetic modifiers of PEV in yeast and flies have proven to be the most efficient and productive strategy to identify protein components of heterochromatin and their modifiers. The list of PEV-modifying genes is large and growing, and for many such modifiers, their gene products are either not known or their role in the silencing mechanism is not understood. For these reasons, we will focus here on a small subset of PEV modifiers whose characterization has yielded major insights into nuclear organization.

1. HETEROCHROMATIN PROTEIN 1

Heterochromatin protein 1 was first described as an antigen recognized by a monoclonal antibody raised to a protein fraction eluted from *Drosophila* embryo nuclei at 1–2 M potassium thiocyanate (84). Its name comes from its predominant (though not exclusive) association with the heterochromatic chromocenter and telomeres of *Drosophila* larval polytene chromosomes (84, 85). Indeed, in interphase and metaphase chromosomes of all eukaryotic cells examined, HP1 family proteins are generally found to be associated with pericentric heterochromatin [exceptions are pchet1 from mealybug (86) and HP1c in *Drosophila* (87)]. Subsequent studies have uncovered HP1 family proteins in fission yeast, nematodes, fish, birds, mammals, and plants [reviewed in (88, 89)].

Independent genetic screens for dominant suppressors of PEV in *Drosophila* (25, 90) identified a strong suppressor locus on the second chromosome called Su(var)2-5, which, upon molecular cloning, was determined to encode HP1 (91–93). A role for HP1 family proteins in PEV silencing has been demonstrated in fission yeast (94, 95), flies (91–92), and mice (96).

The structure of HP1 family proteins consists of two conserved globular domains connected by a relatively nonconserved flexible linker. The globular domains have a characteristic fold of three peptide strands forming a β-sheet, packed against a C-terminal α-helix [reviewed in (97)]. The N-terminal domain, termed the "chromo domain," binds specifically to histone H3 methylated at lysine 9 (67, 68, 98, 99). However, the binding of HP1 family proteins is not obligatorily linked to H3 lysine 9 methylation, as the entire mammalian Barr body X-chromosome is enriched for this methylated H3 isoform, yet the binding of all three mammalian HP1 isoforms is restricted to the pericentric heterochromatin of this and all chromosomes (62). The C-terminal domain of HP1, termed the "chromo shadow domain," mediates self-association (100, 101) and has been implicated in a variety of heterologous interactions with nuclear proteins involved in chromatin structure, DNA replication, and nuclear architecture [reviewed in (88, 102)]. Most of these interactions have been demonstrated by yeast two-hybrid protein assays and/or *in vitro* pulldown assays, and will require *in vivo* interaction tests to determine their physiological significance.

The two-domain structure of HP1 family proteins suggests a bifunctional crosslinker. The specificity of the chromo domain for methylated H3 suggests that HP1 family proteins could use the N-terminal chromo domain to target heterochromatic nucleosomes and anchor heterochromatin protein complexes to nucleosomes through the C-terminal chromo shadow domain and/or through additional interactions via the N-terminal chromo domain. An estimate of HP1 concentration in *Drosophila* is consistent with a roughly stoichiometric binding of two molecules of HP1 per nucleosome in heterochromatin (*21*). It remains unclear how HP1 binding to nucleosomes results in gene silencing; a mechanistic model will require a more complete picture of how HP1-binding proteins interact with HP1, nucleosomes, and DNA.

2. Suppressor of Variegation 3-9 Protein

Su(var)3-9 was first identified genetically in *Drosophila* as a dominant suppressor of PEV (*103*). Molecular cloning and biochemical characterization revealed a structural homology between the C-terminus of SU(VAR)3-9 protein and the proteins encoded by the homeotic silencer *Enhancer of zeste* and the homeotic gene activator *trithorax*, a domain called the SET domain. The structural similarity of the SET domain to the plant Rubisco protein implicated SET domains in protein methylation, and biochemical assays revealed that SU(VAR)3-9 family proteins are histone methyltransferases with specificity for lysine 9 of histone H3 (*63, 65*). As discussed above, methylated lysine 9 of histone H3 represents a high-affinity binding site for the HP1 chromo domain.

The interaction between the HP1 chromo domain and methylated lysine 9 of histone H3, together with the colocalization of lysine 9-methylated H3 and HP1 in yeast and flies (*104, 105*) suggests that *Su(var)3-9* mutations should be epistatic to mutations in HP1. Indeed, mutation in *clr4*, the *S. pombe* homologue of *Su(var)3-9*, results in loss of binding of Swi6p, the *S. pombe* HP1 family protein (*60*). Surprisingly, however, *Su(var)3-9* is not essential in *Drosophila*, though HP1 is essential. This suggests either that HP1 may be targeted to chromatin by other mechanisms, or that the essential function of HP1 does not require chromosome binding. Consistent with the former possibility, HP1 remains associated with the fourth chromosome and telomeres and only partially disappears from the chromocenter in flies lacking SU(VAR)3-9 (*105*).

3. Silent Information Regulator 2 Protein

SIR2 was first identified in a screen for mutations in *S. cerevisiae* that permit a strain with α information at both *HML* and *HMR* and defective **a** information at the *MAT* locus to mate as an α strain [signifying loss of silencing of the *HML* and/or *HMR* cassettes; (*106*)]. Sir2 family proteins are highly conserved, with homologs found from bacteria to humans (*107*). Molecular cloning and biochemical characterization revealed that SIR2 encodes a histone deacetylase (*108*). In yeast, Sir2p is required for telomeric PEV (*109*) and in silencing position effects at the rDNA locus (*110*).

The role of Sir2p in yeast silencing chromatin is thought to be to establish and maintain a domain of deacetylated chromatin as a substrate for silencing factors. At yeast telomeres, Sir2p is recruited by Sir3p–Sir4p complexes, which in turn bind to subtelomeric nucleosomes through N-terminal tails of histones H3 and/or H4 (*111*).

F. RNA as a Component of Heterochromatin

Just as certain DNA sequences and histone modifications are defining characteristics of heterochromatin, recent data suggest that specific RNAs might also be involved in the generation and/or maintenance of heterochromatin. Supporting this theory, RNase A (but not RNase H) digestion of mammalian nuclei destabilizes HP1α binding in pericentric heterochromatin (*112*), suggesting a mechanistic role for RNA in heterochromatin-mediated silencing. RNA transcripts of centric repeats has also been implicated in silencing in fission yeast, where a connection between RNA interference (RNAi) and heterochromatin silencing has emerged from genetic studies (*113*), pointing to the paradoxical idea that transcription of heterochromatin regions actively promotes their silencing. RNAi is the targeted degradation of specific transcripts resulting from the transcription of self-complementary RNA (*114, 115*). In heterochromatic regions, self-complementary RNA could readily arise because of the highly repetitious sequence composition of heterochromatic DNA, together with the tendency of heterochromatic regions to accumulate transposable elements containing promoter elements. The mechanistic relationship between the small double-stranded RNAs that direct sequence-specific mRNA degradation by the RNAi pathway and the pattern of specific chromatin structures that characterize heterochromatin remains obscure. However, it is easy to see that if self-complementary RNA is transcribed from heterochromatin and does not immediately diffuse away from the site of transcription, its accumulation in heterochromatic regions of the genome might serve as a way to target silencing factors specifically to heterochromatic DNA. Precedent for the role of RNA in regulation of gene expression and modulation of chromatin structure is seen for the *Xist* RNA that associates with the silent X-chromosome in female mammals and the *rox* RNAs that associate with the hyperactivated X-chromosome in male *Drosophila* (*116*).

VII. Setting Up Heterochromatin and Euchromatin Domains

Do specific DNA sequences or proteins demarcate the boundaries of heterochromatin and euchromatin? The inverted repeats flanking the silent mating type cassettes in S. *pombe* appear to demarcate a transition point between heterochromatin—rich in Swi6p and histone H3 methylated at lysine 9—and adjacent zones of euchromatin enriched in histone H4 methylated at lysine 4 (*102*). This type of organization is also observed at the chicken β-globin locus, where sharp transitions in histone modifications flank the locus (*117*). Whereas the β-globin genes are embedded within a large domain of acetylated chromatin that contains DNase I hypersensitive sites, the upstream 16 kb is packaged with hypoacetylated histones and is relatively inaccessible to DNase I cleavage. What is responsible for abrupt transitions between different chromatin states? In the case of the β-globin locus, a zinc finger DNA-binding protein termed CTCF associates at the transition site and might be partially responsible for setting up the active domain by recruiting a histone acetyltransferase (*118, 119*).

The idea that specific factors are involved in setting up "barriers" that separate domains of distinctly different chromatin states is supported by several studies in S. *cerevisiae*. First, the termination of telomeric silencing chromatin has been attributed to sequences within subtelomeric X and Y' elements. These elements contain barriers termed STARS (subtelomeric antisilencing regions) that bind the proteins Tbf1p and Reb1p (*120*). These proteins are required to prevent the spread of telomeric silencing chromatin into the adjacent regions containing expressed genes. Second, the termination of silencing chromatin at the telomere-proximal barrier of the silent mating type cassette *HMR* was identified as a tRNA gene. Genetic data indicate that Pol III transcription factors responsible for transcribing this gene, but not the act of transcription itself, are required for barrier activity (*121, 122*). Third, the termination of silencing chromatin from the left of the silent mating type cassette *HML* maps to DNA sequences containing binding sites for three transcriptional activators. It is not yet known whether these activators are required for barrier function (*123*). Overall, the common theme emerging from studies in mammals and yeast is that barriers demarcating transitions from active to inactive chromatin are associated with DNA-binding proteins, many of which play roles in both transcriptional activation and silencing at other locations in the genome. Supporting this hypothesis, a fragment of DNA containing three Rap1 binding sites can act as a barrier when placed within the *HML* silencing region (*124*).

In *Drosophila*, barriers capable of constraining heterochromatin spreading were invoked by Tartof *et al.* (*125*) as part of a model for PEV. The physical reality of such barriers as part of a normal chromosome-organizing mechanism has not been demonstrated. The observation of interdigitated zones of transcriptional activity and silencing does not necessarily imply an underlying organizing principle. It could simply be that, in some cases, transcriptional activators can compete successfully with heterochromatin to overcome silencing (*126*). Such competition has been described for yeast TPE by the yeast transcription factor Pprlp (*36, 127*), and for the yeast activator Gal4p in *Drosophila* (*128*). Competition between silencing and activation could explain how numerous genes on the human Barr body X-chromosome escape silencing. The organization of the *D. melanogaster* fourth chromosome into regions capable of silencing transgenes, alternating with regions of stable transgene expression, suggests that there could be barriers segregating alternating heterochromatic and euchromatic zones on this chromosome (*129*).

VIII. Summary and Future Directions

A general model for heterochromatin in higher eukaryotes that would be applicable to silencing chromatin in yeast implicates distinctive properties of nucleosomes that are propagated thoughout a large domain. Heterochromatin nucleosomes have defining covalent modifications, generally hypoacetylated N-terminal histone tails and (in fission yeast and higher eukaryotes) methylated lysine 9 of histone H3. The patterns of histone modification are thought to recruit specific nonhistone protein complexes. Such complexes could establish and maintain the positioned nucleosome arrays typical of heterochromatin. In the case of the silencing telomeric chromatin of yeast, and perhaps in heterochromatin in higher eukaryotes, nucleoprotein complexes might promote a distinct folded structure. Nucleosome positioning and chromatin folding can serve to exclude one or more DNA-binding factors necessary for transcription of some genes, but it may also facilitate enhancer–promoter communication for other genes—particularly genes that normally reside and function in heterochromatin.

In the future, the histone code will continue to be translated into new meanings. Connections between the code and other nuclear processes such as DNA methylation have recently been reported (*130, 131*). An important goal is to be able to predict and regulate the expression pattern of a gene using the tools identified by the genetic, cytological, and molecular dissection of heterochromatin.

Acknowledgments

We would like to acknowledge Tim Parnell for assistance with Fig. 1. Research in our laboratories is supported by NIH Grant R01GM061513 (LLW), Department of Defense Grant BC010192 (LLW), and NSF Grant MCB 0131414 (JCE).

REFERENCES

1. Heitz, E. (1928). Das Heterochromatin der Moose. *Jb. Wiss. Bot.* **69**, 728–818.
2. Dubinin, N. P., and Siderov, B. N. (1935). [The position effect of the hairy gene]. *Biol. Zh. (Mosk.)* **4**, 555–568.
3. Panshin, I. B. (1935). New evidence for the position-effect hypothesis. *C. R. Acad. Sci. USSR* **9**, 85–88.
4. Karpen, G. H., and Spradling, A. C. (1990). Reduced DNA polytenization of a minichromosome region undergoing position-effect variegation in Drosophila. *Cell* **63**, 97–108.
5. Glaser, R. L., and Spradling, A. C. (1994). Unusual properties of genomic DNA molecules spanning the euchromatic-heterochromatic junction of a Drosophila minichromosome. *Nucleic Acids Res.* **22**, 5068–5075.
6. Wallrath, L. L., Guntur, V. P., Rosman, L. E., and Elgin, S. C. R. (1996). DNA representation of variegating heterochromatic P-element inserts in diploid and polytene tissues of Drosophila melanogaster. *Chromosoma.* **104**, 519–527.
7. Barr, M. L., and Bertram, E. G. (1949). A morphological distinction between neurons of the male and female, and the behavior of the nuclear satellite during accelerated nucleoprotein synthesis. *Nature* **163**, 676–677.
8. Nur, U. (1990). Heterochromatization and euchromatization of whole genomes in scale insects (Coccoidea: homoptera). *Development Suppl.*, 29–34.
9. Nasmyth, K. A., Tatchell, K., Hall, B. D., Astell, C. R., and Smith, M. (1981). A position effect in the control of transcription at yeast mating type loci. *Nature* **289**, 244–250.
10. Klar, A. J. S. (1990). Regulation of fission yeast mating-type interconversion by chromosome imprinting. *Development Suppl.*, 3–8.
11. Lu, B. Y., Ma, J., and Eissenberg, J. C. (1998). Developmental regulation of heterochromatin-mediated gene silencing in Drosophila. *Development* **125**, 2223–2234.
12. Tan, S.-S., Williams, E. A., and Tam, P. P. L. (1993). X-chromosome inactivation occurs at different times in different tissues of the postimplantation mouse embryos. *Nat. Genet.* **3**, 170–174.
13. Roberts, E. (1929). The inheritance of anhidrosis associated with anodontia. *J. Am. Med. Assoc.* **93**, 277–279.
14. Parnell, T. J., Grade, S. K., Geyer, P. K., and Wallrath, L. L. PEV: Lessons from model organisms applied to human genetic diseases. *In* "Encyclopedia of the Human Genome." In press.
15. Weiler, K. S., and Wakimoto, B. T. (1995). Heterochromatin and gene expression in Drosophila. *Annu. Rev. Genet.* **29**, 577–605.
16. Eissenberg, J. C., and Hilliker, A. J. (2000). Versatility of conviction: Heterochromatin as both a repressor and an activator of transcription. *Genetica* **109**, 19–24.
17. Copenhaver, G. P., Nickel, K., Kuromori, T., Benito, M.-I., Kaul, S., Lin, X., Bevan, M., Murphy, G., Harris, B., Parnell, L. D., McCombie, W. R., Martienssen, R. A., Marra Marco, M., and Preuss, D. (1999). Genetic definition and sequence analysis of Arabidopsis centromeres. *Science* **286**, 2468–2474.
18. Hessler, A. Y. (1958). V-type position effect at the *light* locus in Drosophila melanogaster. *Genetics* **43**, 395–403.
19. Wakimoto, B. T., and Hearn, M. G. (1990). The effects of chromosome rearrangements on the expression of heterochromatic genes in chromosome 2L of Drosophila melanogaster. *Genetics* **125**, 141–154.
20. Eberl, D. F., Duyf, B. J., and Hilliker, A. J. (1993). The role of heterochromatin in the expression of a heterochromatic gene, the rolled locus of Drosophila melanogaster. *Genetics* **134**, 277–292.

21. Lu, B. Y., Emtage, P. C. R., Duyf, B. J., Hilliker, A. J., and Eissenberg, J. C. (2000). Heterochromatin protein 1 is required for the normal expression of two heterochromatin genes in Drosophila. *Genetics* **155,** 699–708.
22. de Bruin, D., Zaman, Z., Liberatore, R. A., and Ptashne, M. (2001). Telomere looping permits gene activation by a downstream UAS in yeast. *Nature* **409,** 109–113.
23. Strahl-Bolsinger, S., Hecht, A., Luo, K., and Grunstein, M. (1997). Sir2 and Sir4 interactions differ in core and extended telomeric heterochromatin in yeast. *Genes Dev.* **11,** 83–93.
24. Reuter, G., Werner, W., and Hoffman, H. J. (1982). Mutants affecting position effect heterochromatinization in *Drosophila melanogaster*. *Chromosoma* **85,** 539–551.
25. Sinclair, D. A. R., Mottus, R. C., and Grigliatti, T. A. (1983). Genes which suppress position-effect variegation in *Drosophila melanogaster* are clustered. *Mol. Gen. Genet.* **191,** 326–333.
26. Locke, J., Kotarski, M. A., and Tartof, K. D. (1988). Dosage-dependent modifiers of position effect variegation in Drosophila and a mass action model that explains their effect. *Genetics* **120,** 181–198.
27. Gottschling, D. E., Aparicio, O. M., Billington, B. L., and Zakian, V. A. (1990). Position effect at S. cerevisiae telomeres: Reversible repression of Pol II transcription. *Cell* **63,** 751–762.
28. Lowell, J. E., and Pillus, L. (1998). Telomere tales: Chromatin, telomerase, and telomere function in Saccharomyces cerevisiae. *Cell. Mol. Life Sci.* **54,** 32–49.
29. Hartmann-Goldstein, I. J. (1967). On the relationship between heterochromatization and variegation in Drosophila with special reference to temperature-sensitive periods. *Genet. Res.* **10,** 143–159.
30. Zhimulev, I. F., Belyaeva, E. S., Fomina, O. V., Protopopov, M. O., and Bolshakov, V. N. (1986). Cytogenetic and molecular aspects of position-effect variegation in *Drosophila melanogaster*. I. Morphology and genetic activity of the 2AB region in chromosome rearrangement $T(1;2)dor^{var7}$. *Chromosoma* **94,** 492–504.
31. Belyaeva, E. S., Demakova, O. V., Umbetova, G. H., and Zhimulev, I. F. (1993). Cytogenetic and molecular aspects of position-effect variegation in *Drosophila melanogaster*. V. Heterochromatin-associated protein HP1 appears in euchromatic chromosomal regions that are inactivated as a result of position-effect variegation. *Chromosoma* **102,** 583–590.
32. Henikoff, S., Jackson, J. M., and Talbert, P. B. (1995). Distance and pairing effects on the $brown^{Dominant}$ heterochromatic element in Drosophila. *Genetics* **140,** 1007–1017.
33. Belyaeva, E. S., Koryakov, D. E., Pokholkova, G. V., Demakova, O. V., and Zhimulev, I. F. (1997). Cytological study of the *brown* dominant position effect. *Chromosoma* **106,** 124–132.
34. Cockell, M., and Gasser, S. (1999). Nuclear compartments and gene regulation. *Curr. Opin. Genet. Dev.* **9,** 199–205.
35. Talbert, P. B., and Henikoff, S. (2000). A reexamination of spreading of position-effect variegation in the *white-roughest* region of *Drosophila melanogaster*. *Genetics* **154,** 259–272.
36. Renauld, H., Aparicio, O. M., Zierath, P. D., Billington, B. L., Chhablani, S. K., and Gottschling, D. E. (1993). Silent domains are assembled continuously from the telomere and are defined by promoter distance and strength, and by *SIR3* dosage. *Genes Dev.* **7,** 1133–1145.
37. Csink, A. K., and Henikoff, S. (1996). Genetic modification of heterochromatic associations and nuclear organization in Drosophila. *Nature* **381,** 529–531.
38. Dernburg, A. F., Borman, K. W., Fung, J. C., Marshall, W. F., Philips, J., Agard, D. A., and Sedat, J. W. (1996). Perturbation of nuclear architecture by long-distance chromosome interactions. *Cell* **85,** 745–759.
39. Brown, K. E., Guest, S. S., Smale, S. T., Hahm, K., Merkenschlager, M., and Fisher, A. G. (1997). Association of transcriptionally silent genes with Ikaros complexes at centromeric heterochromatin. *Cell* **91,** 845–854.

40. Andrulis, E. D., Neiman, A. M., Zappulla, D. C., and Sternglanz, R. (1998). Perinuclear localization of chromatin facilitates transcriptional silencing. *Nature* **394,** 592–595.
41. Sass, G. L., and Henikoff, S. (1999). Pairing-dependent mislocalization of a Drosophila *brown* gene reporter to a heterochromatic environment. *Genetics* **152,** 595–604.
42. Brown, K. E., Baxter, J., Graf, D., Merkenschlager, M., and Fisher, A. G. (1999). Dynamic repositioning of genes in the nucleus of lymphocytes preparing for cell division. *Mol. Cell* **3,** 207–217.
43. Gasser, S. M. (2000). Positions of potential: Nuclear organization and gene expression. *Cell* **104,** 639–642.
44. Cryderman, D. E., Morris, I. J., Biessmann, H., Elgin, S. C. R., and Wallrath, L. L. (1999). Silencing at *Drosophila* telomeres: Nuclear organization and chromatin structure play critical roles. *EMBO J.* **18,** 3724–3735.
45. Dorer, D. R., and Henikoff, S. (1994). Expansions of transgene repeats cause heterochromatin formation and gene silencing in Drosophila. *Cell* **77,** 993–1002.
46. Fanti, L., Dorer, D. R., Berloco, M., Henikoff, S., and Pimpinelli, S. (1998). Heterochromatin protein 1 binds transgene arrays. *Chromosoma* **107,** 286–292.
47. Clark, D. V., Sabl, J., and Henikoff, S. (1998). Repetitive arrays containing a housekeeping gene have altered polytene chromosome morphology in Drosophila. *Chromosoma* **107,** 96–104.
48. Luger, K., Mader, A. W., Richmond, R. K., Sargent, D. F., and Richmond, T. J. (1997). Crystal structure of the nucleosome core particle at 2.8 A resolution. *Nature* **389,** 251–260.
49. Jeppesen, P., Mitchell, A., Turner, B. M., and Perry, P. (1992). Antibodies to defined histone epitopes reveal variations in chromatin conformation and underacetylation of centromeric heterochromatin in human metaphase chromosomes. *Chromosoma* **101,** 322–332.
50. Jeppesen, P., and Turner, B. M. (1993). The inactive X chromosome in female mammals is distinguished by a lack of H4 acetylation, a cytogenetic marker for gene expression. *Cell* **74,** 281–289.
51. Boggs, B. A., Connors, B., Sobel, R. E., Chinault, A. C., and Allis, C. D. (1996). Reduced levels of histone H3 acetylation on the inactive X chromosome in human females. *Chromosoma* **105,** 303–309.
52. Suka, N., Suka, Y., Carmen, A. A., Wu, J., and Grunstein, M. (2001). Highly specific antibodies determine histone acetylation site usage in yeast heterochromatin and euchromatin. *Mol. Cell* **8,** 473–479.
53. Turner, B. M., Birley, A. J., and Lavender, J. (1992). Histone H4 isoforms acetylated at specific lysine residues define individual chromosomes and chromatin domains in Drosophila polytene nuclei. *Cell* **69,** 375–384.
54. Braunstein, M., Sobel, R. E., Allis, C. D., Burner, B. M., and Broach, J. R. (1996). Efficient transcriptional silencing in *Saccharomyces cerevisiae* requires a heterochromatin histone acetylation pattern. *Mol. Cell. Biol.* **16,** 4349–4356.
55. Kuo, M.-H., Zhou, J., Jambeck, P., Churchill, M. E. A., and Allis, C. D. (1998). Histone acetyltransferase activity of yeast Gcn5p is required for the activation of target genes in vivo. *Genes Dev.* **12,** 627–639.
56. Utley, R. T., Ikeda, K., Grant, P. A., Côté, J., Steger, D. J., Eberharter, A., John, S., and Workman, J. L. (1998). Transcriptional activators direct histone acetyltransferase complexes to nucleosomes. *Nature* **394,** 498–502.
57. Czermin, B., Schotta, G., Hülsmann, B. B., Brehm, A., Becker, P. B., Reuter, G., and Imhof, A. (2001). Physical and functional association of SU(VAR)3–9 and HDAC1 in Drosophila. *EMBO Rep.* **2,** 915–919.

58. Vaute, O., Nicolas, E., Vandel, L., and Trouche, D. (2002). Functional and physical interaction between the histone methyl transferase Suv39H1 and histone deacetylases. *Nucleic Acids Res.* **30,** 475–481.
59. Peters, A. H. F. M., O'Carroll, D., Scherthan, H., Mechtler, K., Sauer, S., Schoefer, C., Weipoltshammer, K., Pagani, M., Lachner, M., Kohlmaier, A., Opravil, S., Doyle, M., Sibiliaand, M., and Jenuwein, T. (2001). Loss of the Suv39h histone methyltransferases impairs mammalian heterochromatin and genome stability. *Cell* **107,** 323–337.
60. Nakayama, J.-I., Rice, J. C., Strahl, B. D., Allis, C. D., and Grewal, S. I. S. (2001). Role of histone H3 lysine 9 methylation in epigenetic control of heterochromatin assembly. *Science* **292,** 100–113.
61. Boggs, B. A., Cheung, P., Heard, E., Spector, D. L., Chinault, A. C., and Allis, C. D. (2002). Differentially methylated forms of histone H3 show unique association patterns with inactive human X chromosomes. *Nat. Genet.* **30,** 73–76.
62. Peters, A. H. F. M., Mermoud, J. E., O'Carroll, D., Pagani, M., Schweizer, D., Brockdorff, N., and Jenuwein, T. (2002). Histone H3 lysine 9 methylation is an epigenetic imprint of facultative heterochromatin. *Nat. Genet.* **30,** 77–80.
63. Rea, S., Eisenhaber, F., O'Carroll, D., Strahl, B. D., Sun, Z.-W., Schmid, M., Opravil, S., Mechtler, K., Ponting, C. P., Allis, C. D., and Jenuwein, T. (2000). Regulation of chromatin structure by site-specific histone H3 methyltransferases. *Nature* **406,** 593–599.
64. Tachibana, M., Sugimoto, K., Fukushima, T., and Shinkai, Y. (2001). SET domain-containing protein, G9a, is a novel lysine-preferring mammalian histone methyltransferase with hyperactivity and specific selectivity to lysines 9 and 27 of histone H3. *J. Biol. Chem.* **276,** 25309–25317.
65. Czermin, B., Melfi, R., McCabe, D., Seitz, V., Imhof, A., and Pirrotta, V. (2002). *Drosophila* Enhancer of Zeste/ESC complexes have a histone H3 methyltransferase activity that marks chromosomal polycomb sites. *Cell* **111,** 185–196.
66. Müller, J., Hart, C. M., Francis, N. J., Vargas, M. L., Sengupta, A., Wild, B., Miller, E. L., O'Connor, M. B., Kingston, R. E., and Simon, J. A. (2002). Histone methyltransferase activity of a Drosophila Polycomb group repressor complex. *Cell* **111,** 197–208.
67. Bannister, A. J., Zegerman, P., Partridge, J. F., Miska, E. A., Thomas, J. O., Allshire, R. C., and Kouzarides, T. (2001). Selective recognition of methylated lysine 9 on histone H3 by the HP1 chromo domain. *Nature* **410,** 120–124.
68. Lachner, M., O'Carroll, D., Rea, S., Mechtler, K., and Jenuwein, T. (2001). Methylation of histone H3 lysine 9 creates a binding site for HP1 proteins. *Nature* **410,** 116–120.
69. Jenuwein, T., and Allis, C. D. (2001). Translating the histone code. *Science* **293,** 1074–1080.
70. Cao, R., Wang, H., Wang, L., Xia, L., Erdjument-Bromage, H., Tempst, P., Jones, R. S., and Zhang, Y. (2002). Role of histone H3 lysine 27 methylation in Polycomb-group silencing. *Science* **298,** 1039–1043.
71. Jacobson, R. H., Ladurner, A. G., King, D. S., and Tjian, R. (2000). Structure and function of a human $TAF_{II}250$ double bromo domain module. *Science* **288,** 1422–1425.
72. Cheung, P., Tanner, K. G., Cheung, W. L., Sassone-Corsi, P., Denu, J. M., and Allis, C. D. (2000). Synergistic coupling of histone H3 phosphorylation and acetylation in response to epidermal growth factor stimulation. *Mol. Cell* **5,** 905–915.
73. Lo, W. S., Trievel, R. C., Rojas, J. R., Duggan, L., Hsu, J. Y., Allis, C. D., Marmorstein, R., and Berger, S. L. (2000). Phosphorylation of serine 10 in histone H3 is functionally linked in vitro and in vivo to Gcn5-mediated acetylation at lysine 14. *Mol. Cell* **5,** 917–926.
74. Wang, H., Huang, Z.-Q., Xia, L., Feng, Q., Erdjument-Bromage, H., Strahl, B. D., Briggs, S. D., Allis, C. D., Wong, J., Tempst, P., and Zhang, Y. (2001). Methylation of histone H4 at arginine 3 facilitating transcriptional activation by nuclear hormone receptor. *Science* **293,** 853–857.

75. Briggs, S. D., Xiao, T., Sun, Z.-W., Caldwell, J. A., Shabanowitz, J., Hunt, D. F., Allis, C. D., and Strahl, B. D. (2002). Gene silencing: Trans-histone regulatory pathway in chromatin. *Nature* **418**, 498.
76. Sun, Z.-W., and Allis, C. D. (2002). Ubiquitination of histone H2B regulates H3 methylation and gene silencing in yeast. *Nature* **418**, 104–108.
77. Kristjuhanl, A., Walker, J., Suka, N., Grunstein, M., Roberts, D., Cairns, B. R., and Svejstrup, J. Q. (2002). Transcriptional inhibition of genes with severe histone H3 hypoacetylation in the coding region. *Mol. Cell* **10**, 925–933.
78. Vogelauer, M., Wu, J., Suka, N., and Grunstein, M. (2000). Global histone acetylation and deacetylation in yeast. *Nature* **408**, 495–498.
79. Nasmyth, K. A. (1982). The regulation of yeast mating-type chromatin structure by *SIR*: An action at a distance affecting both transcription and transposition. *Cell* **30**, 567–578.
80. Ravindra, A., Weiss, K., and Simpson, R. T. (1999). High-resolution structural analysis of chromatin at specific loci: *Saccharomyces cerevisiae* silent mating-type locus HMRa. *Mol. Cell. Biol.* **19**, 7944–7950.
81. Wallrath, L. L., and Elgin, S. C. R. (1995). Position effect variegation in Drosophila is associated with an altered chromatin structure. *Genes Dev.* **9**, 1263–1277.
82. Sun, F.-L., Cuaycong, M. H., and Elgin, S. C. R. (2001). Long-range nucleosome ordering is associated with gene silencing in *Drosophila melanogaster* pericentric heterochromatin. *Mol. Cell. Biol.* **21**, 2867–2879.
83. Cryderman, D. E., Tang, H. B., Bell, C., Gilmour, D. S., and Wallrath, L. L. (1999). Heterochromatic silencing of Drosophila heat shock genes acts at the level of promoter potentiation. *Nucleic Acids Res.* **27**, 3364–3370.
84. James, T. C., and Elgin, S. C. R. (1986). Identification of a nonhistone chromosomal protein associated with heterochromatin in Drosophila and its gene. *Mol. Cell. Biol.* **6**, 3862–3872.
85. James, T. C., Eissenberg, J. C., Craig, C., Dietrich, V., Hobson, A., and Elgin, S. C. R. (1989). Distribution patterns of HP1, a heterochromatin-associated nonhistone chromosomal protein of Drosophila. *Eur. J. Cell Biol.* **50**, 170–180.
86. Epstein, H., James, T. C., and Singh, P. B. (1992). Cloning and expression of Drosophila HP1 homologs from a mealybug, *Planococcus citri*. *J. Cell Sci.* **101**, 463–474.
87. Smothers, J. F., and Henikoff, S. (2001). The hinge and chromo shadow domain impart distinct targeting of HP1-like proteins. *Mol. Cell. Biol.* **21**, 2555–2569.
88. Eissenberg, J. C., and Elgin, S. C. R. (2000). The HP1 protein family: Getting a grip on chromatin. *Curr. Opin. Genet. Dev.* **10**, 201–210.
89. Jones, D. O., Cowell, I. G., and Singh, P. B. (2000). Mammalian chromodomain proteins: Their role in genome organisation and expression. *BioEssays* **22**, 124–137.
90. Wustmann, G., Szidonya, J., Taubert, H., and Reuter, G. (1989). The genetics of position-effect variegation modifying loci in *Drosophila melanogaster*. *Mol. Gen. Genet.* **217**, 520–527.
91. Eissenberg, J. C., James, T. C., Foster-Hartnett, D. M., Hartnett, T., Ngan, V., and Elgin, S. C. R. (1990). Mutation in a heterochromatin-specific chromosomal protein is associated with suppression of position-effect variegation in *Drosophila melanogaster*. *Proc. Natl. Acad. Sci. USA* **87**, 9923–9927.
92. Eissenberg, J. C., Morris, G. D., Reuter, G., and Hartnett, T. (1992). The heterochromatin-associated protein HP-1 is an essential protein in Drosophila with dosage-dependent effects on position-effect variegation. *Genetics* **131**, 345–352.
93. Eissenberg, J. C., and Hartnett, T. (1993). A heat shock-activated cDNA rescues the recessive lethality of mutations in the heterochromatin-associated protein HP1 of *D. melanogaster*. *Mol. Gen. Genet.* **240**, 333–338.

94. Lorentz, A., Ostermann, K., Fleck, O., and Schmidt, H. (1994). Switching gene *swi6*, involved in repression of silent mating-type loci in fission yeast, encodes a homologue of chromatin-associated proteins from Drosophila and mammals. *Gene* **143**, 139–143.
95. Allshire, R. C., Nimmo, E. R., Ekwall, K., Javerzat, J. P., and Cranston, G. (1995). Mutations derepressing silent centromeric domains in fission yeast disrupt chromosome segregation. *Genes Dev.* **9**, 218–233.
96. Festenstein, R., Sharghi-Namini, S., Fox, M., Roderick, K., Tolaini, M., Norton, T., Saveliev, A., Kioussis, D., and Singh, P. (1999). Heterochromatin protein 1 modifies mammalian PEV in a dose- and chromosomal-context-dependent manner. *Nat. Genet.* **23**, 457–461.
97. Eissenberg, J. C. (2001). Molecular biology of the chromo domain: An ancient chromatin module comes of age. *Gene* **275**, 19–29.
98. Jacobs, S. A., Taverna, S. D., Zhang, Y., Briggs, S. D., Li, J., Eissenberg, J. C., Allis, C. D., and Khorasanizadeh, S. (2001). Specificity of HP1 chromo domain for methylated N-terminus of histone H3. *EMBO J.* **20**, 5232–5241.
99. Jacobs, S. A., and Khorasanizadeh, S. (2002). Structure of HP1 chromodomain bound to lysine 9-methylated histone H3 tail. *Science* **295**, 2080–2083.
100. Smothers, J. F., and Henikoff, S. (2000). The HP1 chromo shadow domain binds a consensus peptide pentamer. *Curr. Biol.* **10**, 27–30.
101. Cowieson, N. P., Partridge, J. F., Allshire, R. C., and McLaughlin, P. J. (2000). Dimerisation of a chromo shadow domain and distinctions from the chromodomain as revealed by structural analysis. *Curr. Biol.* **10**, 517–525.
102. Li, Y., Kirschmann, D. A., and Wallrath, L. L. (2002). Does heterochromatin protein 1 always follow code? *Proc. Natl. Acad. Sci. USA* **99**(Suppl. 4), 16462–16469.
103. Tschiersch, B., Hofmann, A., Krauss, V., Dorn, R., Korge, G., and Reuter, G. (1994). The protein encoded by the Drosophila position-effect variegation suppressor gene Su(var)3–9 combines domains of antagonistic regulators of homeotic gene complexes. *EMBO J.* **13**, 3822–3831.
104. Noma, K., Allis, C. D., and Grewal, S. I. S. (2001). Transitions in distinct histone H3 methylation patterns at the heterochromatin domain boundaries. *Science* **293**, 1150–1155.
105. Schotta, G., Ebert, A., Krauss, V., Fischer, A., Hoffmann, J., Rea, S., Jenuwein, T., Dorn, R., and Reuter, G. (2002). Central role of Drosophila SU(VAR)3–9 in histone H3-K9 methylation and heterochromatic gene silencing. *EMBO J.* **21**, 1121–1131.
106. Rine, J., and Herskowitz, I. (1987). Four genes responsible for a position effect on expression from *HML* and *HMR* in *Saccharomyces cerevisiae*. *Genetics* **116**, 9–22.
107. Brachmann, C. B., Sherman, J. M., Devine, S. E., Cameron, E. E., Pillus, L., and Boeke, J. D. (1995). The SIR2 gene family, conserved from bacteria to humans, functions in silencing cell cycle progression and chromosome stability. *Genes Dev.* **9**, 2888–2902.
108. Imai, S. I., Armstrong, C. M., Kaeberlein, M., and Guarente, L. (2000). Transcriptional silencing and longevity protein Sir2 is an NAD-dependent histone deacetylase. *Nature* **403**, 795–800.
109. Aparicio, O. M., Billington, B. L., and Gottschling, D. E. (1991). Modifiers of position effect are shared between telomeric and silent mating-type loci in *S. cerevisiae*. *Cell* **66**, 1279–1287.
110. Smith, J. S., Brachmann, C. B., Pillus, L., and Boeke, J. D. (1998). Distribution of a limited Sir2 protein pool regulates the strength of yeast rDNA silencing and is modulated by Sir4p. *Genetics* **149**, 1205–1219.
111. Hecht, A., Laroche, T., Strahl-Bolsinger, S., Gasser, S. M., and Grunstein, M. (1995). Histone H3 and H4 N-termini interact with *SIR3* and *SIR4* proteins: A molecular model for the formation of heterochromatin in yeast. *Cell* **80**, 583–592.
112. Maison, C., Bailly, D., Peters, A. H. F. M., Quivy, J.-P., Roche, D., Taddei, A., Lachner, M., Jenuwein, T., and Almouzni, G. (2002). Higher-order structure in pericentric heterochromatin

involves a distinct pattern of histone modification and an RNA component. *Nat. Genet.* **30**, 329–334.
113. Volpe, T. A., Kidner, C., Hall, I. M., Teng, G., Grewal, S. I. S., and Martienssen, R. A. (2002). Regulation of heterochromatic silencing and histone H3 lysine-9 methylation by RNAi. *Science* **297**, 1833–1837.
114. Cogoni, C., and Macino, G. (2000). Post-transcriptional gene silencing across kingdoms. *Genes Dev.* **10**, 638–643.
115. Matzke, M., Matzke, A. J. M., and Kooter, J. M. (2001). RNA: Guiding gene silencing. *Science* **293**, 1080–1083.
116. Kelley, R. L., and Kuroda, M. I. (2000). The role of chromosomal RNAs in marking the X for dosage compensation. *Curr. Opin. Genet. Dev.* **10**, 555–561.
117. Mutskov, V. J., Farrell, C. M., Wade, P. A., Wolffe, A. P., and Felsenfeld, G. (2002). The barrier function of an insulator couples high histone acetylation levels with specific protection of promoter DNA from methylation. *Genes Dev.* **16**, 1540–1554.
118. Litt, M. D., Simpson, M., Gaszner, M., Allis, C. D., and Felsenfeld, G. (2001). Correlation between histone lysine methylation and developmental changes at the chicken beta-globin locus. *Science* **293**, 2453–2455.
119. Litt, M. D., Simpson, M., Recillas-Targa, F., Prioleau, M. N., and Felsenfeld, G. (2001). Transitions in histone acetylation reveal boundaries of three separately regulated neighboring loci. *EMBO J.* **20**, 2224–2235.
120. Fourel, G., Boscheron, C., Revardel, E., Lebrun, E., Hu, Y.-F., Simmen, K. C., Muller, K., Li, R., Mermod, N., and Gilson, E. (2001). An activation-independent role of transcription factors in insulator function. *EMBO Rep.* **2**, 124–132.
121. Donze, D., Adams, C. R., Rine, J., and Kamakaka, R. T. (1999). The boundaries of the silenced HMR domain in Saccharomyces cerevisiae. *Genes Dev.* **13**, 698–708.
122. Donze, D., and Kamakaka, R. T. (2001). RNA polymerase III and RNA polymerase II promoter complexes are heterochromatin barriers in Saccharomyces cerevisiae. *EMBO J.* **20**, 520–531.
123. Bi, X., and Broach, J. R. (2001). Chromosomal boundaries in *S. cerevisiae*. *Curr. Opin. Dev. Biol.* **11**, 199–204.
124. Bi, X., and Broach, J. R. (1999). UASrpg can function as a heterochromatin boundary element in yeast. *Genes Dev.* **13**, 1089–1101.
125. Tartof, K. D., Hobbs, C., and Jones, M. (1984). A structural basis for variegating position effects. *Cell* **37**, 869–878.
126. Martin, D. I. K. (2001). Transcriptional enhancers—on/off gene regulation as an adaptation to silencing in higher eukaryotic nuclei. *Trends Genet.* **17**, 444–448.
127. Aparicio, O. M., and Gottschling, D. E. (1994). Overcoming telomeric silencing: A transactivator competes to establish gene expression in a cell cycle-dependent way. *Genes Dev.* **8**, 1133–1146.
128. Ahmad, K., and Henikoff, S. (2001). Modulation of a transcription factor counteracts heterochromatic gene silencing in Drosophila. *Cell* **104**, 839–847.
129. Sun, F.-L., Cuaycong, M. H., Craig, C. A., Wallrath, L. L., Locke, J., and Elgin, S. C. R. (2000). The fourth chromosome of *Drosophila melanogaster*: Interspersed euchromatic and heterochromatic domains. *Proc. Natl. Acad. Sci. USA* **97**, 5340–5345.
130. Tamaru, H., and Selker, E. U. (2001). A histone H3 methyltransferase controls DNA methylation in *Neurospora crassa*. *Nature* **414**, 277–283.
131. Gendrel, A.-V., Lippman, Z., Yordan, C., Colot, V., and Martienssen, R. (2002). Dependence of heterochromatic histone H3 methylation patterns on the *Arabidopsis* gene *DDM1*. *Science* **297**, 1871–1873.

Index

A

Abiotic stress resistance, PM H$^+$-ATPase and, 211
Acetylated histone isoforms, heterochromatin and, 283–284
Active repression/passive repression, CDP/Cux/Cut proteins and, 24
ADAR1 proteins
 adenosine deaminase (editase) domain and, 140
 antiviral defense and, 140
 dsRBM(s) (double-stranded binding-motif) proteins and, 140
ADAR2 proteins
 adenosine deaminase (editase) domain and, 140
 dsRBM(s) (double-stranded binding-motif) proteins and, 140
Adenosine deaminase (editase) domain
 ADAR1/ADAR2 proteins and, 140
 properties of, 140
Adenosine triphosphate. *See* ATP (adenosine triphosphate)
AID (activation-induced cytidine deaminase), DSB(s) (double-strand breaks) formation and, 161
Akt, 15, 17
 cancer and, 15
 IKK and, 15
 NF-κB activation/regulation and, 15
Analogue hypothesis
 evidence supporting, 91–93
 Moffat, J. G./Tate, W. P. and, 91
 molecular mimicry and, 83
 prediction of, 99
 RNA molecules and, 83, 91
 tRNA and, 83, 91
Angiogenesis
 chemokines associated with, 2
 ELR motif and, 2
Anticodon, bacterial RF, 103–105
Anticodon loops, tRNA and, 89–90, 104

Antiviral defense, PKR (protein kinase, RNA dependent) and, 140–141
Aquifex aeolicus, RNase III domain and, 139
Arabidopsis, heterochromatin and, 278
Archebacteria, dsRBM(s) (double-stranded RNA-binding motif) proteins and, 124
Artemis, DNA-PKcs interaction with, 188–189, 190
Asthma, chemokines and, 9
Atherosclerosis, chemokines and, 9
ATP (adenosine triphosphate)
 homologous DNA pairing/strand exchange and, 171
 PM H$^+$-ATPase and, 204
 RAD54 and, 175–176
 reverse gyrase and, 49–50
 Walker mutants and, 175

B

Bacillus stearothermophilis, ribosomes and, 85
Bacterial termination complex
 RF, bacterial decoding and, 106
 specific interactions within, 106–111
Bacterial/viral infections, chemokines and, 9
Barr body X chromosome
 heterochromatin and, 276
 lysine 9-methylated H3 enriched, 285
Baumann, P., NHEJ (nonhomologous endjoining) and, 187
BIR (break-induced replication)
 D-loop (DNA joint) and, 165
 DNA and, 177
BLAST tool
 CK2α gene, 251
 CK2α' gene (CSNK2A2), 251
 CK2β gene (CSNK2B) and, 251
 dsRBM search and, 126
 human protein kinase CK2 genes and, 251
Blue light receptors, plant PM H$^+$-ATPase and, 216

Break-induced replication. *See* BIR
    (break-induced replication)
Brown^{Dominant} (bw^D) mutation
    Drosophilia and, 281

C

CAAT boxes
    CK2α gene and, 253
    CK2α' gene (CSNK2A2) and, 257
    CK2β gene (CSNK2B) and, 257
Caenorhabditis elegans (nematode worm),
    dsRBM(s) (double-stranded
    binding-motif) proteins and, 125, 126
CAF proteins, Dicer proteins and, 136
Cancer cells, pathways for survival/
    proliferation of, 15, 16
Cancers, Akt, up-regulation in, 15
CAP (catabolite gene activator protein),
    DNA binding and, 39
Casein kinase II, protein kinase CK2m,
    replaced by, 240
Catabolite gene activator protein. *See* CAP
    (catabolite gene activator protein)
CBP
    human, 18
    transcription induction and, 18, 19
    transcriptional activation and, 19
CDKs (cyclically activated protein kinases),
    protein kinase CK2 and, 244
CDP/Cux/Cut proteins, 23
    active repression/passive repression of, 24
    CXCL1 transcription and, 23
    DNA-binding and, 23
    down-modulation of, 24
    PARP-1 (Poly(ADP-ribose) polymerase-1),
        displaced by, 25
    repression activity of, 24–25
Cell elongation, PM H$^+$-ATPase and, 211
Cell-cycle
    engine control of, 264–266
    phases of, 262–263
    progression.exit genes and, 264–266
    protein kinase CK2 and, 262–263, 264–266
Cellular Tyrosine Recombinases: The
    Thermophilic Xer Proteins, 69–70, 73
    chromosomal metabolism and, 69
    XerC/XerD protein, homologues of, 69

Chemokines
    angiogenesis and, 2
    asthma and, 9
    atherosclerosis and, 9
    bacterial/viral infections and, 9
    binding of, 5, 6
    CXC, 6, 7, 8, 9, 10, 11–12
    cytokines and, 5, 7
    definition of, 2, 7
    differential expression of, 10–11
    diseases/infections/viruses associated
        with, 8–10
    G-proteins and, 5, 6
    leukocyte recruitment/activation
        and, 7
    receptors of, 5–7
    regulation of, 5
    selectivity of, 6
    structure of, 5
    tumors and, 9–10
    types of, 2–4
    wound healing/diseases and, 7–10
Cis-Spreading, heterochromatin and, 280–282
CK2. *See* Protein kinase CK2
CK2α'gene
    BLAST tool and, 251
    CAAT boxes, presence of in, 253
    Ets1 motif in, 258, 259, 260
    GC boxes, occurrence of in, 253
    human protein kinase CK2 genes and,
        244–245, 247, 251
    introns/exons of, 248
    protein kinase CK2, overexpression and,
        244–245
    single transcription start site of, 257
    TATA box, lack of and, 253
    theileriosis (East coast fever) and, 253
    transcription start sites of, 253
CK2α' gene (CSNK2A2)
    BLAST tool and, 251
    CAAT boxes, presence of in, 257
    Ets1motif in, 261
    GC boxes, occurrence of in, 257
    human protein kinase CK2 genes and,
        247–248, 251
    introns of, 248
    Sp1 motif in, 261
    TATA boxes, lack of, 257
    transcriptionally relevant upstream
        sequence of, 257

INDEX

303

CK2β gene (CSNK2B)
  BLAST tool and, 251
  CAAT boxes and, 257
  Ets1 motif in, 258, 260
  GC boxes, occurrence of in, 257
  human protein kinase CK2 genes and, 244–245, 247, 248
  introns of, 248
  theileriosis (East coast fever) and, 261
  transcription sites of, 257–258
ClustalX tool, dsRBM(s) (double-stranded RNA-binding motif) and, 126
Codon recognition, NIK (NF-κB-inducing kinase) and, 99
COPII (coat protein complex II)
  ER-to-Golgi vesicle formation and, 219
  PM H^+-ATPase and, 219–220
  Saccharomyces pombe (SpPMA1) and, 223
  ScPMA1-DIG association and, 223
Cox, M. M., RAD51-mediated homologous DNA pairing/strand-exchange reaction and, 173
Cryoelectron/cryoimmunoelecton microscopy, ribosomes and, 85, 90–91
C-terminal domain/"chromo shadow domain"
  HP1 (heterochromatin protein 1) and, 288
  reverse gyrase and, 288
C-terminus, plant PM H^+-ATPase and, 214–215
CXC, chemokines and, 6, 7, 8, 9, 10, 11–12
CXCL1 protein, 1, 2, 3, 6, 7, 8, 9, 10, 11, 12
  melanoma cells and, 17
  NF-κB and, 13
  transcription induction and, 13
  transcription regulation and, 11–12
CXCL1 transcription
  CBP in, 18–20
  CDP/Cux/Cut proteins in, 23
  PARP-1 (Poly(ADP-ribose) polymerase-1) in, 20–22
CXCL8 protein, transcription regulation and, 11
Cyclically activated protein kinases. *See* CDKs (cyclically activated protein kinases)
Cytokines, 5
  chemokines and, 5, 7
  interleukin-1 (IL-1)[1], 5
  modulation of transcription of, 12
  NF-κB dependent, 10
  proinflammatory (TNF-α, IL-1), 7, 13

  transcription factors of, 10–11
  tumor necrosis factor-α (TNF-α), 5

D

DEAD/DEAH helicase domains, 139
  RHA proteins and, 139
Detergent-insoluble glycolipid-enriched domain. *See* DIG (detergent-insoluble glycolipid-enriched domain)
Detergent-resistant membrane. *See* DRM (detergent-resistant membrane)
Diacidic motif, plant PM H^+-ATPase and, 220, 221
Dicer proteins, 134, 136, 139
  CAF proteins, plant homologue of, 136
  domains of, 136
DIG (detergent-insoluble glycolipid-enriched domain), 222, 223
  ScPMA1 incorporated into, 223
Dimer formations
  Hjr (Holliday junction resolving and), 67
  RuvC and, 67
D-loop (DNA joint). *See also* HR (homologous recombination)
  BIR (break-induced replication) and, 165
  homologous DNA pairing and, 163
  HR (homologous recombination) and, 163, 164–166
  Orr-Weaver, T. L. and, 164
  processing/recombination models of, 164–166
  RAD 54 and, 175
  Rothstein, R. J. and, 164
  SDSA (DNA synthesis-dependent single-strand annealing), 164
  Stahl, F. W. and, 164
  Szostak, J. W. and, 164
Dnl4/Lif1, Pol4 and, 184–185
DNA
  BIR (break-induced replication) of, 177
  differing cutting patterns of, 67
  Drosophilia and, 283
  heterochromatin and, 283
  HJR (holiday junction resolvases) and, 59
  HR (homologous recombination) and, 170
  Hyperthermophilic Topoisomerase (Type IA, Archaea) and, 45, 46

DNA (cont.)
  Ku70/Ku80, topological linking to, 187
  PARP-1 (Poly(ADP-ribose) polymerase-1) and, 20, 21, 22
  RAD52, end-processing of, 167, 168
  replication forks, stalling of, 160
  reverse gyrase and, 46, 49, 50, 51
  Saccharomyces cervisiae (bakers' yeast) and, 160
  single-stranded, HR (homologous recombination) and, 170
DNA, homologous pairing. *See* Homologous DNA pairing/strand exchange
DNA joint. *See* D-loop (DNA joint)
DNA ligase IV. *See also* DSB(s) (double-strand breaks)
  DNA-PK and, 186
  Ku70/Ku80 and, 187
  NHEJ (nonhomologous endjoining) and, 181
  NHEJ genes and, 181
DNA synthesis-dependent single-strand annealing. *See* SDSA (DNA synthesis-dependent single-strand annealing)
DNA-binding
  CAP, 39
  CDP/Cux/Cut proteins and, 23
  PARP-1 and, 20, 21, 22
DNA-PK
  DNA ligase IV and, 186
  NHEJ genes and, 181
DNA-PKcs
  Artemis, interaction with, 188–189, 190
  autophosphorylation of, 188
  end-bridging by, 188
  hRad50/hMre11/NBS1 complex and, 190
DNA-PKcs/Artemis complex, 188
Domain I/II, tip of
  eRFI and, 104
  GGQ motif and, 103
  NIKS motif and, 104
Domain IV models
  EF-G and, 98
Double-stranded RNA-binding motif proteins. *See* dsRBM(s) (double-stranded RNA-binding motif) proteins
Down-modulation, CDP/Cux/Cut proteins and, 24
Down-regulation, PM H$^+$-ATPase and, 212
DRBD. *See* dsRBM(s) (double-stranded RNA-binding motif) proteins

DRBM. *See* dsRBM(s) (double-stranded RNA-binding motif) proteins
DRM(s) (detergent-resistant membrane), 222
  unassigned dsRBM-containing protein families and, 145–146
Drosophilia
  brown^{Dominant} (bw^D) mutation of, 281
  DNA, repetitious nature and, 283
  dsRBM(s) (double-stranded binding-motif) proteins and, 125, 126, 129
  Henikoff, S. and, 281
  heterochromatin and, 278, 279, 280, 281, 282, 283, 287–288
  HP1 (heterochromatin protein 1) and, 287–289
  PEV and, 276, 278, 281, 288
  Sass, G. L. and, 281
  SU(VAR)3-9 protein and, 289
Drosophilia melanogaster (fruit fly)
  dsRBM(s) (double-stranded binding-motif) proteins and, 125, 126, 129, 145
DSB(s) (double-strand breaks), 159. *See also* HR (homologous recombination); RAD52
  AID (activation-induced cytidine deaminase) and, 161
  Hdf1/Hdf2 and, 183
  HR (homologous recombination) and, 159, 162
  immunoglobin gene switching and, 161
  introduction to, 160
  meiotic breaks and, 161–162
  NHEJ (nonhomologous endjoining) and, 159, 178, 186
  recombination events induced by, 162, 163
  RSSs (recombination signal sequences) and, 161
  Saccharomyces cerevisiae (bakers' yeast) and, 161, 162
  XRCC (X-Ray Cross Complementing), repair of, 185
dsRBD. *See* dsRBM(s) (double-stranded RNA-binding motif) proteins
dsRBM(s) (double-stranded RNA-binding motif) proteins. *See also* Dicer proteins; RNase III domain; RNase IIIA domain; RNase IIIB domain
  archebacteria and, 124, 146
  biochemical/cytological properties of, 125
  Caenorhabditis elegans (nematode worm) and, 125, 126

completely sequenced genomes of, 125
domain structures of, 129, 131
DRMs and, 145–146
Drosophilia melanogaster (fruit fly) and, 125, 126, 129
E. coli (escherichia coli), 125, 126
EBI (European Bioinformatics Institute) and, 125
Homo sapiens (human) and, 125, 126
human genome and, 131–132
introduction to, 124–125
metazoans and, 129
NRF (NF-κB repressing factor), 144
occurrence, frequency, conservation of, 126–129
origin/evolution of, 148–150
other names for, 124
PKR (protein kinase, RNA dependent) and, 140
plants and, 146–147
protein function of, 150–151
RHA proteins and, 139
RNase IIIA and, 134, 136
Saccharomyces cervisiae (bakers' yeast) and, 125, 126, 145
scope of, 151–152
similarities among, 133
Staufen proteins and, 141
viruses and, 147–148
DT40 chicken cells, Ku70/Ku80 and, 187
Dynan, W. S., HeLa extracts and, 189

E

E. coli (escherichia coli), 39, 44
alignment of Xer-like proteins with, 70
biochemical/structural date on, 54
dsRBM(s) (double-stranded binding-motif) proteins and, 125, 126
GGQ motif and, 100, 101
RNase III domain and, 139
RuvC and, 63, 65
T. maritima and, 44
Xer site-specific recombination system and, 69
XerC/XerD protein and, 70
EBI (European Bioinformatics Institute), 125
dsRBM(s) (double-stranded binding-motif) proteins documented by, 125
Interpro database and, 125, 126

Pfam database and, 125
PRINTS database and, 125
ProDom database and, 125
SMART database and, 125
EF-G, 94, 95, 97
domain IV models of, 98
RF3, homologous to, 109, 113
EF-Tu, 94, 97
RF3, homologous to, 109
Eggler, A. L., RAD51-mediated homologous DNA pairing/strand-exchange reaction and, 173
Elongation factors
EF-G, 94, 95, 97
EF-Tu, 94, 97
ELR motif, chemokines containing, 2
End-bridging
DNA-PKcs and, 188
Ku70/Ku80 and, 189
Rad50/Mre11/Xrs2 complex and, 188
Endonucleolytic activity, RNase III domain and, 134
ER export motifs, PM H^+-ATPase and, 221
ER quality control, PM H^+-ATPase and, 219
eRF1 structure
bacterial polypeptide release factor 2/eukaryotic, 101
Domain I/II, tip of and, 104
features of, 99
NIK (NF-κB-inducing kinase) and, 100
RF (polypeptide chain release factors) and, 99
RF, structure of and, 99
RF2 function and, 99
ribosomes, active center and, 99
three-dimensional coordinates of, 100
trinucleotide codons in, 112
tRNA and, 99, 101
ER-to-Golgi vesicle formation, COPII (coat protein complex II) and, 219
Escherichia coli. See *E. coli (escherichia coli)*
E-site/P-site, tRNA and, 107
Ets1 motif
CK2β gene(CSNK2B), 258, 260
CK2α'gene (CSNK2A2) and, 261
human protein kinase CK2 genes and, 258, 259, 260, 261
TATA-less genes and, 259
Eubacterial/eukaryotic decoding, RF (polypeptide chain release factors) and, 111

Euchromatin domains, heterochromatin and, 290–291
Eukaryotes
   heterochromatin and, 283
   NHEJ, mechanism/function in, 179–191
   single-cell, RAD52 epistasis group and, 178
European Bioinformatics Institute. *See* EBI (European Bioinformatics Institute)

F

Fink, G. R., *Saccharomyces cerevisiae* (bakers' yeast) and, 205
14-3-3 proteins, plant PM H^+-ATPase, 214, 215, 216, 217
Fusicoccin binding site, plant PM H^+-ATPase and, 214, 215

G

GC boxes
   CK2α gene and, 253
   CK2α′ gene (CSNK2A2) and, 257
   CK2β gene (CSNK2B) and, 257
Gene activation, heterochromatin and, 278–279
Gene silencing, heterochromatin and, 276, 278
GGQ motif
   domain II and, 103
   E. coli and, 100, 101
   radicals generated close to, 108
   RF (polypeptide chain release factors) and, 92, 99, 100
   SPF motif and, 102
G-proteins
   chemokines and, 5, 6
   translational, 109
GTF, RF3, form of, 113
GTP hydrolysis, 113
Guard cells, PM H^+-ATPase and, 210, 216
Gyrase (Type IIa, Eubacteria and Some Archaea)
   enzymatic properties of, 54, 55
   hypotheses on physiological role of, 56–57
   phylogenic position/sequence organization of, 53–54, 55, 56
   T. maritima and, 56
   temperature and, 57

H

Haloarcula marismortui, ribosomes and, 85, 86
HDAC corepressor proteins, NF-κB/NF-κB activation/regulation and, 14
Hdf1/Hdf2
   DSB(s) (double-strand breaks) and, 183
   Ku70/Ku80 and, 187
   RAD50/Mre11/Xrs2 complex and, 183
HeLa extracts, Huang, J./Dynan, W. S. and, 189
Helix-hairpin-helix motif. *See* HhH motif
Henikoff, S., *Drosophilia* and, 281
Heterochromatin, 275
   acetylated histone isoforms, 283–284
   activator/silencer function of, 278
   Arabidopsis and, 278
   cis-Spreading and, 280–282
   DNA structure and, 283
   Drosophilia and, 278, 279, 280, 281, 282, 283, 287–288
   euchromatin domains and, 290–291
   eukaryotes and, 283
   gene activation and, 278–279
   gene silencing and, 276, 278
   H3 enriched, 285
   histone code hypothesis and, 286
   histone modifications of, 283, 285–287
   nonhistone proteins and, 287–288
   nucleosome-DNA interactions in, 283, 287
   PEV (position-effect variegation) and, 276, 278
   Rag-1/Tdt genes next to, 282
   setting up domains of, 290–291
   trans-Inactivation and, 280–282
Heterochromatin protein 1. *See* HP1 (heterochromatin protein 1)
HhH motif, 52, 53
   Topoisomerase V (Type IB, Archaea) and, 51, 52, 53
Histone
   heterochromatin and, 283, 285–287
   histone code hypothesis and, 286
   modifications of, 283, 285–287
   trans-histone regulation and, 286
Hjc (Holliday junction cleavage)
   HJR, archael activity and, 61–62
   mutants of, 65–66
   recombinant, 65
   RuvC and, 66
   sequence alignments of, 65

INDEX

Hje (Holliday junction endonuclease)
  HJR, archael activity and, 62
HJR (Holliday junction resolvases)
  biochemical/enzymatic properties of, 63–67
  classes/superfamilies of, 61, 63
  DNA repair and, 59
  Hjc/Hje/Hjr, activity of, 61–62
  P. furiosus, 62, 65, 66, 67
  phylogeny of, 61–63
  S. solfataricus, 62, 65, 66, 67
  structure of archaeal, 67
Hjr (Holliday junction resolving)
  dimer formations in, 67
  HJR, archaeal activity and, 62
HMM tool, dsRBM search and, 126
Holliday junction cleavage. *See* Hjc (Holliday junction cleavage)
Holliday junction endonuclease. *See* Hje (Holliday junction endonuclease)
Holliday junction resolvases. *See* HJR (Holliday junction resolvases)
Holliday junction resolving. *See* Hjr (Holliday junction resolving)
Homo sapiens (human), dsRBM(s) (double-stranded binding-motif) proteins and, 125, 126
Homologous DNA pairing/strand exchange, 170
  ATP (adenosine triphosphate) and, 171
  DNA branch migration (strand exchange) and, 173
  presynaptic phase of, 171–172
  RAD51 and, 171
  synapsis and, 172
Homologous recombination. *See* HR (homologous recombination)
HP1 (heterochromatin protein 1), 287–288
  C-terminal domain/"chromo shadow domain" of, 288
  Drosophilia and, 287–288
  two-domain structure of, 288
HR (homologous recombination), 159. *See also* Homologous DNA pairing/strand exchange
  D-loop (DNA joint) and, 163, 164–166
  DSB(s) (double-strand breaks) and, 159, 162
  initiation of, 162–164
  single-stranded DNA and, 170
hRad50/hMre11/NBS1 complex
  DNA-PKcs and, 190

NHEJ (nonhomologous endjoining) and, 189–191, 192
Huang, J., HeLa extracts and, 189
Human protein kinase CK2 genes, 245. *See also* CK2α *gene;* CK2α′ *gene;* CK2β *gene*
  BLAST tools and, 251
  CK2 loci of, 245
  CK2α gene and, 247, 251
  CK2α′ gene (CSNK2A2) and, 247–248, 251
  CK2β gene (CSNK2B) and, 244–245, 247, 248
  distribution/transcripts of, 250–251
  Ets1 motif in, 258, 259, 260, 261
  introns/exon of, 245–246, 248
  mRNA and, 250, 251
  NF-κB and, 256
  Sp1 motif in, 258, 259, 260
  transcript-based phylogeny of, 251, 253
  transcriptional coordination of, 258–261
Hyperthermophilic Topoisomerase III (Type IA, Archaea), 42, 45–46
  DNA and, 45, 46

I

IF1 (initiation factor 1)
  molecular mimicry and, 95, 96, 97
  tRNA and, 95
IκB proteins
  NF-κB and, 13
  types of, 13
IKK, NF-κB and, 14, 15
Immunoglobin gene switching, DSB(s) (double-strand breaks) and, 161
Inflammatory response, NF-κB activation/regulation in, 13
Initiation factor 1. *See* IF1 (initiation factor 1)
Inman, R. B., RAD51-mediated homologous DNA pairing/strand-exchange reaction and, 173
Interferon, PKR (protein kinase, RNA dependent), induced by, 140
Interleukin-1 (IL-1)[1], cytokines and, 5
Internal pH homeostasis, PMH$^+$-ATPase and, 211
Interpro database, EBI and, 125, 126
Introns/exons, human protein kinase CK2 genes and, 245–246, 248
IUR element, transcription regulation and, 12

K

Kanadaptin (kidney anion exchanger adaptor protein), 144–145
Keratinocytes, 8
Kidney anion exchanger adaptor protein. *See* Kanadaptin (kidney anion exchanger adaptor protein)
Kielland-Brandt, M. C., *Saccharomyces cerevisiae* (bakers' yeast) and, 205
KRKFEEFKK motif, RF3 and, 113
Ku70/Ku80
 DNA Ligase IV/XRCC4, interaction with, 187
 DNA, linked to, 187
 DT40 chicken cells and, 187
 end-bridging and, 189
 Hdf1/Hdf2 and, 187
 NHEJ (nonhomologous endjoining) and, 189
 Rad50/Mre11/Xrs2 complex and, 183

L

Last universal common ancestor (LUCA), 48
Leukocytes
 chemokines and, 7
 migration of, 7
 recruitment/activation of, 7
 wound healing and, 7–8
Lysine 9-methylated H3
 Barr body X chromosome and, 285
 di-/tri-methylated, 285–286

M

Melanoma cells
 CXCL1 proteins expressed in, 17
 NF-κB-inducing kinase (NIK) in, 15, 17–18
Meristemic regions, PM H$^+$-ATPase and, 211
Metazoans, dsRBM(s) (double-stranded RNA-binding motif) proteins and, 129
Moffat, J. G., analogue hypothesis and, 91
Molecular mimicry, 83
 analogue hypothesis and, 83
 class I RFs and, 97
 class II RFs and, 97
 IF1 (initiation factor 1) and, 95, 96, 97
 ribosome, active center and, 94
 tRNA and, 86, 88, 89, 90, 94
MRE11
 multifunctional nature of, 167
 RAD50/Mre11/Xrs2complex and, 182
mRNA
 human protein kinase CK2 genes and, 250, 251
 nucleotides of, 107
 RF, bacterial decoding and, 106
 ribosomes and, 86, 107
 Staufen proteins and, 141

N

N. plumbaginifolia, PM H$^+$-ATPase and, 221
NEJ1, NHEJ genes and, 182
Neurospora crassa, PM H$^+$-ATPase and, 208, 209
NF90 (nuclear factor 90)
 properties of, 143
 spermatid perinuclear protein (SPNR) and, 143–144
 spermatogenesis and, 143
NF-κB. *See also* NF-κB activation/regulation
 acetylation of, 20
 activation/regulation of, 13–15
 composition of, 13
 CXCL1 protein and, 13
 cytokines and, 10
 HDAC corepressor proteins and, 14
 human protein kinase CK2 genes and, 256
 IκB proteins and, 13
 IKK and, 14, 15
 inflammation in, 13
 melanoma cells and, 15, 17–18
 pathogenic stimuli of, 14
 transactivation function of, 14
 transcription induction and, 13
 transcription regulation and, 11, 12
NF-κB activation/regulation
 Akt and, 15
 HDAC corepressor proteins in, 14
 IκB proteins, 13
 IKK, 14, 15
 inflammatory response in, 13
 PARP-1 (Poly(ADP-ribose) polymerase-1) in, 22
 Ras genes and, 17

INDEX

NF-κB repressing factor. *See* NRF (NF-κB repressing factor)
NF-κB-inducing kinase. *See* NIK (NF-κB-inducing kinase)
NHEJ (nonhomologous endjoining), 159
  Baumann, P. and, 187
  DNA ligase IV and, 181
  DNA-PK and, 181
  DSB(s) (double-strand breaks) and, 159, 178, 186
  eukaryotes and, 179–191
  hRad50/hMre11/NBS1, role in, 189–191, 192
  Ku70/Ku80 and, 189
  mammals, molecular mechanisms of, 187–191
  NEJ1 and, 182
  Pol4, role of in, 183–184
  Saccharomyces cerevisiae (bakers' yeast) and, 179, 182–185
  V(D)J recombination and, 161, 186, 189
  West, S. C. and, 187
NHEJ genes
  DNA ligase IV and, 181
  DNA-PK and, 181
  mammalian, 185–186
  NEJ1 and, 182
  Saccharomyces cerevisiae, 181–182
  XRCC4 and, 181, 185
NIK (NF-κB-inducing kinase), 17–18, 100, 104
  codon recognition and, 99
  Domain I/II, tip of and, 104
  eRFI and, 100
  melanoma and, 15, 17–18
  up-regulation of NF-κB activity and, 18
Nonhistone proteins, heterochromatin and, 287–288
Nonhomologous endjoining. *See* NHEJ (nonhomologous endjoining)
NRF (NF-κB repressing factor), dsRBM(s) (double-stranded binding-motif) proteins and, 144
Nuclear factor 90. *See* NF90 (nuclear factor 90)
Nucleosome-DNA interactions, heterochromatin and, 283, 287
Nucleotide-binding (N) domain, PM H$^+$-ATPase and, 206

O

Oligomerization
  PM H$^+$-ATPase and, 209, 224–225
Orr-Weaver, T. L., D-loop (DNA joint) and, 164

P

P. furiosus, HJR (Holliday junction resolvases) and, 62, 65, 66, 67
PARP-1 (Poly(ADP-ribose) polymerase-1), 20
  CDP/Cux/Cut proteins displacement of, 25
  CXCL1 transcription and, 20
  DNA-binding and, 20, 21, 22
  functions of, 21
  NF-κB activation/regulation and, 22
  PIC (preinitiation complex) and, 22
  RNAPII (RNA polymerase II) and, 22
Peptidyl-tRNA hydrolysis, RF (polypeptide chain release factors)/RF2 and, 112
PEV (position-effect variegation)
  dominant, 281
  Drosophilia and, 276, 278, 281, 288
  enhancer of, 279
  heterochromatic, 276, 278
  recessive, 280
  Su(VAR)3-9, dominant suppressor of, 289
  suppressor of, 279
Pfam database, EBI (European Bioinformatics Institute) and, 125
PIC (preinitiation complex), PARP-1 (Poly(ADP-ribose) polymerase-1) and, 22
PKR (protein kinase, RNA dependent), 140–141
  antiviral defense and, 140–141
  dsRBM(s) (double-stranded binding-motif) proteins and, 140
  interferon induced, 140
  substrate, 141
Plant PM H$^+$-ATPase, 214–216
  blue light receptors and, 216
  C-terminus of, 214–215
  diacidic motif and, 220, 221
  14-3-3 proteins and, 214, 215, 216, 217
  fusicoccin binding site and, 214, 215

Plant PM H$^+$-ATPase (cont.)
    phosphorylation of, 216
    posttranscriptional modification of, 217
    Saccharomyces pombe (ScPMA1) and, 218
PM H$^+$-ATPase. *See also* Plant
        PM H$^+$-ATPase; PM H$^+$-ATPase,
        trafficking
    abiotic stress resistance and, 211
    ATP (adenosine triphosphate) and, 204
    cell elongation in, 211
    !COPII (coat protein complex II) and,
        219–220
    down-regulation in, 212
    enzymology of, 207–209
    ER export motifs in, 221
    ER quality control and, 219
    five subfamilies of, 206
    14-3-3 proteins and, 214, 215, 216, 217
    guard cells and, 210, 216
    internal pH homeostasis and, 211
    introduction to, 204–205
    meristemic regions and, 211
    N. plumbaginifolia and, 221
    Neurospora crassa and, 208, 209
    nucleotide-binding (N) domain of, 206
    oligomerization and, 209, 224–225
    organization/expression, 205–206
    other regulatory mechanisms of, 217–218
    phosphorylation (P) domain of, 206
    posttranslational regulation of, 212–218
    P-type ATPase superfamily and, 205
    rafts and, 222–223, 224–225
    regulation, plant, 214–216
    Saccharomyces cerevisiae and, 205, 217
    Saccharomyces pombe (SpPMA1)
        and, 204
    secretory pathway of, 218
    topology/structure of, 206–207
    translational regulation of, 212
    transport of, 209–211
PM H$^+$-ATPase, trafficking, 218–225
Pol4
    Dn14/Lif1 and, 184–185
    NHEJ (nonhomologous endjoining)
        and, 183–184
Polypeptide chain release factors. *See* RF
    (polypeptide chain release factors)
Poly (ADP-ribose) polymerase-1. *See* PARP-1
    (Poly (ADP-ribose) polymerase-1)

Position-effect variegation. *See* PEV
    (position-effect variegation)
Preinitiation complex. *See* PIC
    (preinitiation complex)
Pretranslocation/posttranslocation state,
    ribosomes, active center and, 88, 90
PRINTS database, EBI (European
    Bioinformatics Institute) and, 125
ProDom database, EBI (European
    Bioinformatics Institute) and, 125
Protein kinase CK2. *See also* Human protein
        kinase CK2 genes
    casein kinase II, replacing with, 240
    CDKs (cyclically activated protein kinases)
        and, 244
    cell division and, 266
    cell-cycle engine control and, 264–266
    cell-cycle phases of, 262–263
    cell-cycle progression/exit genes
        and, 264–266
    CK2α overexpression and, 244–245
    deviation patterns of, 266
    discovery of, 241
    flexible signal-mediated cellular localization
        dynamics of, 243
    functional links of, 262
    gene control and, 243
    human genome of, 239, 242
    metabolic pathway/nutrition supply genes
        and, 263–264
    pathophysiological implications of,
        244–245, 267
    physiological role of, 242–245, 267
    Saccharomyces cerevisiae (bakers' yeast)
        and, 241, 242, 244, 262–263
    survival factor function of, 264
    tetrameric structure of, 242
    theileriosis (East coast fever) and, 244
    transcription, global role of, 267
    transcription regulation and, 240
    Tyr phosphorylating activity of, 242
Protein kinase, RNA dependent. *See* PKR
    (protein kinase, RNA dependent)
Protein synthesis
    genetic code and, 84
    RF (polypeptide chain release factors)/RF3
        and, 93
P-type ATPase superfamily, PM H$^+$-ATPase
    and, 205, 207

R

Rad50/Mre11/Xrs2 complex, 167, 169
  end-bridging mediated by, 188
  Hdf1/Hdf2, required by, 183
  Ku70/Ku80 and, 183
  mRE11, subunit of, 182
  NHEJ (nonhomologous endjoining) and, 181
RAD51, 167, 169–170
  homologous DNA pairing/strand exchange and, 171
  RAD54 interacts with, 172
RAD51-mediated homologous DNA pairing/strand exchange reaction, RPA and, 173–174
RAD51-mediated homologous DNA pairing/strand-exchange reaction, 171, 173–177
  Cox, M. M. and, 173
  Eggler, A. L. and, 173
  Inman, R. B. and, 173
  RAD52 in, 174–175
  RAD55 and, 175
  RAD57 and, 175
RAD52
  DNA end processing and, 167, 168
  group, components of, 166
  RAD51-mediated homologous DNA pairing/strand-exchange reaction and, 174–175
  single-cell eukaryotes of, 178
  SSA (single-strand annealing) and, 178
RAD54, 167, 172
  ATP (adenosine triphosphate) and, 175–176
  D-loop formation and, 175
  RAD51 interacts with, 172
  RAD51-mediated homologous DNA pairing/strand-exchange reaction and, 175–177
  RDH54 (RAD Homologue 54) and, 177
RAD55, RAD51-mediated homologous DNA pairing/strand-exchange reaction and, 175
RAD57, RAD51-mediated homologous DNA pairing/strand-exchange reaction and, 175
RAD59, SSA (single-strand annealing) and, 178
Rafts, PM H$^+$-ATPase and, 222–223
Rag1/Rag2 endonuclease, V(D)J recombination and, 189
Rag-1/Tdt genes, heterochromatin and, 282

Ras genes, NF-κB activation and, 17
RDH54 (RAD Homologue 54), RAD54 and, 177
Recombinases. *See also* Cellular Tyrosine Recombinases: The Thermophilic Xer Proteins; HJR (Holliday junction resolvases); Viral Tyrosine Recombinases: The Fuselloviradae Integrases
  families of, 61
  thermophilic representatives of, 60
  tyrosine, 68
Recombination signal sequences (RSSs), DSB(s) (double-break strands) and, 161
Release factors, polypeptide chain. *See* RFs (release factors)
Reverse gyrase, 42, 46–51
  ATPase activity of, 49–50
  C-terminal domain of, 47
  DNA and, 46, 49, 50, 51
  functions of, 50–51
  origin/evolution of, 48–49
  sequence organization of, 46–48
  structure of, 49–50
  T. maritima and, 49
  temperature and, 48
RF (polypeptide chain release factors), 91
  bacterial, 100, 106
  class I, 95, 96
  class II, 95, 96
  decoding, 106, 107
  eRF1 structure and, 99
  eubacterial/eukaryotic decoding, 111
  GGQ and, 92, 99, 100
  peptidyl-tRNA hydrolysis and, 112
  protein synthesis termination and, 93
  RF, bacterial, "anticodon" and, 103–105
  RF1, 95, 96
  RF2, 95, 96
  RF3, 95, 96, 97
  SPF motif and, 102
  structure of, 99–103
RF (polypeptide chain release factors) rRNA and, 107
RF, bacterial, "anticodon," RF (polypeptide chain release factors) and, 103–105
RF, bacterial coding, mRNA and, 106
RF1, 95, 96
  RF (polypeptide chain release factors) and, 95, 96

RF1 (cont.)
  rRNA and, 108
  tripeptide motifs in, 104
RF2, 95, 96, 102
  eRFI structure and, 99
  peptidyl-tRNA hydrolysis domain of, 108, 109
  RF (polypeptide chain release factors) and, 95, 96
  rRNA and, 108
  tripeptide motifs in, 104
RF3, 95, 96, 97
  bacterial termination of protein synthesis and, 109
  EF-G and, 109, 113
  Ef-Tu and, 109
  function of, 99
  GTF form of, 113
  KRKFEEFKK motif and, 113
  RF (polypeptide chain release factors) and, 95, 96, 97
  ribosomes, active center, interaction with, 109, 113
  ribosomes, interaction with, 109
  Zavialov, A., study of, 111, 113
RHA proteins
  DEAD/DEAH helicase domains and, 139
  dsRBM(s) (double-stranded binding-motif) proteins, domain and, 139
  RHA1 (RNA helicase A) and, 139, 140
  RHA2 and, 139, 140
RHA1 (RNA helicase A), 139, 140
  retroviruses and, 139
  RHA proteins and, 139, 140
RHA2, 139
  dsRBM(s) (double-stranded binding-motif) proteins of, 139
  RHA proteins and, 139, 140
Ribosomes. See also Ribosomes, active center; tRNA
  atomic level structures/subunits of, 85–87
  Bacillus stearothermophilus and, 85
  Cryoelectron/cryoimmunoelecton microscopy and, 85, 90–91
  Haloarcula marismortui and, 85, 86
  mRNA and, 86, 107
  RF3 interaction with, 109
  RNA molecules, components of, 83, 84, 86, 87
  rRNA and, 86, 87, 88, 90, 107

Thermus thermophilis and, 85
  tRNA and, 86, 88, 89, 90
Ribosomes, active center
  biochemical structure of, 88
  characteristic forms of, 88
  eRFI interacts with, 99
  interactions of, 88–89
  molecular mimicry and, 94
  pretranslocation/posttranslocation state of, 88, 90
  RF3 interaction with, 109, 113
  RNA molecules and, 87–90
  rRNA lining in, 90
  three-dimensional space of, 90–93
  tRNA and, 88, 89, 90, 91, 92
RNA helicase A proteins. See RHA proteins
RNA helicase A. See RHA1
RNA helicase. See RHA1, 139
RNA molecules
  analogue hypothesis and, 83, 91
  ribosomes, active center and, 87–90
  ribosomes and, 83, 84, 86, 87
RNAPII (RNA polymerase II), PARP-1 (Poly (ADP-ribose) polymerase-1) and, 22
RNase III domain, 134, 136, 139
  Aquifex aeolicus and, 139
  archaea and, 149
  classes of, 134
  E. coli (escherichia coli) and, 139
  endonucleolytic activity of, 134
  functions of, 134
RNase IIIA domain, dsRBM(s) (double-stranded binding-motif) proteins and, 134, 136
RNase IIIB domain, 136
Rothstein, R. J., D-loop (DNA joint) and, 164
RPA
  RAD51-mediated homologous DNA pairing/strand exchange reaction and, 173–174
  SSA (single-strand annealing) and, 178
rRNA
  RF (polypeptide chain release factors) and, 107
  RF1/RF2 and, 108
  ribosomes, active center and, 90
  ribosomes and, 86, 87, 88, 90, 107
RSSs (recombination signal sequences), DSB(s) (double-strand breaks) and, 161, 162

INDEX

RuvC
  acidic residues of, 65
  dimer formations in, 67
  E. coli, 63, 65
  Hjc, functional homolog of, 66
  T. maritima and, 63

S

S. solfataricus
  HJR (Holliday junction resolvases) and, 62, 65, 66, 67
  SSV1 integrase and, 69
Saccharomyces cerevisiae (bakers' yeast)
  chromatin structure, studies of, 280
  DNA and, 160
  DSB(s) (double-strand breaks) and, 161, 162
  dsRBM(s) (double-stranded binding-motif) proteins and, 125, 126, 145
  Fink, G. R. and, 205
  Kielland-Brandt, M. C. and, 205
  mating type switching and, 162
  NHEJ (nonhomologous endjoining) and, 179, 182–185
  NHEJ genes and, 181–182
  PM H$^+$-ATPase and, 205, 217
  Protein kinase CK2 and, 241, 242, 244, 262–263
  Saccharomyces pombe (ScPMA1) and, 217
  ScPMA1/ScPMA2 and, 217
  Serrano, R. and, 205
Saccharomyces pombe (ScPMA1)
  COPII (coat protein complex II) and, 223
  DIG (detergent-insoluble glycolipid-enriched domain) and, 223
  plant PM H$^+$-ATPase and, 218
  PM H$^+$-ATPase and, 204, 209
  Saccharomyces cerevisiae (bakers' yeast) and, 217
  sphingolipids and, 222, 223
  yeast PM H$^+$-ATPase and, 204, 209
Sass, G. L., *Drosophilia* and, 281
SDSA (DNA synthesis-dependent single-strand annealing), D-loop and, 164
Serrano, R., *Saccharomyces cerevisiae* (bakers' yeast) and, 205
Single-strand annealing. See SSA (single-strand annealing)

SIR2 (silent information regulator 2 protein), 289
SMART database, EBI (European Bioinformatics Institute) and, 125
SON protein, 144
Sp1 motif
  CK2α' gene (CSNK2A2) and, 261
  human protein kinase CK2 genes, 258, 259, 260
  phosphorylation of, 259
Spermatid perinuclear protein (SPNR), NF90 (nuclear factor 90) and, 143–144
Spermatogenesis
  NF90 (nuclear factor 90) and, 143
  Staufen proteins and, 141
  TRBP (TAR RNA-binding protein) and, 143
SPF motif
  GGQ motif and, 102
  RF (polypeptide chain release factors) and, 102
Sphingolipids, *Saccharomyces pombe* (ScPMA1) and, 222, 223
SSA (single-strand annealing), recombination by, 166
  RAD52/RAD59 and, 178
  RPA and, 178
SSV1 integrase
  S. solfataricus and, 69
  Viral Tyrosine Recombinases: The Fuselloviradae Integrases and, 68
Stahl, F. W., D-loop (DNA joint) and, 164
Staufen proteins, 141
  dsRBM(s) (double-stranded RNA-binding motif) proteins and, 141
  mRNA and, 141
  spermatogenesis and, 141
SU(VAR)3-9,
  Drosophilia and, 289
  PEV (position-effect variegation) and, 289
  Survival factor function, protein kinase CK2, 264
Szostak, J. W., D-loop (DNA joint) and, 164

T

T. maritima
  E. coli and, 44
  Gyrase (Type IIa, Eubacteria and Some Archaea) and, 56

T. maritima (cont.)
  properties of, 44
  reverse gyrase and, 49
  RuvC, 63
TAR RNA-binding protein (TRBP),
    spermatogenesis and, 143
TATA boxes
  CK2α′ gene (CSNK2A2) and, 257
  CK2α gene and, 253
TATA-less genes, Ets1 motif and, 259
Tate, W. P., analogue hypothesis and, 91
Termination mechanism, 111–113
Theileriosis (East coast fever)
  CK2β (CSNK2B), lack of and, 261
  CK2α, high levels of and, 261
  protein kinase CK2 and, 244
Thermus thermophilis, ribosomes and, 85
Topoisomerase I, thermophilic
    (Type IA, Eubacteria)
  E. coli and, 39, 44
  enzymatic properties of, 44–45
  phylogenic position / sequence organization,
      39, 43–45
  T. maritima and, 44
Topoisomerase V (Type IB, Archaea)
  enzymatic properties of, 53
  HhH motif and, 51, 52, 53
  phylogeny of, 51–53
  temperature and, 52
  transcription regulation and, 52
Topoisomerase VI (Type IIb, Archaea)
  biochemical properties / structural data
      of, 58–59
  phylogenic position / sequence organization
      of, 58
Topoisomerases. *See* Gyrase (Type IIa,
    Eubacteria and Some Archaea);
    Hyperthermophilic Topoisomerase III
    (Type IA, Archaea); Reverse Gyrase;
    Thermophilic Topoisomerase III (Type IA,
    Eubacteria); Topoisomerase V (Type IB,
    Archaea); Topoisomerase VI
    (Type IIb, Archaea)
Transcription induction
  CXCL1 protein in, 13
  NF-κB in, 13
Transcription regulation
  CXCL1 in, 11–12
  CXCL8 in, 11
  IUR element in, 12

NF-κB in, 11, 12
protein kinase CK2 and, 240
topoisomerase, Type V (Type IB, Archaea)
    and, 52
Transcriptional activation, CBP and, 19
Trans-histone regulation, 286
  histone and, 286
Trans-Inactivation, heterochromatin and,
    280–282
Trinucleotide codons, eRF1 structure
    in, 112
Tripeptide motifs, RF1 / RF2 and, 104
tRNA. *See also* Analogue hypothesis;
    Ribosomes; Ribosomes, active center
  analogue hypothesis of, 83, 91
  anticodon loops of, 89–90, 104
  eRF1 structure and, 99, 101
  E-site / P-site, 107
  IF1 (initiation factor 1) and, 95
  molecular mimicry of, 86, 88, 89, 90, 94
  ribosomes, active center and, 88, 89,
      90, 91, 92
  ribosomes and, 86, 88, 89, 90
  translocation of, 89
Tumor necrosis factor-α (TNF-α), cytokines
    and, 5
Tumors, chemokines and, 9–10
Tyr phosphorylating activity, Protein kinase
    CK2 and, 242
Tyrosine, recombinases and, 68

V

Viral Tyrosine Recombinases: The
    Fuselloviradae Integrases, 68–69
  enzymatic properties of, 68–69
  phylogeny of, 68
  SSV1 integrase and, 68
V(D)J recombination
  NHEJ (nonhomologous endjoining) and,
      161, 186, 189
  Rag1 / Rag2 endonuclease and, 189

W

Walker mutants
  ATP (adenosine triphosphate) and, 175

INDEX

West, S. C., NHEJ (nonhomologous endjoining) and, 187
Wound healing, 7–10
  leukocyte migration in, 7–8

X

Xer site-specific recombination system
  E. coli (escherichia coli) and, 69
XerC/XerD protein
  Cellular Tyrosine Recombinases: The Thermophilic Xer Proteins and, 69, 71, 72, 73
  E. coli and, 70
XRCC (X-Ray Cross Complementing), DSB(s) (double-strand breaks) repair and, 185
XRCC4, NHEJ genes and, 181, 185

XRN1 proteins, dsRBM(s) (double-stranded binding-motif) proteins and, 145
XRS2, multifunctional nature of, 167

Y

Yeast PM H^+-ATPase
  regulation of, 213–214
  ScPMA1 and, 218

Z

Zavialov, A.
  RF3 study and, 111, 113

ISBN 0-12-540074-8